ABOUT ISLAND PRESS

Island Press is the only nonprofit organization in the United States whose principal purpose is the publication of books on environmental issues and natural resource management. We provide solutions-oriented information to professionals, public officials, business and community leaders, and concerned citizens who are shaping responses to environmental problems.

In 2002, Island Press celebrates its eighteenth anniversary as the leading provider of timely and practical books that take a multidisciplinary approach to critical environmental concerns. Our growing list of titles reflects our commitment to bringing the best of an expanding body of literature to the environmental community throughout North America and the world.

Support for Island Press is provided by The Nathan Cummings Foundation, Geraldine R. Dodge Foundation, Doris Duke Charitable Foundation, Educational Foundation of America, The Charles Engelhard Foundation, The Ford Foundation, The George Gund Foundation, The Vira I. Heinz Endowment, The William and Flora Hewlett Foundation, Henry Luce Foundation, The John D. and Catherine T. MacArthur Foundation, The Andrew W. Mellon Foundation, The Moriah Fund, The Curtis and Edith Munson Foundation, National Fish and Wildlife Foundation, The New-Land Foundation, Oak Foundation, The Overbrook Foundation, The David and Lucile Packard Foundation, The Pew Charitable Trusts, The Rockefeller Foundation, The Winslow Foundation, and other generous donors.

The opinions expressed in this book are those of the author(s) and do not necessarily reflect the views of these foundations.

OCEANS 2020

OCEANS 2020

Science,

Trends, and

the Challenge

of Sustainability

THE INTERGOVERNMENTAL OCEANOGRAPHIC COMMISSION
THE SCIENTIFIC COMMITTEE ON OCEANIC RESEARCH
THE SCIENTIFIC COMMITTEE ON PROBLEMS OF THE ENVIRONMENT

John G. Field

Gotthilf Hempel

Colin P. Summerhayes

ISLAND PRESS
Washington • Covelo • London

Copyright © 2002 Island Press

All rights reserved under International and Pan-American Copyright Conventions. No part of this book may be reproduced in any form or by any means without permission in writing from the publisher: Island Press, 1718 Connecticut Avenue, N.W., Suite 300, Washington, DC 20009.

ISLAND PRESS is a trademark of The Center for Resource Economics.

Library of Congress Cataloging-in-Publication Data
Oceans 2020 : science, trends, and the challenge of sustainability / edited by J. G. Field, G. Hempel, C. P. Summerhayes
   p. cm.
Includes bibliographical references and index.
  ISBN 1-55963-469-3 (hardcover : alk. paper) — ISBN 1-55963-470-7 (pbk. : alk. paper)
  1. Marine sciences—Research. I. Intergovernmental Oceanographic Commission. II. Field, J. G. III. Hempel, Gotthilf. IV. Summerhayes, C. P.
GC57 .O29 2002
551.46—dc21   2002005950

British Cataloguing-in-Publication Data available.

Printed on recycled, acid-free paper

Manufactured in the United States of America
09 08 07 06 05 04 03 02   8 7 6 5 4 3 2 1

This book is dedicated to Elizabeth Gross, who was until October 2000 Executive Director of the Scientific Committee on Oceanic Research (SCOR). Liz decided to retire from full-time SCOR commitment (and commitment it certainly was) after exactly twenty years of enthusiastic hard work at the helm. For many years, with an eagle eye for important detail, she acted as the nerve center of a network of marine scientists, her broad knowledge of marine science enabling her to keep her finger on the pulse. She encouraged and mentored many with good advice. She was (and still is) a tower of strength to a succession of new SCOR presidents.

This book is just one of the many examples of the effects of her efforts, both as an individual and as a team player. Without her contribution to organizing the preparatory meetings, and the piles of administrative labor involved in shipping chapters back and forth between editors and authors, compiling the bibliography, and keeping track of all the figures, this book would probably never have got to the press.

Liz, for your dedicated twenty years, and for this book, we salute you!

CONTENTS

List of Boxes   xi

Foreword   xiii

Preface   xvii

1. Introduction   1
   Colin P. Summerhayes, John G. Field, and Gotthilf Hempel

2. Ocean Studies   9
   John A. McGowan and John G. Field

PART I. ISSUES

3. The Coastal Zone: An Ecosystem Under Pressure   49
   Han Lindeboom

4. Climate Change and the Ocean   85
   Gerald R. North and Robert A. Duce

5. Fisheries and Fisheries Science in Their Search for Sustainability   109
   Gotthilf Hempel and Daniel Pauly

6. Ocean Studies for Offshore Industry   137
   Colin P. Summerhayes and Karin Lochte

7. Marine Information for Shipping and Defense   163
   David P. Rogers, Mary G. Altalo, and Richard W. Spinrad

PART II. TOOLS AND APPROACHES

8. Operational Oceanography   187
   Colin P. Summerhayes and Ralph Rayner

9. A Vision of Oceanographic Instrumentation and Technologies in the Early Twenty-first Century   209
   Tommy D. Dickey

PART III. SOCIAL/INSTITUTIONAL CHALLENGES

10. Framework of Cooperation   257
    Geoffrey Holland and Patrico Bernal

11. Capacity Building   283
    Miguel Fortes and Gotthilf Hempel

12. The Vision to 2020   309
    John G. Field, Colin P. Summerhayes, and Gotthilf Hempel

References   321

About IOC, SCOR, and SCOPE   349

List of Contributors   351

Index   355

LIST OF BOXES

2.1. The Joint Global Ocean Flux Study, by Hugh Ducklow 26
2.2. The Ocean Drilling Program, by Colin P. Summerhayes 34
2.3. Hydrothermal Systems, by Nils Holm 36
2.4. The Earth System, by John G. Shepherd 38
3.1. Mangroves and Aquaculture: Malaysia Case Study, by Jin E. Ong and Wooi K. Gong 54
3.2. Great Barrier Reef Case Study: Wise Use from Partnerships with Science, by Chris Crossland 56
3.3. The Dutch Coast: A Case Study, by Henk De Kruik and Frank Van der Meulen 61
3.4. The Caspian Sea Problem, by Vladimir E. Ryabinin 64
3.5. Nitrogen, Phosphates, and the Eutrophication of Coastal Waters, by Han Lindeboom 69
4.1. The World Ocean Circulation Experiment and the Ocean's Role in Climate, by W. John Gould 86
4.2. The South Atlantic and Climate, by Ilana Wainer and Edmo Campos 88
5.1. The Role of Science in an Unsteady Market-Driven Fishery: The Patagonian Case, by Ramiro Sánchez 114
5.2. Deep-Water Coral Ecosystems, by Jan H. Fosså 121
5.3. The North Sea Herring: A Case Study for Single Species Fisheries Management, by Christopher Zimmermann 123
5.4. Demands on Fishery Science and Technology: A Fishery Perspective, by Richard Ball 130
6.1. Contribution of Oceanography to Offshore Oil and Gas Activities: A Case Study from Brazil, by Leonardo F. Souza, Jose Antonio M. Lima, M. E. R. Casneiro, and Waldemar T. Junior 143
6.2. Effects of Oil Pollution from Oil Drilling in the North Sea, by Karl-Heinz Van Bernem 145
6.3. Effects of Manganese Nodule Mining on the Deep-Sea Ecosystem, by Karin Lochte 153

6.4. Ocean CO_2 Sequestration, by Peter Brewer   158
7.1. The Role of Marine Science: A U.S. Navy Perspective, by Richard W. Spinrad   170
7.2. Ship Routing and Real-Time Weather Forecasting, by David P. Rogers and Mary G. Altalo   175
7.3. Ballast Water, by Julie Hall   177
8.1. A Definition of Operational Oceanography, by Colin P. Summerhayes and Ralph Rayner   188
8.2. Matters of Public Concern About the Oceans, by Ralph Rayner   189
8.3. Broad Classes of Users and Customers, by Ralph Rayner   190
8.4. Case Study: North Brazil Current Rings Experiment, by Ralph Rayner and David Szabo   190
8.5. The Atlantic Margin Metocean Project, by Ralph Rayner   194
9.1. Autonomous Underwater Vehicles, by Gwynn Griffiths   221
9.2. NEPTUNE: Telescope to Inner Space, by John Delaney   243
9.3. Information Technology in Marine Science, by Eric Wolanski   251
10.1. The Needs for Organization and Cooperation in Ocean Science, by Rana A. Fine   260
10.2. Fostering a Coherent Approach to the Development of the Marine Information Business: The Marine Foresight Process, by Ralph Rayner   262
10.3. An International Forum for Intergovernmental and Industrial Cooperation in the Application of Science and Engineering to the Maritime Transportation Industry, by David P. Rogers   262
10.4. Research Ship Sharing Scheme of the European Union, by Colin P. Summerhayes   265
10.5. The Regional Seas Programme, by Stjepan Keckes   269
10.6. Partnership for Observation of the Global Oceans (POGO), by Lisa Shaffer   278
11.1. Capacity Building: A Thai Perspective, by Manuwadi Hungspreugs   287
11.2. Elements of Assistance in Capacity Building in Marine Science   290
11.3. Capacity Building in Marine Sciences in South America: A Plea for Networking and General Education, by Eduardo Marone and Paulo Lana   290
11.4. A Pacific Regional Perspective, by Russell Howorth   298
11.5. MADAM: Research and Capacity Building through an Integrated Long-Term Project, by Ulrich Saint-Paul   302

FOREWORD

This book is the outcome of the third assessment of Ocean Sciences, which was conducted jointly by the Intergovernmental Oceanographic Commission (IOC) of UNESCO working with two organizations of the International Council of Science (ICSU)—the Scientific Committee on Oceanic Research (SCOR) and the Scientific Committee on Problems of the Environment (SCOPE).

The purpose of these assessments of the status of ocean sciences, conducted by IOC about every ten years, is to identify the opportunities opened by new knowledge and technologies and to highlight the outstanding gaps and societal needs that call for further work. This prospective aspect of the exercise has been particularly valuable and useful in the past.

The assessment was conducted by active scientists nominated by their peers and selected by the three sponsoring organizations. Participants prepared synthesis papers in the different fields of ocean sciences and gathered in a workshop (Potsdam, Germany, October 1999) to discuss them with other invited presenters. The visions and opinions expressed here are based on their collective best judgment as experts.

This is not a "pure" assessment of the internal trends of ocean sciences, considered in isolation of its context of development. On the contrary, the terms of reference given to each of the participants made an explicit appeal to conduct their appraisal while keeping in mind the ever-increasing number of questions for which society is turning to science for answers.

The relationship between science and the institutions of science and society is complex. There are many-faceted means of mutual influence, some of which are quite unapparent, and many others are outside the sphere of influence of classical scientific institutions. There is no simple way to orient science in new directions and impress new acceleration to its development. On the other hand, the benefits that society obtains directly and indirectly from the development of science are proven and many, though difficult to quantify according to standard economic practices.

Planning and funding can and do make the difference. Plans are made for the foreseen science, the planned experiments. The funding is usually tailored to the requirements to build and run a new special instrument or large facility to develop and conduct the experiments. Investments in large science facilities, such as the complex of southern hemisphere observatories, have had a distinct impact on the rate of development of astrophysics.

Recent progress in ocean sciences has been achieved through international coordination and planning. The sequence of modern large-scale experiments of physical oceanography related to weather and climate prediction, from the Global Atmospheric Programme (GARP) to the World Ocean Circulation Experiment, are good examples of this. Another successful case is the Deep Sea Drilling Project, the sustained effort of which made possible the observations that confirmed the theory of plate tectonics. Although the Deep Sea Drilling Project is an example of the financing of a common international "facility"—the building and operation of the series of Glomar Challenger expeditions—it is also relevant from the point of view of signaling a close cooperation with an industry—the oil industry—that could benefit directly and indirectly from these scientific operations and from the applications of the results.

A key new trend toward planning and funding, which is very well reflected in the text, moves ocean sciences from the single experiment using research vessels toward the more permanent establishment of large distributed facilities. These instrumental facilities, such as the Tropical Atmosphere Ocean array for moored instruments and the Argo project for drifting profiling floats, integrate by means of what is truly a new emerging engineering of distributed systems thousands of self-operated instruments that can deliver observations in real time and quasi real time. This promise of instrumentation of the planet's ocean surface extends toward the chemistry and biology of the ocean, disciplines that are closely following physics in the development of their transducers and sensors. Essential components of these new large distributed facilities are the numerical models that integrate satellite observations with the in situ systems and that can use optimal assimilation techniques to improve forecasting ability.

Numerical models of the complexity necessary to resolve eddies in the upper layers of the ocean were not possible ten or so years ago because computers were not powerful enough. Faster computers have permitted realistic modeling of the physics of the ocean and will permit the attack on more complex, coupled systems that incorporate the chemistry, biology, and geology of the ocean. Forecasting the future states of fisheries and of coastal water quality will involve some degree of ecosystem modeling. In the text, there is a reiterated and tantalizing mention of this trend of ocean science toward increased integration. We take this to signal the development of a new discipline generating the conceptual framework to deal with the real level of complexity of the natural world. It is not too risky to say that in the next decade we will see the initial development of this disciplinary integration. The effective protection of the ocean environment and ecosystems, a high priority in the international agenda, will directly benefit from this exciting new development.

PATRICIO BERNAL

PREFACE

This book is based on the deliberations of a large group of experts in ocean science and ocean management who met in Potsdam, Germany, from October 2 to 6, 1999, to analyze the state of marine science, identify key scientific issues for sustainable development, and evaluate the capability of scientific, governmental, and private sector communities in different parts of the world to respond to these issues. The experts considered what has been achieved in recent years, defined the problems, outlined solutions, and determined the needs for and directions of ocean science in support of sustainable development for the next twenty years.

Scientists from a broad range of ages and from many different countries mixed with other stakeholders representing the management and user community to ensure that both the science and the societal issues were addressed in a realistic, credible, and comprehensive manner. During the workshop the participants considered a set of specially prepared background papers and then broke up into discussion groups to tackle the main issues and crosscutting factors. The authors of this book were participants and used the background papers and records of discussions when subsequently writing their chapters. The aim is to guide countries in developing their marine science and technology strategies and priorities in support of sustainable development to 2020.

We would like to thank sincerely the many people who helped to create this book. We thank the Intergovernmental Oceanographic Commission (IOC), Scientific Committee on Oceanic Research (SCOR), and Scientific Committee on Problems of the Environment (SCOPE) for their sponsorship and their invaluable work behind the scenes, including the efforts of the experts who developed the idea for the Potsdam meeting. We thank the

IOC, the German government, the German Science Foundation, and the U.S. National Oceanic and Atmospheric Administration (NOAA) for their generous financial support. We thank the local organizing committee, especially Sabine Luetkemeier and her staff from the Potsdam Institute for Climate Impact Research, and Dr. Christiane Schnack and her colleagues from the Centre for Tropical Marine Ecology in Bremen for their hospitality, efficiency, and enthusiasm. And last but not least we thank the many busy people who took the time to come to the workshop. Many experts unselfishly contributed sections to the different chapters or helped in the process of reviewing and editing, especially Ken Brink and Ken Mann. S. Krishnaswami, Richard Ball, and Eduardo Marone provided helpful comments on early drafts of the introductory chapter.

JOHN G. FIELD
GOTTHILF HEMPEL
COLIN P. SUMMERHAYES

EDITORIAL BOARD FOR PEER REVIEW

Peer review comments on each chapter were provided by an editorial board comprising John Field (University of Cape Town), Gotthilf Hempel (University of Bremen), Elizabeth Gross (SCOR), Véronique Plocq-Fichelet (SCOPE), Patricio Bernal (IOC), Robert Duce (Texas A&M University), Brian Rothschild (University of Massachusetts, Dartmouth), Ilana Wainer (University of São Paulo), Geoff Holland (Consultant, Canada), and Colin Summerhayes (IOC). In addition, the board requested external reviews of all chapters from Ken Brink (Woods Hole Oceanographic Institution) and Ken Mann (Bedford Institute of Oceanography). Science writer Peter Coles (Paris, France) was commissioned to edit the text of draft chapters to ensure that they followed a common format and were written in an easily readable style.

Chapter 1

Introduction

Colin P. Summerhayes, John G. Field, and Gotthilf Hempel

> The living ocean drives planetary chemistry, governs climate and weather, and otherwise provides the cornerstone of the life-support system for all creatures on our planet, from deep-sea starfish to desert sagebrush.
> —Sylvia Earle (1995)

THE OCEANS MATTER

The oceans cover 72 percent of the Earth's surface and, with an average depth of several kilometers, provide an enormous and varied living space that is mainly hidden. They drive our climate and weather, controlling the global deliveries of heat and freshwater. They provide a livelihood for many millions of people through fishing, exploitation of energy and mineral resources, shipping, defense, and leisure activities. The oceans contribute enormously to the biodiversity of the planet. Sediments from the ocean floor contain a record of life's evolution, the changing position of the continents, and the past variability of the Earth's climate. The picture is not always rosy. The oceans pose threats to human life and property through floods, storms, sea-level change, and coastal erosion.

Inevitably, humans and their effects on the marine environment threaten the sea's natural bounty. Overfishing has led to dramatic declines in many fish stocks. Humankind's influence on the sea is changing the patterns of biodiversity, probably irrevocably. The seas are being used increasingly for oil and gas exploitation and world trade. Population is rising inexorably, especially in the coastal zone where more than half the world's population lives. Mounting pressure on fragile coastal systems and coastal seas increases environmental damage. The sea is directly and indirectly used for waste disposal. Most waste eventually ends up in the oceans, affecting marine water quality and the health of the environment and of humans. Runoff from land pollutes coastal seas with fertilizers, pesticides, insecticides, and a growing sheaf of complex chemicals. Some of these appear to disrupt the endocrine systems of organisms by mimicking hormones. Toxic algal blooms are also on the increase, and some coastal seas are suffering from a combination of nutrient enrichment, excessive productivity, and oxygen starvation, which leads to eutrophication. Exotic species are being introduced into coastal waters through the discharge of ballast waters. Global warming is causing widespread bleaching and death of corals. There is growing concern that we are not proving as successful as might be wished in protecting our planet and sustaining our future (Watson et al. 1998).

The picture is not all gloom and doom. For instance, the introduction of effective controls on tanker activity by the International Maritime Organisation has reduced the spillage of oil into the sea from tankers by a substantial amount in recent years. New management principles have evolved in fisheries. Nations have agreed to control the runoff of pollutants from land. This book suggests that we can manage the marine environment in a sustainable way by effectively addressing environmental problems at local, regional, and global levels. Effective management means equipping managers with the best understanding and tools that marine science and technology can provide. Part of that understanding is that we cannot continue to use a sectoral approach to environmental questions. In nature there are complex physical, chemical, and biological linkages between what people commonly see as different environmental issues. These interlinked issues have to be addressed in the future in an integrated manner. In preparing this book we addressed that need for integration by bringing together different science groups dealing with the ocean, fostering dialogue between ocean scientists and ocean managers, and considering broad themes—such as climate or coasts—rather than single scientific disciplines or narrow sectoral issues.

WHAT THIS BOOK IS ABOUT

This book suggests what can be done about major marine environmental issues through the better development and application of marine science and technology. It is for policymakers, government officials, resource managers, scientists, the media, and the public. It provides a clear message about the oceans and gives advice on how to gather and use ocean information efficiently and cost-effectively for a multitude of purposes. It also suggests what to invest in to get the best results. It addresses increasing public concern about the direction, magnitude, and consequences of environmental change, helping to answer such questions as, Is the ocean changing? What is the evidence? Why is it happening? And, So what? The book is designed around a limited set of socially important and scientifically exciting issues addressed through chapters illustrated by case histories conveying important messages.

In keeping with the drive for integration, the book focuses on major topics of interest to broad and fairly well-defined groups of users. The book is not a comprehensive study of all marine science and technology. Marine geosciences, for instance, get short shrift. Based on a broad appreciation of societal concerns, the core of the book comprises chapters on basic ocean sciences, coastal research, climate, fisheries, ocean industries, and shipping and navigation. These are complemented by chapters on crosscutting issues, including operational oceanography, ocean instrumentation, the framework of cooperation, and capacity building. All these are sandwiched between an introduction and the book's final chapter, A Vision to 2020. In focusing on particular issues, we have had to leave out several important topics; for example, we have deliberately made little mention of polar regions, ice-covered seas, and the problems that go with them.

Why did we write this book? Now and then organizations need to pause, to take stock of where they are, to consider where they would like to be, and to plan a way forward. To meet this basic human need for orientation in a complex world, the Intergovernmental Oceanographic Commission (IOC) of the United Nations Education, Scientific and Cultural Organization (UNESCO) and its sister ocean organization, the Scientific Committee on Oceanic Research (SCOR), a component of the International Council for Science (ICSU), have twice worked together to assess the progress of marine science and technology. The first time was in Ponza, Italy, in 1969 (SCOR 1969), and the second in Villefranche, France, in 1982 (IOC 1984). By calling on the expertise of a large number of scientists, the reports of those meetings helped in their own small way to shape the course of the subject toward the end of the twentieth century. As the new millennium approached, the IOC and SCOR considered that it was time to take stock again, with the aid of the Scientific Committee on Problems of the Environment (SCOPE), another member of the ICSU family. This time, however, the emphasis was to be different. The stocktaking was to be carried out in the context of meeting societal needs and the challenge of sustainable development. For the purposes of this exercise we used the Bruntland Commission definition of sustainable development: "Development that meets the needs of the present without compromising the ability of future generations to meet their own needs" (UNCED 1987).

Our stocktaking is intended to complement the same kind of exercise that national governments undertake. It will provide an international perspective viewed from three angles: *political* (advice to governments) through the IOC; *environmental* (the needs to protect the marine environment) through SCOPE; and *scientific* (the needs of the international science community) through SCOR. One question raised within this perspective is how developed and developing nations alike can address important ocean issues. That question leads naturally to the issue of the transfer of knowledge and technology to help build the capacity of developing nations to carry out marine science and technology in support of their own sustainable development.

This book is not intended to be an assessment of the present health of the ocean; that would require a much longer and more technical volume. Readers wanting such an assessment are referred to recent publications of the Group of Experts on the State of Marine Pollution (GESAMP 2001a, 2001b). For readers searching for an in-depth assessment of progress in ocean science over the past two decades, we recommend the results of a recent comprehensive review on that topic sponsored by the United States's National Science Foundation (NSF) and fully reported on the Internet as *Ocean Sciences at the New Millennium* (http://www.geo-prose.com/decadal/). In contrast to the GESAMP and NSF assessments, *Oceans 2020* fills a hitherto unoccupied niche by taking a look at ocean science issues that are most closely related to human and sustainable development.

FORECASTING OCEAN SCIENCE

In writing the science and technology chapters in this book, the authors have peered into the future, extrapolating recent trends to see what might be possible in 2020. The authors were chosen because they are in a good position to see with a fair degree of confidence where present scientific trends are leading. They have considered what we already know and what we still need to know in marine science and technology, bearing in mind the natural variability of the ocean and the Earth system of which it is part. They have also taken account of the many human expectations and pressures on the marine environment and its resources.

How accurate are the forecasts reported in this book likely to be? A similar study of major trends in ocean research up to the year 2000 was held in Villefranche, France, in April 1982 (for more details, see IOC 1984). Unlike the largely issue-based studies in the present book, the previous study was discipline-based. When we compare its predictions with what actually emerged, we find that, on the whole, they were remarkably accurate, not least in singling out two areas of interdisciplinary research with significant potential impact on society: *climate research* and *ecosystem studies*.

In ocean physics, the previous workshop forecast the need, met through the World Ocean Circulation Experiment (WOCE), for a global scale hydrographic survey. The workshop foresaw that ocean physics research would increasingly underpin the work of the World Climate Research Programme (WCRP), a prediction fulfilled through the Tropical Ocean Global Atmosphere (TOGA) experiment in the equatorial Pacific, which showed that El Niño events could be forecast. TOGA ended in 1995, leaving us with the operational Tropical Atmosphere Ocean array of buoys, whose measurements underpin these forecasts today. The workshop recognized that the world's ocean needs to be studied as a whole if we are to understand the global climate system, foreshadowing the development of the WCRP's new program on decadal Climatic Variability (CLIVAR). And it recognized that a global monitoring system would be required for climate studies, a concept realized with the creation of the Global Ocean Observing System (GOOS) in 1991. Several trends were recognized that continue today, including increased measurements of ocean properties by moored arrays of instruments, nonrecoverable instruments, freely drifting floats, Doppler current profilers on ships, acoustic tomography, and ocean-observing satellites. The workshop foresaw increased demands to model and forecast ocean behavior and properties to improve ocean services for marine operations as well as increased demands for collaboration between oceanographers and meteorologists to improve weather forecasting. Missed were the development of profiling floats, now in widespread use, and of the autonomous underwater vehicles that are just becoming available as research tools.

The previous workshop foresaw a need in ocean chemistry now being met through the Land-Ocean Interactions in the Coastal Zone (LOICZ) project to focus on the fluxes of materials across the coastal zone to establish the influence of rivers on ocean chemical budgets. It saw a need later addressed through the Joint Global Ocean Flux Study (JGOFS) to map the transport, fates, and effects on the environment of carbon dioxide. It saw a need met through major national projects coordinated through the international InterRIDGE Programme to examine the influence on ocean chemistry of the (then) recent discovery of hydrothermal vents on the deep sea floor. And it saw the need for purposeful tracer experiments, like those later used in WOCE, to supply valuable information on mixing in the oceans.

The workshop recognized the importance of technology in facilitating chemical discovery, foreseeing advances in instruments from outside oceanography that would shape future chemical oceanographic research, especially improvements in (1) gas chromatography/mass spectrometry, enabling the identification of individual organic compounds; and (2) airborne lasers, advancing those chemical studies of the oceans requiring synoptic data. It also foresaw the need for improvements in sampling devices, to enable the collection of progressively larger and uncontaminated samples, and in sensors, to enable the collection of chemical data on the same time and space scales as physical data, a development essential for understanding the control of ocean chemistry by ocean physics.

In ocean biology the previous workshop recognized several key trends that continue today, including giving more attention to the properties and functions of ecosystems and quantifying the role of microorganisms on the flux of organic carbon through marine ecosystems. The structure, dynamics, and cycling of matter in marine ecosystems are now better understood, not least through the efforts of JGOFS and GLOBEC, the Global Ocean Ecosystem Dynamics program. Much research on large-scale processes, aimed at understanding the functioning and structure of the ocean ecosystem, is currently being undertaken through GLOBEC.

The workshop saw a need to relate biological processes to the behavior of the physical system in which the processes take place (now being met, for example, through GEOHAB, the Global Ecology and Oceanography of Harmful Algal Blooms project). A need was seen to study the ecology of communities inhabiting hydrothermal vents. Vent organisms have now been studied using in situ experiments, and some have been brought to the surface at ambient pressure for physiological studies. The workshop called for monitoring processes at the benthic boundary, a need met by the development of long-term observing stations in coastal waters.

In addition, the need was seen for interdisciplinary approaches to improve management of living resources in coastal seas and to apply the findings of biological oceanography to fisheries science. Progress is being made in both areas, too, though there is yet more to be done to develop an ecosystem-based approach to fisheries management.

The workshop attributed the (still ongoing) bottleneck in descriptive biological oceanography to two main factors: the time-consuming process of sorting and identifying catches and the rapidly diminishing number of trained taxonomists.

Many major new discoveries or developments were inevitably not foreseen, including:

- Broad areas of the open ocean are iron-limited.
- Organisms use chemical signals extensively.
- Microbes live buried deep beneath the deep-sea floor.
- Ocean color satellites capable of mapping plankton distributions would become widespread in the late 1990s.
- Acoustical and archival tags would enable the collection of environmental and behavioral data on pelagic animals.
- Molecular probes and DNA technology would be used to unravel the genetic structure of marine organisms and populations.
- Marine biotechnology would grow so rapidly.

It will be a pleasant surprise if the predictions made in this book are as good as those made by our predecessors. What our predictions will do is help to set priorities for the investment of effort in the relatively near term.

MARINE SCIENCE IN A CHANGING POLITICAL ARENA

Developments in science or technology are not the only influences on the way in which our science evolves. Social forces play a part too. The United Nations Convention on the Law of the Sea (UNCLOS), which became a treaty in November 1994, gave nations rights over huge extensions of their territories in the shape of exclusive economic zones, the existence of which will have an increasing effect on the ways in which countries study and monitor the oceans.

Since the early 1980s there has been a significant increase in political emphasis on the environment. The United Nations Conference on Environment and Development in Rio de Janeiro in 1992 led to publication of Agenda 21, an agenda for the twenty-first century that calls for improved management and sustainable use of oceans and seas, including the development and implementation of GOOS. Progress against Agenda 21 will be measured at the World Summit on Sustainable Development in Johannesburg in September 2002.

Other changes unforeseen in 1982 included the ending of the cold war, which had a number of unexpected spin-offs for marine science. These included the use of nuclear submarines by civilian scientists under the Arctic ice; the release of formerly confidential ocean data; use of the U.S. Navy's underwater sound surveillance system (SOSUS) to monitor earthquakes, listen to animals communicating, and monitor acoustic signals for detecting climate change through acoustic thermometry of ocean climate (ATOC); and more widespread development and use of remotely operated and autonomous underwater vehicles.

Societal changes led to positive changes in the climate for science and stimulated more marine scientific and technological advances than our predecessors had dreamed possible. Similar changes, which we cannot foresee, may be expected to generate yet more advances in the next twenty years.

Chapter 2

Ocean Studies

John A. McGowan and John G. Field

Earth is a blue water planet, which satellite photographs so clearly and beautifully demonstrate. This fact has had a powerful influence on the tempo and mode of biological evolution as well as on the development of civilization. Water has special physical properties that facilitate the transport of heat and momentum. The ocean absorbs heat and releases it over decades or even centuries, whereas the atmosphere's energy release has time delays of only a few weeks. This very slow release allows the ocean to act as a sort of flywheel or governor on climatic variability. The enormous volume of seawater, with its large heat capacity, regulates our climate and can greatly affect climate variability. Geological evidence shows that circulation changes in the past have accompanied large climatic variations. Another important property of water is its power as a solvent. It is a mixture of most of the elements and many compounds. The result is that it is not only salty but denser than fresh water, its density varying from place to place because of large differences in temperature, precipitation, and evaporation.

Life undoubtedly began in a watery medium. Many animal and plant groups left the sea for land during the course of evolution, but only a few have been successful in maintaining themselves out of water. Many millions of years of evolution took place before there were any terrestrial plants or animals. So, the principal mechanisms of natural selection and speciation must have originated in the ocean. It is even reasonably certain, now, that the earth's atmosphere was once heavily laden with carbon dioxide. The activities of primeval, green, photosynthesizing marine organisms must have removed much of this gas from the air and supplied the oxygen upon which most life depends today. In the process of changing the earth's atmosphere, marine plankton deposited vast amounts of organic carbon byproducts, including what we now call fossil fuel (Schlesinger 1991). This biologically driven change in the chemistry of the atmosphere also changed the heat balance of the earth and, with it, the wind systems and patterns of precipitation and evaporation. This, in turn, changed the circulation of the oceans. The study of this complex interaction of multiple systems on a vast scale has become the interdisciplinary field of oceanography.

THE SCIENCE OF OCEANOGRAPHY

Academic oceanography was founded as a separate field because of the importance of the oceans to human welfare, the richness of scientific questions, the specialized knowledge required to investigate them, and the sheer scale of processes and events in the ocean. Today most large research universities, and many small ones, include courses in oceanography, and it is possible to obtain advanced degrees in this field. But because the subject matter is so diverse, ranging from atmosphere-ocean heat balance to the physiology of marine microorganisms, there is a lot of pressure to specialize in one of the subdisciplines, while simultaneously recognizing the linkages between them.

The purpose of academic oceanography is not only to transfer knowledge and train practitioners, but also to develop new facts and insight into the structure and functioning of the global atmosphere-ocean system: its chemistry, biology, and long-term history of the lithosphere, hydrosphere, and biosphere.

While the study of living creatures is essential to oceanography, the heart of the science is large-scale physics: the movement of water, current patterns, and water chemistry. *Physical oceanography* is the study of the physical processes that govern the way the oceans work and interact with the atmosphere. It uses theoretical approaches, modeling, and observational techniques, such as measuring the distribution of properties and currents.

Marine chemistry looks at chemical and geochemical processes, including the physical and inorganic chemistry of seawater and ocean circulation based on stable chemical and isotopic tracers; organic chemistry and natural products chemistry; and the geochemical cycles of carbon, sulfur, nitrogen, and other elements.

Biological oceanography is concerned with the interactions of organisms with one another and with their physical and chemical environment. Research in this field is conducted on a broad spectrum of space and time scales and includes nutrient regeneration, population and community dynamics, primary and secondary productivity, biogeography, and the consequences of climate change to biological systems.

Marine geology and geophysics use both direct observation and theoretical methods to understand processes that alter the earth's crust and to analyze the long-term history of the ocean and its contents.

Paleoceanography interprets changes in the fossils and chemical composition of deep-sea sediment to reconstruct past conditions in the oceans and atmosphere.

Recently, *climate sciences* have also come to play a much more intimate role in oceanographic research and teaching. The growing realization of the powerful connections between atmosphere and ocean have made it essential to reach a better understanding of these interactions and their importance to the biogeochemistry of the seas. Studies focus on large atmosphere-ocean perturbations such as El Niño, La Niña, decadal shifts, and global warming.

Academic oceanography has seen much real progress in the past twenty years or so. We know much more about why the ocean is the way it is. But, now we must find out whether the ocean is changing; the direction, rate, and magnitude of the change; and the consequences to the planet and its living systems, including humans. This organizing conception is very rapidly evolving into a quest to understand very large-scale earth systems.

What follows is a summary of the larger basic research programs of the past twenty years. Although the programs focus on very different questions, they have a coherent theme: climate-ocean variability and change.

OCEAN CIRCULATION

Thermohaline Circulation

The movement of water in the oceans is influenced by the earth's rotation, by winds driving the ocean surface, and by the internal distribution of density. In many ways, it is difficult to distinguish cause and effect in the density patterns. However, the consequences of the ocean's variations in density from place to place are profound. In addition to being linked to the horizontal flow of water in the surface currents, the density distribution is linked to a globally connected system of deeper currents known as the thermohaline circulation (because of its dependence on temperature and salinity) (figure 2.1). The thermohaline circulation involves a system in which warm surface water is cooled at high latitudes and sinks to fill the deep basins of the global ocean with water at close to 0°C (see chapter 4). The key to this circulation is the salt content of seawater, which allows the water density to increase before it freezes.

The main sites for this conversion of shallow salty, warm water to deep, cold water are the Northern Atlantic, the Arctic Ocean, and the Weddell Sea in the Antarctic. This deep and bottom cold water is carried to the Indian Ocean and on to the North Pacific, where it slowly mixes upward to shallow depths, to be transported by the wind-driven upper-level circulation back to the major downwelling regions in the North Atlantic and Antarctica (see figure 2.1 in the color section). This system has been described as a "conveyor belt." However, it is in fact a complex system of interlinked and variable currents, whose overall result is to transport heat, water, and other properties around the globe and to maintain the balance of the earth's climate system.

The sinking water carries with it all of the properties it acquired when it was in contact with the atmosphere, such as gases, atmospheric particles, and the byproducts of biological production (Rahmstorf 1999) (figure 2.2). It also transports huge quantities of dissolved carbon, both as carbon dioxide and organic carbon. The concentration of dissolved carbon dioxide and oxygen in the upper layer of the oceans is on average in equilibrium with the atmosphere. Photosynthesis by phytoplankton in the upper layer removes CO_2 in order to synthesize organic compounds. Some phytoplankton also secrete little calcium carbonate ($CaCo_3$) shells or platelets, and these fall out of the upper layer when these plants die. Some of these platelets dissolve in mid-water, but others form vast amounts of calcareous sediment (billions of tons) on the floor of the oceans. Thus, this is one mechanism for sequestering CO_2 from the atmosphere to a long-term sink. The phytoplankton also produce dissolved and particulate organic matter. Some of this material simply diffuses downward. This is a second mechanism for the sequestration of CO_2 in the deep. Much of the organic matter synthesized by phytoplankton is eaten by small animals, the zooplankton. These animals also produce dissolved and particulate organic detritus. Many animals perform diel vertical migrations, feeding in the upper layers at night and respiring at depths during the day. This is another mechanism for the removal of CO_2 from the surface layers. As the upper layer waters move northward, they become chilled and therefore more dense and, in certain areas, sink to great depths. As this water sinks, it carries with it dissolved CO_2, organic matter, detritus, and material derived from the atmosphere through fallout or solution. In the case of the formation of North Atlantic Deep Water, the rate of sinking has been estimated to be 15 to

20 million cubic meters per second. Other areas of sinking transport water and materials to intermediate depths. Sinking is the major mechanism for removal of CO_2 from the atmosphere via the ocean. The organic material is fed upon at mid-depths by microorganisms and through the process of respiration; they remove much of the O_2 and add CO_2 to the mid-depth intermediate waters. These waters are greatly oversaturated with CO_2 and undersaturated with oxygen. Thus, the intermediate waters of all oceans, the North Atlantic Deep Water, the Antarctic Bottom Water, and the sediment are the greatest reservoirs of carbon on earth. We are still trying to determine the residence times of these sinks, but deeper waters are known to be several hundreds of years old and the sediments much older. The water that sinks must be replaced by upward mixing. Much of this occurs in the interior of the ocean but eventually by upwelling within the upper few hundred meters. Thus, "old" CO_2 is returned to the atmosphere.

Because of its huge volume and the long time it takes to circulate, this

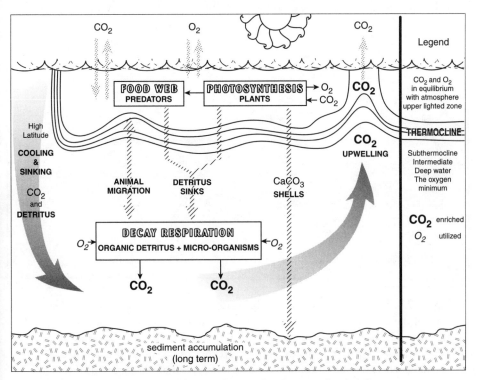

Figure 2.2. The ocean-atmosphere carbon cycle. See text for details.

system is very important to the climate, especially because it redistributes heat, water, and carbon dioxide on a global scale. The annual flow of the "conveyor belt" is equivalent to more than 100 Amazon rivers, and the heat it delivers to the North Atlantic is equal to about 25 percent of that received from the sun.

We now know that there are additional places around the Southern Ocean where great volumes of water sink, but only to intermediate depths. This intermediate water also carries with it vast quantities of dissolved inorganic and organic carbon. Thus, there is a second great reservoir in the subsurface waters for heat, atmospheric gases such as carbon dioxide, and byproducts of biological production.

Wind-Driven Circulation

In addition to the thermohaline circulation is the horizontal, wind-driven circulation system of the shallow layers of the world's ocean (figure 2.3). This system also plays a role in the horizontal transport of heat, salt, and other properties, also on a grand scale. Thus, by redistributing heat, gases,

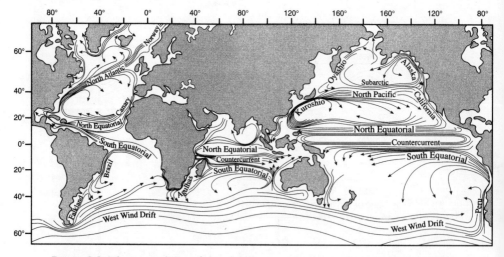

Figure 2.3. The streamlines of the global ocean average surface circulation. This map of the current systems is based on the accumulated knowledge of many decades. Corrections for ship drift from navigational fixes, inferences from temperature, salinity patterns, winds, and sea level heights are all used to construct them. There is no indication here of the relative speed of these currents. From Charnock 1996.

and chemical compounds, the oceans exert a profound influence on the climate, geochemistry, and biology of the earth. This influence is so pervasive that the geological history of the earth, its atmosphere, and the evolutionary history of its biota are recorded in deep-sea sediments. The motion of ocean waters is fundamental to the biology, chemistry, and physics of the earth.

Paleoceanography

To help us understand future climate change, paleoceanographers examine ancient periods of known extreme warmth and rapid climate change. There are variations in fossils of North Atlantic, surface-living, climate-sensitive plankton, in the layers of sediment beneath the zone where dense water sinks. The variations reveal that the conveyor belt has varied in intensity on decadal scales during the past 13,000 years (figure 2.4). In the deeper, older parts of both ice and sediment cores, both isotopically measured temperatures and indicator fossils agree that there have been large climatic changes that must be related to changes in the rate of North Atlantic Deep Water formation and the thermohaline heat conveyor belt. However, much more needs to be learned about the entire climate-ocean-biology system before predictions can be made.

Influence of Observation Techniques and Instruments

Although many basic ocean processes have been known for many years, they vary on scales of centimeters to thousands of kilometers and seconds to thousands of years. We know these processes exist, but we are only beginning to know about the rate at which these variations take place. It is the objective of basic research in this field to resolve this variability in space and time, understand its causes, and predict its consequences (Martinson et al. 1995).

Our understanding of upper-layer ocean circulation has increased greatly in the past two decades, chiefly through the introduction of new technologies such as automated instruments, satellite observations, the rapid transmission of data, global positioning systems, and high-speed computers. These have allowed the deployment of large-scale, long-term systems of measurements that were not possible before (figure 2.5). The training of dedicated young researchers by academic institutions has also contributed greatly to this expansion of knowledge.

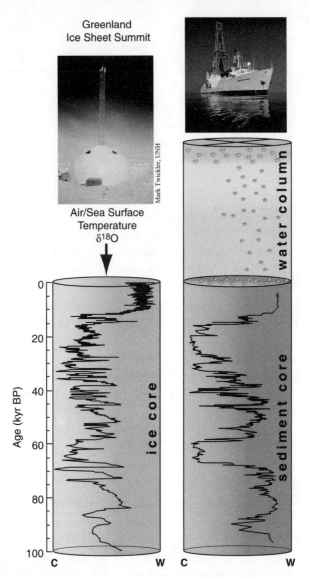

Figure 2.4. The oxygen isotope content of air bubbles in long ice cores from drill holes in the Greenland ice sheet (left) have been used to estimate past temperature variations in the North Atlantic atmosphere and ocean. Although there are rapid decadal sea surface temperature variations, the past 11,000 years have been rather stable. Fossil remains in bottom sediment cores taken near the area of sinking North Atlantic Deep Water (right) also show a fairly constant faunal assemblage, particularly of North Atlantic surface water foraminifera. In the deeper, older parts of both ice and sediment cores, isotopically measured temperatures and indicator fossils demonstrate large climatic changes that must be related to changes in the rate of NADW sinking and the thermohaline heat conveyor belt. Upper left drilling rig for ice cores; upper right the drill ship *Joides Resolution* and its derrick, at sea, on station. Shells of *Neogloboquadrina pachyderma,* an indicator plankton species, are shown falling out of the water column to the sediment on the sea floor (After Severinghaus and Brook, 1999; Bond, et al., 1992; and Bond and Lotti, 1995).

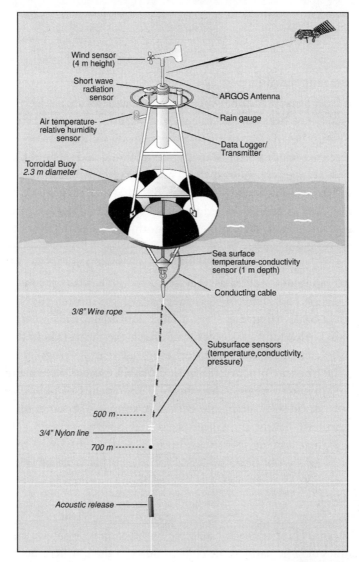
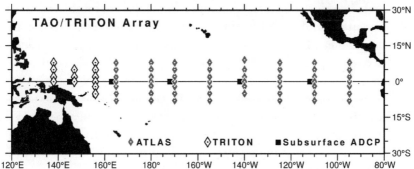

Figure 2.5. The autonomous temperature line acquisition system (ATLAS) environmental sensing buoys are anchored in the equatorial Pacific Ocean to continually monitor variations in the atmospheric and oceanographic conditions. This array evolved from the original Tropical Ocean Global Atmosphere program.

Meanders and Eddies

Both the Gulf Stream and the Kuroshio, two of the world's largest current systems, have now been studied intensively. We now know that rather than the ponderous, broad river-like streams previously shown in many illustrations, these two major currents are highly dynamic and changeable. Both of them, on average, transport over 50 million cubic meters of warm water per second from low latitudes to high. But they are each made up of a complex system of meanders, eddies, and ephemeral countercurrents so that at any one instant in time they look nothing like broad rivers (see figure 2.6 in the color section).

One of the more spectacular characteristics they share is the shedding of very large circular eddies as they bend to the east off Japan or New England. These eddies are of the order of 300 km in diameter and may persist for months (figure 2.7). If a large meander bends to the south of the main current it traps cooler water from the north side of the main body of the current. These meanders eventually form "rings" with cold centers. The rings are pinched off from the main current system and tend to drift off to the southwest, forming isolated islands of physically and biologically distinct water in the ocean. A similar phenomenon occurs on the northern side of the currents, so that both warm and cold core rings are regularly being formed and transported to the south or north. These rings affect the heat, salt, and biological balances of large areas by the introduction of "exotic" allochthonous, components (Ring Group, 1981).

We also know now that there are many other large eddies in all of the world's oceans. These features, called mesoscale eddies, dominate the physical dynamics of large parts of the globe. Eddies in the northern hemisphere that rotate in a clockwise direction (anticyclonic) tend to cause water to sink in their centers, while cyclonic eddies rotating in the opposite direction have upwelling centers. Eddies cause vast amounts of heat, salts, gases (including carbon dioxide), and biological products to move up or down the water column.

The result of the activity of many thousands of such simultaneous mesoscale eddies is very large heat, gas, and material exchanges between the atmosphere, the upper layers, and the intermediate depths of the oceans. Satellite-tracked floats drifting with the currents for many months have provided detailed information about these mesoscale movements (figure 2.8) (Niiler 2000).

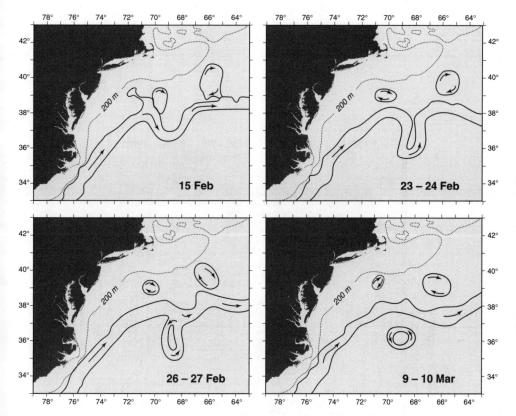

Figure 2.7. Gulf Stream cold core rings. This diagram, based on sea surface temperature measurements, shows the mechanism of formation. The Gulf Stream (arrows) commonly has many meanders. Sometimes these trap cold northern waters in their centers, are pinched off, and drift, often toward the warm southwest. There may be many such cold core rings of different ages present at the same time. There are also warm core rings moving to the north. These rings transport exotic biological and chemical properties, as well as water, into systems with different environmental histories. Such rings or eddies have been seen in many other parts of the world's oceans.

Ocean-Atmosphere Coupling and El Niño

It wasn't until the late 1960s that it became clear that the coupled ocean-atmosphere interaction was essential to the tropical El Niño phenomenon. This interaction involves great atmospheric features: the South Pacific high-pressure zone in the east and the Indonesian low-pressure zone in the west.

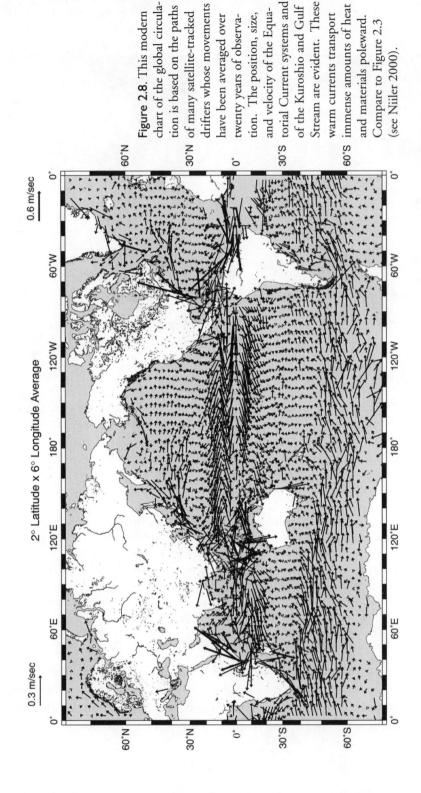

Figure 2.8. This modern chart of the global circulation is based on the paths of many satellite-tracked drifters whose movements have been averaged over twenty years of observation. The position, size, and velocity of the Equatorial Current systems and of the Kuroshio and Gulf Stream are evident. These warm currents transport immense amounts of heat and materials poleward. Compare to Figure 2.3 (see Niiler 2000).

The pressure gradient between these two is responsible for the great trade wind belt of the Pacific. But there are large year-to-year differences in the slope of this gradient. These variations are called the Southern Oscillation.

The trade winds drive the equatorial surface current system, causing the upwelling of cool water along the equator (see chapter 4). When the trade winds periodically relax (the Southern Oscillation), equatorial waters get warmer, eventually affecting the great fisheries of the Peru and California Current systems. This ocean-atmosphere interaction is called ENSO: El Niño–Southern Oscillation. This much was known by the 1980s.

The 1982–83 El Niño caught the scientific community off guard, so a response was called for. In 1985, a program called TOGA (Tropical Ocean Global Atmosphere) was initiated in order to develop a more complete description of the variation in the tropical Pacific and global atmosphere. The aim was to understand, on the one hand, how predictable this system can be on time scales of months to years and, on the other hand, to understand the mechanisms and processes underlying that predictability (Anderson et al. 1998). The idea was to study the feasibility of modeling a coupled ocean-atmosphere system in the hopes of eventually being able to predict the shift.

The TOGA researchers conducted analytic and diagnostic studies of ENSO by relating it to variability in other regions, especially in the monsoon regions of the earth. It is now clear that there are teleconnections between ENSOs and extensive areas at mid-latitudes. That is, climatic conditions vary, in concert, at areas far removed from one another. El Niño thus has virtually global effects.

The actual TOGA plan of research consisted of the installation of a series of monitoring buoys in the Pacific, either side of the equator (see figure 2.5). The buoys recorded ocean temperatures, wind speed and direction, and other variables. The array of buoys was designed to provide a virtually continuous record for ten years—enough time to resolve the cycle of ENSOs. The program resulted in great gains in understanding and predicting ENSOs. It showed that ENSO is a self-sustaining cycle in which anomalies of sea surface temperature cause the trade winds to strengthen or slacken. Essentially, the ocean, with its slower time scale of adjustment, provides the "memory" that carries the oscillations from phase to phase.

The program led to forecasts of El Niños, with the lead times of six to twelve months. But little is known of the limits of these predictions or how forecasts can be made outside the tropics. The program was just a beginning. ENSO was the first climate phenomenon to be shown to depend essentially on coupled interactions of both ocean and atmosphere. TOGA can now serve as a prototype for future studies (McPhaden et al. 1998; Neeland et al. 1998).

Researchers from many parts of the world have also come together in another major program to look at ocean circulation. The multinational World Ocean Circulation Experiment (WOCE) is a research program designed to help us understand the variability of coupled global ocean-atmosphere interactions, thereby improving our ability to predict their behavior. While these interactions occur on many scales, WOCE was designed to investigate the ocean's role on the decadal scale. This is why a field program on a global scale was needed (see box 4.1).

Water Mass Tracers, Nutrients, and the Carbon Cycle

Large-scale stirring, mixing, and sinking occur constantly in the ocean. In order to improve our understanding of these processes, geochemists have developed a series of tracers that help to show how rapidly the oceans remove atmospheric gases, pollutants, and particles, sequestering them in the deeper water, out of contact with the atmosphere. Among the first tracers were carbon 14 (^{14}C) and tritium (hydrogen3). Natural radiocarbon 14 is produced in the atmosphere at an approximately constant rate during the cosmic ray bombardment of stable nitrogen. This radiocarbon begins to decay immediately from ^{14}C back to nitrogen, with a decay rate or half-life of about 5730 years. If the ^{14}C is removed from contact with the atmosphere, for example, by dissolving it in the ocean, the amount of ^{14}C left can provide a good estimate of the time since removal (figure 2.9).

Using these measures, it is possible to estimate the age of the carbon dioxide dissolved in the intermediate and deep water as well as the dissolved and particulate organic detritus. This tells us a great deal about the rate of turnover of the atmosphere-ocean-biology-carbon cycle. The atmospheric testing of thermonuclear weapons during the early 1960s produced a great deal of additional ^{14}C. This global isotopic spike has also been used as a tracer for exchanges on time scales of years to decades. The smaller the ratio of ^{14}C to ^{12}C, the greater the age of the water mass. Tritium, with its much shorter half-life (about twelve years), is used for the same purpose in the shallower layers of the upper ocean where ^{14}C, because of its much longer half-life, is less useful (Schlosser and Smethie 1995).

Chlorofluorocarbons (CFCs or Freon) are used worldwide in refrigeration and aerosol sprays. They make especially valuable tracers because they

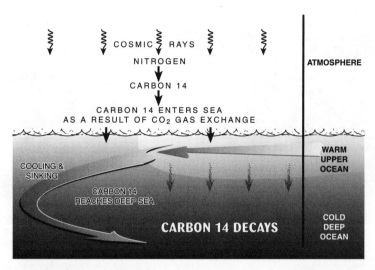

Figure 2.9. The ^{14}C age dating of deep seawater. Cosmic rays in the upper atmosphere bombard nitrogen atoms, turning them into an unstable isotope, ^{14}C, at a fairly constant rate. ^{14}C combines with oxygen and enters the CO_2 pool. This CO_2 equilibrates with the CO_2 in the surface water. But when the surface water is chilled and sinks, the CO_2 is out of contact with any newly formed ^{14}C. ^{14}C decays to nitrogen at a known rate. The time since a body of water has left the surface may be measured by its content of ^{14}C.

escape into the atmosphere and eventually dissolve into the ocean. Unlike ^{14}C, there is no natural source to complicate matters. Furthermore, we know very well when different formulations or types of CFCs were first made and when they entered the atmosphere. They are also rather easy to measure at sea and remain stable, since they are not metabolized by organisms. Because different types of CFCs were produced at different times, the presence of different CFCs in deep and intermediate depth water samples can be used to identify when the intermediate and deep water left the surface (for a deep water example, see Smethie 1993 and figure 2.10).

The ability to determine the age of entire masses of intermediate and deep waters and to calculate their rates of movement and residence times is of great importance (Fine et al. 2001). Although the ocean can sequester vast amounts of materials such as gases and pollutants, there are no permanent stores. If water sinks in one place it must eventually raise, somehow, somewhere else (see figure 2.1). For example, average residence time for the intermediate water (below 1500 m) in the Pacific is about 500 years. The deep and bottom waters are much older. These water masses are the second largest

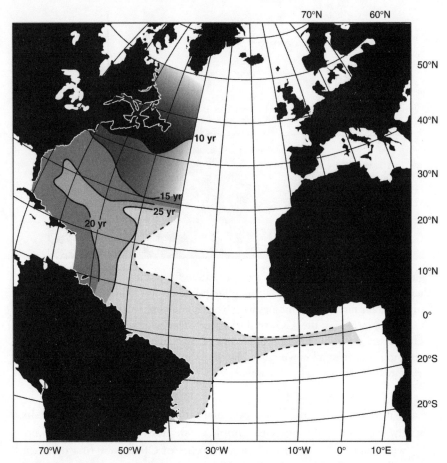

Figure 2.10. Estimates of the age of North Atlantic Deep Water from the concentrations of CFC (Freon) measure the nature of the thermohaline circulation in the North Atlantic and nicely confirm earlier studies using different methods. After Smethie et al. 2000.

reservoirs of carbon dioxide on the planet (deep sea carbonate sediment being the largest). But this water, too, will some day be back in contact with the atmosphere, exchanging carbon dioxide and other gases. The question is: how long will it take?

The water in the surface layers is almost always low in the plant fertilizers phosphate and nitrate. Because this layer receives sunlight, these nutrients are rapidly consumed by plant growth. But below this zone, huge amounts of these and other plant nutrients are dissolved in the water. The high concentrations of nutrients found in the deep ocean are the accumulation of eons of fallout and sinking of detritus from the surface layer. The detritus is broken down by microorganisms, releasing mineral nutrients at depths. It is generally agreed that global ocean production is regulated by the rate at which these nutrient-rich, subsurface waters are mixed up into the sunlit zone, where planktonic plants are fertilized, photosynthesize, and produce food for the rest of the food web. In doing so, these plants utilize not only the mineral nutrients brought up from below, but also the carbon dioxide dissolved in the water. This carbon dioxide is replaced by drawdown from the atmosphere (see figure 2.2). The rate at which photosynthesis occurs, often called primary productivity, is highly variable from place to place and from time to time. Studies of vertical stirring, mixing, diffusion, and upwelling are therefore critical for a second reason: the rate at which plant nutrients are supplied from below has considerable control on the rate of production. Thus, the cycle of nutrients and carbon is completed (Sarmiento 1993).

In 1984, a "meeting of experts" was held to "observe and understand the biogeochemical cycles of the ocean sufficiently well to predict the interaction between the oceanic, atmospheric and sedimentary cycles of the biologically active elements such as carbon, nitrogen, oxygen and sulfur" (Brewer et al. 1986). This meeting arose from concern over the impact of rising carbon dioxide levels and climate change. This and subsequent meetings resulted in an international research program called the Joint Global Ocean Flux Study, or JGOFS (see box 2.1). Its goal was "to focus on the global interactions of cycles of organic carbon, nutrients and biological productivity."

We have learned that most of the carbon released from the burning of coal, oil, and gas will end up in the oceans, where a complex cycle of circulation and other processes controls its fate. Since the ocean has a very large buffering capacity, about 95 percent of all the carbon is present in the ocean as carbon dioxide. Some of it is utilized by photosynthesis, but much more is removed from the surface by the thermohaline sinking of vast quantities of water to intermediate, deep, and bottom depths.

During this cyclic process, oxygen is introduced into the surface waters by photosynthesis and consumed in the deep water by respiration and decay. This cycle has gone on for many millions of years, and the net result is a gradual drawdown of carbon dioxide from the atmosphere into sediment, limestone, fossil fuels, and the vast reservoirs of dissolved carbon dioxide and dissolved organic carbon in the depths of the oceans. Roger Revelle and

Box 2.1. The Joint Global Ocean Flux Study
Hugh Ducklow

The Joint Global Ocean Flux Study (JGOFS) was formed in 1987 under Scientific Committee on Oceanic Research (SCOR) sponsorship and subsequently became a core project in the International Geosphere-Biosphere Global Change Research Program (IGBP). The objectives are

1. To determine and understand on a global scale the processes controlling the time-varying fluxes of carbon and associated biogenic elements in the ocean, and to evaluate the related exchanges with the atmosphere, sea floor, and continental boundaries, and
2. To develop a capacity to predict on a global scale the response to anthropogenic perturbations, in particular those related to climate change.

JGOFS includes twenty-one member nations participating in one or more of the four large JGOFS elements:

1. A series of process studies in key ocean biogeochemical provinces chosen for their sensitivity to climate change and/or because they exhibit large carbon flux signals.
2. A network of time series observatories for documenting seasonal to decadal scale variations in key biogeochemical, physical, and ecological parameters (figure BX2.1).

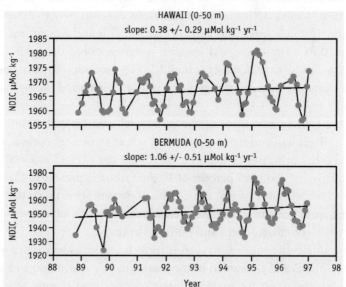

Figure BX2.1. The average concentration of dissolved inorganic carbon (DIC) at the (top) Hawaiian Ocean Time Series and (bottom) Bermuda Atlantic Time Series sites, 1988–97, in the upper 50 m of the ocean. This layer actively exchanges CO_2 with the atmosphere. The trend lines show the increasing CO_2 content of the ocean, caused by rising atmospheric CO_2, and ocean uptake. After Bates et al. 1996; Winn et al. 1998.

3. A large-scale survey of CO_2 and related properties, conducted with WOCE (see chapter 4). The CO_2 survey has realized the first comprehensive, high-precision global scale maps of surface pCO_2, allowing new estimates of carbon uptake by the oceans (figure BX2.2).
4. A synthesis and modeling program aimed at analyzing JGOFS and related observations over the past decade and formulating insights in a hierarchy of mathematical models. An early synthesis activity is the Ocean Carbon Model Intercomparison Project (OCMIP), whose goal is to identify the key elements of coupled 3-D models leading to convergence and divergence of global estimates of carbon uptake.

JGOFS therefore aims at understanding and predicting the role of the ocean in the global carbon cycle, clarifying the roles of the physical and biological carbon pumps. Together, these store about 2 gigatons (Gt; 10^9 metric tons) of carbon in the ocean annually. Figure BX2.1 shows the first direct observations of carbon uptake by the surface ocean. These observations have been made possible by establishing long-term time series observatories and by developing high-precision CO_2 analysis. The longer-term signal can now be detected against a background of great seasonal variability even in these oligotrophic sites. Figure BX2.2 shows the global distribution of the net flux of CO_2 between the atmosphere and surface ocean. The map shows that the Equatorial Pacific is the major global source for atmospheric CO_2, and the North Atlantic and Southern Oceans are the principal sinks (removal areas). The annual global net flux is 2.2 GtC, a figure about 0.7 GtC greater than previous estimates, as a result of new data from JGOFS and an increase in the atmospheric pCO_2 of about 7 ppm since 1995. The Southern Ocean is an efficient CO_2 sink. It has 10 percent of the global ocean area, but accounts for 29 percent of the total ocean CO_2 uptake. Cold temperature and moderate photosynthesis are responsible for the large uptake by the Southern Ocean. The Atlantic Ocean is the largest net sink for atmospheric CO_2 (39 percent); the Southern Ocean (22 percent), Indian Ocean (22 percent), and Pacific (11 percent) follow. The large sink in the northern oceanic areas is attributed to the intense biological drawdown of CO_2 in the North Atlantic and Arctic seas during the summer months. The equatorial Pacific CO_2 source flux may be totally or partly eliminated during El Niño events. This effect alone could increase the global ocean uptake flux up to 0.6 GtC per year (25 percent) during an El Niño year.

others have pointed out that humans are now engaged in an uncontrolled geophysical experiment on a massive scale, where within a few centuries by burning fossil fuel we are returning to the atmosphere carbon that was stored in sediments over millions of years.

We know from the long time-series of direct measurements at Mauna

Loa mountain observatory in Hawaii that carbon dioxide has been increasing in the atmosphere. So the question arises, Can the ocean absorb this excess? We are not yet able to tell if there has been an increase in carbon dioxide in the deep ocean, chiefly because we lack the essential baseline data to reveal such a change. But we know that large quantities of CO_2 are continually being removed from the surface ocean by the large scale sinking of surface water and through the action of the global biological "pump" being studied by JGOFS. Indeed, it is the slow rate of supply of plant nutrients, with its resulting limitation on productivity, that prevents the biological pump from removing all of the "excess" anthropogenic carbon dioxide.

Certain observations have led some oceanographers to believe that, in addition to the plant nutrients nitrate (NO_3) and phosphate (PO_4), the scarcity of the trace element iron (Fe) may further limit plant growth in certain oceanic areas. In iron enrichment experiments performed in the Pacific and Southern Oceans, large amounts of iron, added to a small body of surface water, resulted in blooms of phytoplankton. But since there is no known, natural reservoir of dissolved iron in the subsurface waters, we still do not know the extent to which this element can regulate productivity on a large scale. One hypothesis is that it does so through aeolian fallout (Falkowski et al. 1998).

The issue of whether additions of iron by man can cause extensive phytoplankton blooms has significant implications for environmental policy. It has been suggested that fertilization of the entire Southern Ocean with Fe could sequester substantial quantities of atmospheric carbon dioxide, delaying the predicted greenhouse warming of the earth. Debates about the feasibility, wisdom, and ethics of such climate engineering have been greatly handicapped by our limited knowledge of the response of the ocean to nutrient enrichment (Carpenter et al. 1995).

The great size of the oceans and the variability of production make it difficult to come up with reliable estimates of global production or how much it varies over time. Satellite oceanography could help to provide the answers. Ocean color, caused mostly by plant chlorophyll concentrations, can be monitored frequently by satellite sensing on very large scales, giving an indication of biological activity. The relationship of these remote measurements to the actual rate of photosynthesis, that is, the rate of uptake of carbon dioxide and the biological pump, is a topic of current research.

Plankton, Fisheries, and Ocean Ecology

PLANKTON

The very small living organisms floating and drifting in the sea, collectively known as plankton, are the basis of the entire food web. Although life in the sea ranges in size by eight orders of magnitude, tiny organisms account for a large fraction of the total. Some estimates of the total production of organic matter by the plant portion (phytoplankton) suggest that the oceans produce as much as the land in tons of organic material per year (Field et al. 1998). But these estimates are highly tenuous. Global maps of this ocean primary productivity are based on a single map made originally in 1970—and that original had many caveats (Koblentz-Mishke et al. 1970). Since the 1980s, satellite ocean color scanners have surveyed globally for chlorophyll, the chief plant pigment. But satellites can see only the ocean's surface layer, so the resulting maps are of standing crops of the upper few tens of meters and not of the entire column of water, nor of productivity.

It is one of the outstanding tasks of modern biological oceanography to produce global ocean estimates of primary production (i.e., the rate at which carbon dioxide is converted to organic carbon by photosynthesis, per unit time) that are accurate enough to be used in future global carbon cycle models. In some parts of the world, oceanic productivity is highly seasonal, with regular blooms of planktonic algae.

A great deal of effort has gone into trying to understand the mechanisms behind these blooms. Seasonal overturn or water column mixing and the resulting changes in the rate of supply of nutrients from below the lighted zone are frequently involved. But there are large ecosystems where seasonal blooms do not seem to occur, although there is seasonal mixing of the water column and probably an adequate supply of the standard nutrients. This is the case in the sub-Arctic North Pacific. One explanation (supported by evidence) is that although plant growth does increase in the spring, the grazing animals are so efficient that, as the plants grow, they are grazed down by zooplankton, so that there is no apparent change in plant biomass. Much the same effect may be seen in a pasture full of young cattle in the spring that grow fat by autumn, while the pasturage height and color stay the same to the eye.

Another explanation is that a micronutrient such as iron, rather than nitrate or phosphate, may play a limiting role. The iron enrichment experiments also support this explanation in some areas.

OCEAN ECOSYSTEMS

One of the larger research programs in biological oceanography is the Global Ocean Ecosystem Dynamics program (GLOBEC), a component of the International Geosphere-Biosphere Global Change Research Program (IGBP), co-sponsored by the Scientific Committee on Oceanic Research (SCOR). The basic mission of GLOBEC is to understand the structure and function of ocean ecosystems and their response to physical forcing. In the Georges Bank area off the northeastern United States, the combination of 3-D simulations of regional circulation with models of population dynamics for zooplankton, fish eggs, and larvae has contributed to new understanding of how plankton populations persist in a complex physical milieu. This has been worked out for the dominant copepod *Calanus finmarchicus*, which is a principal source of food for commercial fish stocks.

One of GLOBEC's aims is to use physical simulations to determine the distributions of fish eggs and larvae and the effects of physical variability on fish stock recruitment. In the Antarctic, this approach is helping researchers to understand the consequences of ocean current variability on the life cycle of the largest crustacean biomass in the world ocean—krill, *Euphausia superba*. The populations of krill at South Georgia are crucial for the fisheries there.

The synthesis of GLOBEC's worldwide studies will provide general proof of a basic postulate of biological oceanography: that the regularities in the physical dynamics of the ocean define the first-order explanations for biological distributions and dynamics of populations. This approach can be categorized as "bottom-up," and the successes from GLOBEC are a measure of the validity of this assumption. At the same time, the short and longer term variability in these populations, especially fish stocks, forces us to look at the second order—or complementary—processes of top-down or community control that regulate or ameliorate this induced variability.

FISHERIES

The term *regime shifts* is a convenient shorthand for the relatively large and rapid shifts in plankton and fish communities that appear to separate periods of relative stasis. It is often difficult to separate the two disparate causes, climate and overfishing. The implicit assumption in this concept of regime shifts is that the community response to forcing is essentially nonlinear; that makes it difficult to disentangle the causes or forcings by normal statistical analysis.

The open ocean away from the continental shelf is effectively pristine, but it is also subject to forcing at the top and the bottom. Climate change is likely to have an increasing bottom-up impact in these waters. The evidence indicates that physical regime shifts not only may result in changes in productivity, but also may induce changes in species structure that can have their largest effects and most dramatic impacts at the top trophic levels. At the same time, commercial fishing is moving off the shelf into deeper water, to the continental slope, sea mounts, and mid-ocean ridges, thereby affecting open ocean pelagic stocks through top-down control.

The California Cooperative Oceanic Fisheries Investigations (CalCOFI) is a time-series approach to understanding the consequences of climatic variability for both plankton and the fish that feed on them. Since it was established in 1949, this program has continuously monitored a large sector of the California Current. The original objective was to clarify the extent to which natural climatic variations are responsible for the large variations in harvestable fish, as opposed to the harvesting itself. In other words, are the variations in fish availability caused by humans (a result of overfishing) or do they occur naturally? The study used a time-series methodology to track populations and environmental and climatic variables through time. The process requires enough time to measure the variations, identify those that are most important, and then compare the spectra of variability of different entities with one another in a search for a common thread. This work was supplemented with numerous process-oriented studies. On the one hand, it has shown that the abundances of both plankton and larval fish vary greatly over long periods of time, but are not well correlated (if at all) (McGowan et al. 1996). On the other hand, the large, low-frequency plankton variations are well correlated with large-scale climatic changes.

There are two important scales of change in the California Current: the interannual or El Niño scale and the interdecadal or regime-shift scale. El Niños appear along the west coast of the United States at about a twelve-month lag after they occur along the equator. During El Niños, plankton, larval fish, seabirds, and kelp bed populations decline. Cool La Niña periods have been less studied, but it appears that these populations increase then; i.e., they recover from El Niños.

The larger, decadal scale warming of regime shifts shows a different picture. From around 1977 there was a massive (over 70 percent) long-term decline of plankton in the California Current and in many other elements of the food web (Roemmich and McGowan 1995) (figure 2.11). As yet, populations have not recovered from this collapse. It is not entirely clear whether the regime shift is due to a generalized warming or an increased frequency of

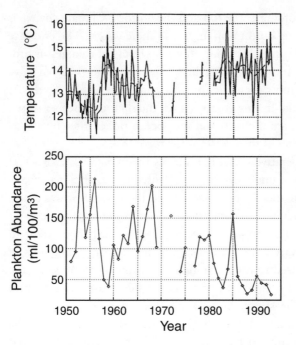

Figure 2.11. Top: The average temperature of the upper 100 m of a large sector of the California Current. Each point represents hundreds of individual measurements. Bottom: The annual average plankton abundance in the same area as above. As ocean temperatures increased, plankton decreased over time. From Roemmich and McGowan 1995.

warm episodes due to tropical El Niños. The regime shift was not limited to the California Current, but involved the entire North Pacific and was clearly linked to a major change in atmospheric circulation. There are not enough plankton baselines from other sectors of this ocean to establish the spectrum of variability or to determine whether there has been Pacific Ocean–wide biological change. The decades-long Continuous Plankton Recorder timeseries has also produced convincing evidence of strong linkage between climate-ocean variations and plankton variations in the North Atlantic (Aebischer et al. 1990; Reid et al. 1998).

So, far, the weight of the evidence indicates that overharvesting is a major factor in the decline of many stocks of fish. However, in the case of the California sardine, where a moratorium on fishing was called decades ago, the population did not respond until a significant change in ocean climate occurred. Indeed, the response, to date, has been weak. Thus, the changes in sardine population in this area are not controlled just by fishing; climate also plays an important role.

Part of the evidence that climatic variations do play a role in regulating fish population sizes comes from the examination of bottom sediment cores where fish scales have been preserved in annual layers for hundreds of years.

Off California, these show that there were great variations in the number of sardine scales deposited—and, presumably, the number of sardines in the water column—from decade to decade, long before any harvesting took place. This probably reflected changes in the environment caused by climate change.

The Sea Floor and the History of Change

Our understanding of the ocean floor has undergone a spectacular transformation resulting from the well-known demonstration of plate tectonics and our new appreciation of the rearrangement of continents. Although, properly speaking, this was not an oceanographic study, most of the original observations were made at sea from oceanographic ships. Another geological study, the Ocean Drilling Program, however, very much concerns the ocean, its history, and the biogeochemical changes that have occurred over long time periods. As such it has given us great insight into what kind of climate-ocean changes have occurred and the circumstances of those changes. The ocean floor provides an ideal location to explore Earth history because deep sediments and rock layers are generally much more continuous and less disturbed than those found on continents. The Ocean Drilling Program (ODP) (see box 2.2) has recovered sediments ranging in age from the last decade all the way back to the Triassic Period some 227 million years ago. The chemistry of these sediments and the fossil microorganisms they contain are indicators of environmental conditions in the ocean at the time of deposition (Norris 1997).

This program has provided evidence for drastic changes in the Earth surface environments. The most well known is the "impact event" about 65 million years ago, when a large extraterrestrial object slammed into the Earth and caused widespread extinction of perhaps 70 percent of all species. Further study of ODP cores will reveal the postapocalyptic repopulation of the ocean and give us some new insights into the rates of evolution and the conditions affecting those rates (Norris 1997). The next event occurred more than 55 million years ago when the temperature of sea water increased by 5 to 8°C at the surface and by 4°C at the bottom within a thousand or so years. This abrupt warming (the "Late Paleocene thermal maximum") led to a mass extinction of benthic fauna (Kennett and Stott 1991) and is interpreted as a result of input of carbon dioxide into the atmosphere and ocean.

Box 2.2. The Ocean Drilling Program (ODP)
Colin P. Summerhayes

In March 1996 the ODP published its long-range plan, a ten-year look ahead to 2006. The ODP's approach to scientific ocean drilling parallels the evolving recognition that the Earth is a large, complex, interactive, and dynamic system (see figure BX2.2 in the color section).

The plan identified two major research themes: dynamics of the Earth's environment and dynamics of the Earth's interior. Both are linked to studies of natural resource potential, global environmental change, and natural hazards. Crossing the two themes are three frontier initiatives that capitalize on new drilling and logging technologies and on advances in scientific understanding, and where ocean drilling is either the best or the only means of solving outstanding fundamental scientific questions:

1. Understanding natural climate variability and the causes of rapid climate change.
2. In situ monitoring of geological processes.
3. Exploring the deep structure of continental margins and oceanic crust up to 6 km beneath the seabed in waters deeper than 4 km. This depends on the development of new technology—deep-water riser drilling—combined with blowout prevention to seal off boreholes where there is a risk of an oil or gas escape. The technology will not be available until after 2003.

DYNAMICS OF EARTH'S ENVIRONMENT

Key objectives here include understanding Earth's changing climate; determining the causes and effects of sea-level change; and examining sediments, fluids, and bacteria as agents of change.

Major achievements include

- reconstructing the ocean's deep-water chemistry and circulation and their links to climate;
- relating the exchange of water between the Arctic and other ocean basins to climate;
- improved understanding of the magnitude, age, and mechanisms of Cenozoic sea-level change;
- a highly precise geologic time scale based on past variations in Earth's orbital geometry;
- demonstrating that the Great Barrier Reef is much younger than supposed; and
- finding live bacteria thriving hundreds of meters deep in the sediment column.

New goals, several requiring riser drilling, include

- addressing the role of greenhouse gases in climate change;
- advancing our understanding of the spatial patterns and timing of biological extinctions;

- expanding documentation of the timing, rates, and magnitude of sea-level change;
- improving understanding of the variability in the carbon cycle;
- exploring gas hydrates;
- studying the role of fluids in biogeochemical transformations; and
- exploring the nature and extent of the sub-seabed biosphere.

DYNAMICS OF EARTH'S INTERIOR

Key objectives here include exploring the transfer of heat and materials to and from the Earth's interior and investigating deformation of the lithosphere and earthquake processes.

Major achievements include

- understanding global geochemical cycles and evaluating Earth's resources;
- understanding the development of massive, sediment-hosted sulfide mineral deposits;
- ascertaining the role of fluids in faulting on accretionary wedges where plates collide;
- revealing some of the fundamental processes responsible for early rifting and subsidence; and
- documenting the pathways, sources, rates and effects of fluids in the ocean crust.

New goals include

- borehole observatories for high resolution tomographic images of the Earth's mantle;
- how mantle circulation has changed through time and how it relates to crustal dynamics;
- establishing the fluxes of heat, mass, and volatiles from the Earth's interior;
- investigating the processes occurring in the root zones of hydrothermal systems;
- shedding light on the role of low-angle normal faulting in continental break-up; and
- improving understanding of deformational processes and earthquake mechanisms.

Hydrothermal Activities

Recent deep-sea investigations have discovered intensive energy and material exchanges between Earth's surface and interior through the sea floor. Hydrothermal activities were found in the 1970s at the mid-ocean ridges. There seawater cycles through the newly formed igneous rocks, extracts heat and large quantities of sulfur and base metals from them, and transports these components to the sea floor, giving rise to hydrothermal vents of hot water associated with large polymetallic sulfide deposits and peculiar fauna. New evidence shows that fluid cycling through passive margins may exceed the amount that is cycled through mid-ocean ridges, in terms of volume and significance (see box 2.3).

Geological processes may lead to reorganization of the global ocean-water circulation. Some marine "gateways," for example, influence the thermohaline circulation and may act as thresholds in the global climate system, and plate migration may open or close the gateways. The onset of the Antarctic Glaciation was related to opening of the Drake and Tasmanian passageways, which enabled the formation of the Antarctic Circumpolar Current. Similarly, the closure of the Panama gateway may have resulted in the onset of a permanent Northern Hemisphere Glaciation. Sea-level fluctuations can also cause changes in marine gateways and in the routes of ocean currents. During the last glaciation, such changes forced a deflection of the western boundary current in northwest Pacific, the Kuroshio, which in turn enhanced aridity and cooling in East Asia (Wang 1999).

Box 2.3. Hydrothermal Systems
Nils Holm

Hydrothermal exchange at mid-ocean ridges has a profound effect on the chemistry of both the oceans and the underlying crust. In addition, hydrogen sulfide and other energy-rich substances present in hydrothermal fluids provide the fuel for the synthesis of organic matter in the deep sea. Plant life is impossible in the total darkness of the deep sea, and food resources and animal life are rare. Most deep-sea food chains are nourished by organic debris that sediments down from surface waters where phytoplankton carry out photosynthesis. The deep sea floor has often been described as a desert-like environment. So the discovery of luxuriant oases of giant tube worms, clams, and mussels clustering around hydrothermal vents at depths exceeding 2000 m came as a complete surprise to biologists. Since their first discovery in 1977, more than 100 sites of either active or extinct hydrothermal activity have been found around the global ridge-crest system.

During hydrothermal circulation seawater percolates down into fractured oceanic crust toward a heat source that may be an active magma chamber or a zone of partially melted rock that is referred to as *crystal mush*. As pressure and temperature increase downward in this hydrothermal circulation cell, a series of progressive seawater-rock reactions takes place that fundamentally alters the composition of the circulating fluid, while modifying the host rock. The net effect of these reactions is oxidation of the rocks and release of metals and reduced sulfur, in the form of hydrogen sulfide, into the fluid.

After the first discovery of active hydrothermal systems in 1977, biologists scrambled to identify the food source of the unusual ecosystems found around the vents. The presence of hydrogen sulfide in hydrothermal fluids and an abundance of sulfide-oxidizing bacteria were the first clues that led to the development of the hypothesis that biological productivity at hydrothermal vents is sustained by the chemosynthesis of organic matter by microorganisms. They use energy from chemical oxidation to produce organic matter from carbon dioxide. This was a conceptual challenge to the long-held view that Earth's ecosystems require sunlight and photosynthesis to supply food for animal food chains. In addition, the concept of the deep biosphere was pushed forward by the discovery of chemoautotrophic organisms in the fractured oceanic crust.

Bacterial life is now thought to be pervasive, possibly even reaching the Earth's mantle. Such features are also relevant to planetary science and the origin of life. There is a strong school of thought that believes that life on earth originated in such hydrothermal vent systems, based on chemosynthesis. The deep biosphere is generally considered to be a topic that is ripe for major breakthrough in a new area of fundamental scientific importance (see figure BX2.3 in the color section).

CONCLUSIONS AND RECOMMENDATIONS

The theme of the basic research over the past twenty years reported here is ocean variability and change. In order to understand large-scale variations it is necessary to understand the ocean's interactions with the atmosphere. The variability of the ocean, in turn, affects its biology and chemistry. These feed back into atmospheric variations. Resolving the nature of these global-scale interactions is the necessary first step in the effort to predict future rates, magnitudes, and causes of change. The geological evidence of the past suggests that future large magnitude changes may occur on time scales of decades to centuries. Understanding how this variability occurs demands a new kind of scientific research, one that requires global-scale and interdisciplinary observation programs encompassing long periods in both time and space.

The organizing principles of the earth-ocean atmosphere system (see box 2.4) clearly involve the mechanisms of transport of heat and momentum and the nature of the carbon cycle. We emphasize the source of the water, its contents, and the rates of its transport to the intermediate, deep, and bottom

Box 2.4. The Earth System
John Shepherd

The Earth system is made up of the solid Earth, the oceans, the atmosphere, the cryosphere, and the biosphere (both terrestrial and marine). These components are not independent. They interact with one another over an enormous range of scales in space and time to form a coupled system that functions as an almost closed system ("Spaceship Earth"). It interacts with the rest of the universe mainly through gravity (including tidal forces) and electromagnetic radiation (incoming ultraviolet and visible light, and outgoing infrared), apart from perturbations caused by major but infrequent collisions with asteroids and large meteorites.

This self-sustaining, almost autonomous system exhibits many interesting phenomena, such as the formation and movement of continents, the opening and closing of ocean basins, the formation and erosion of mountain ranges, the waxing and waning of massive ice sheets, and the inception and evolution of life in all its marvelous diversity. The processes involved in the functioning of the Earth system, and thus our conception of the system itself, depend greatly on the time-scale of interest, as summarized in the table below.

Time-scale (years)	Key Aspects and Processes
Billions	Continental accretion, solar evolution, origin of life, creation and oxygenation of atmosphere and ocean.
Hundreds of millions	Continental drift, ocean basin formation, evolution of phyla, mass extinctions.
Tens of millions	Mountain building, major volcanic episodes.
Millions	Crustal weathering, ocean chemistry (calcium), sediment accumulation, evolution and extinction of individual species.
Hundreds of thousands	Repeated glacial-interglacial cycles, insolation (eccentricity), sea-level changes.
Tens of thousands	Last glacial to current interglacial, last change of sea-level, insolation (precession), ocean chemistry (phosphorous).
Thousands	Millennial climate changes, ocean circulation/mixing, migration and modification of terrestrial ecosystems.
Hundreds	Anthropogenic climate change, abrupt (natural) climate variations.
Tens	Ocean-atmosphere coupled (decadal) variability.

Each of the different aspects and processes of the Earth system involves interactions between physical, biological, chemical, and geological processes, and so demands an interdisciplinary approach. Their study requires observations based on technology and engineering and modeling using mathematical and computational techniques.

A good example is the Earth's climate system (profiled in more detail in chapter 4), which is a complex nonlinear system involving multiple feedbacks, both positive and negative. Fundamentally, it is controlled by the balance between incoming sunlight (insolation) and outgoing infrared radiation, as illustrated in figures BX2.3 (in the color section) and BX2.4. But it also involves living things, as recognized by James Lovelock's Gaia paradigm.

The major factors include:

- the spatial and seasonal distribution of insolation, which is modulated by small changes in the Earth's orbital parameters;
- the Earth's albedo (its reflectivity to sunlight), which is affected by clouds and by atmospheric dust, and by changes in the coverage of land and sea surfaces by vegetation, snow, and ice;
- the transmission of infrared radiation through the atmosphere, which is affected by greenhouse gases such as atmospheric water vapor, carbon dioxide, and methane;
- the transport (redistribution) of heat and water between the equatorial regions and the poles, by the circulation and turbulent mixing of the ocean and the atmosphere; and
- the concentration of carbon dioxide in the atmosphere, which is determined naturally by biological and physicochemical processes in the oceans, but is now being substantially increased by humans.

Massive Northern Hemisphere ice sheets, which grow and decay with time constants of tens of thousands of years, cause major time lags and inertia in the system.

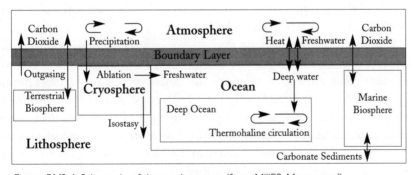

Figure BX2.4. Schematic of the earth system (from MillESyM proposal).

(continues)

Box 2.4. Continued

The depression of bedrock in response to ice loading and iceberg calving, melting, and consequential sea-level rise are also involved and may cause surprisingly rapid deglaciation. The consequences of interactions between many of these processes are well illustrated by data from ice cores drilled from the polar ice caps (Petit et al. 1999) (see figure 2.4).

In coming years we face the challenge of incorporating the biological and chemical controls on carbon cycling in the oceans, along with the behavior of carbon dioxide in the atmosphere, into dynamic global simulations of climate, from the seasonal scale right through to the scale of glacial-interglacial cycles. This calls for the development of models of intermediate complexity, because climate models based on normal GCMs (general circulation models) of the ocean and the atmosphere are computationally much too expensive for such long simulations.

---

waters because these water masses are gigantic reservoirs of carbon dioxide, nutrients, and dissolved organic carbon. It is now clear that atmospheric events not only drive this system but also are driven by it. The atmosphere, the ocean, and its inhabitants are intimately linked in what must be the world's greatest chicken and egg question.

We have made considerable progress in understanding the large-scale movement of water and the cycles of substances in it. The history of these mechanisms is becoming clearer, as are the conditions under which both short-term variations and long-term changes occur. We are slowly perceiving the role of the biology of the carbon cycle in influencing climate-ocean variations. But much needs to be done.

There is also a larger issue that we are only beginning to deal with, namely, the biological response to changes in the atmosphere-ocean system. Sediment cores show that variations in climate must have had significant impacts on the structure of ecosystems in the past, but at least three contemporary studies have shown large biological responses to rather small physical perturbations. Disturbance theory is at the very heart of modern concepts of the evolution and maintenance of diversity and of the regulation of organized ecosystems, and intensified disturbances of the state of the ocean can be expected in the future. We cannot, at present, predict which physical perturbations will disturb the existing biological structure in terms of direction, magnitude, or persistence.

Learning more about the likely effects of ocean-atmosphere variations on the structure and function of ecosystems is of more than academic importance. There is still no unambiguous resolution to the question of whether the great variations we see in marine fisheries are caused by humans (overharvesting, pollution, habitat disruption) or climate variations. We now know that unharvested populations, such as plankton and seabirds, are strongly affected by climatic variations. But the relationship of climate variability to harvested populations is not yet clear, except locally.

There are two rather disparate concepts of how to design research on what the consequences of climate variability may be to biological systems. The most popular is the traditional "scale-up" approach where observations made at small space-time scales are used to determine possible, fundamental, mechanistic attributes of biological communities or "laws of nature." These are then extrapolated upward and outward in models to predict large-scale responses. However, events observed at small scales do not necessarily reveal dominant processes that generate large-scale patterns. Further, the complexity of biological systems is so great that what we see depends on the scale of observations (Goldenfeld and Kadanoff 1999). It is difficult to assess which of the many intermediate and changeable mechanistic pathways are the ones to incorporate in an ecological model. The alternative, or "scale-down" design of research, is based on determining the large-scale time-space patterns of a set of indicator variables (biomass, for example), which, given enough observations, may then be cross correlated with physical and chemical measurements to identify possible causal relationships and to test hypotheses (Root and Schneider 1995). Short-term "scale-up" studies cannot detect the effects of climate change, and long-term "scale-down" usually cannot specify the intermediary mechanisms should change occur.

Behind both of these research philosophies is the concept of "change." This word is generally used to mean the fact of becoming different. But natural environmental properties vary on many time-space scales (figure 2.12) so that further definition is necessary. With regard to climate-ocean-biology change, we mean a shift from one pattern of variability to another, or a transformation from one state to another. In order to detect the direction, magnitude and frequency of such deviations, we must have a norm or a baseline from which to compare the departures. Since very large spatial changes happen only rarely and slowly (i.e., are of low frequency), very long baseline sets of rather frequent measurements are necessary. There are few such time-series in marine biology, chemistry, or physics. An objective for the future is to establish such monitoring systems, because serial measurements of these properties are essential to our understanding of the word "change."

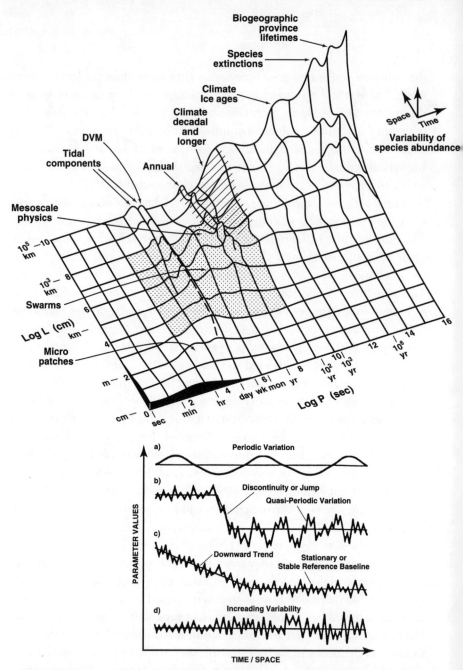

Figure 2.12. Because of large-scale studies done chiefly in the past twenty years, we now know that there are variations in all oceanic properties on many time and space scales (bottom) and how time-space scales are related (top). We also know that different time-space peaks in the spectrum of biological changes are due to different causes. Thus, extrapolation from studies done on one time-space scale to other scales is problematic. The stippling in the top Figure shows the typical scales for most biogeochemical studies, and the cross-hatching shows the decade to century scale for comparison. DVM = diurnal vertical migration. From Haury and McGowan 1998; and Martinson et al. 1995.

Perhaps the greatest lesson to be learned from the experience of the past two decades is that of the importance of scale in the design of research programs that aim to improve our understanding of global physical-chemical-biological change. Extrapolations outward in space and onward in time from limited local observations are not often successful in providing this understanding. One obvious reason is that large, and therefore important, changes are rather slow to develop, making them difficult to detect, because of local high-frequency variability due to local causes. So the question of what temporal or spatial scales to use for measurement depends on the objectives of the research—whether it is to resolve a local problem or a general one.

There has been a clear improvement in our understanding of how the atmosphere and ocean interact, especially in the tropics. El Niño and La Niña forecasts are now a distinct possibility. Further, the decadal regime shift of the Pacific is an observed phenomenon with a great impact on marine ecosystems and fisheries. Now that the nature of this event is becoming clearer it may be studied with the aim of eventual prediction.

The entire question of how much anthropogenic carbon dioxide the ocean can sequester depends strongly on the rate of sinking of dense water and the residence times of intermediate, deep, and bottom water. The rates of formation and eventual upwelling of these water masses (the thermohaline circulation) is of critical importance and is being intensively studied. But upper layer horizontal circulation is important as well, because it is from the upper layers that rapid exchanges of heat, momentum, and gases occur with the atmosphere. The role of mesoscale eddies, meanders, jets, and episodic countercurrents in the current systems of the world has become much clearer and has led to better predictions.

Biological productivity, the subject of diligent study for decades, has become more complex by virtue of the discovery that much of the basic photosynthesis is carried out by very small chlorophyll-bearing microorganisms. Little is known of their natural history or what may regulate their populations. Biological production is responsible for the uptake and sequestering of much carbon dioxide, but its significance in the total carbon cycle is still somewhat uncertain. The controls on production clearly involve the rate of input of essential chemical nutrients. While most of these kinds of input are known, their sources and rates of supply are not well understood. Part of this problem of rates of supply has to do with the circumstances under which vertical mixing and upwelling of deep, nutrient-rich water occur, so knowledge of ocean-atmosphere variation is of essence here. El Niños and the Pacific regime shift resulted in warming and a less dense upper layer. This, in turn, led to changes in the upwelling mixing regime and a decline in biological production across the food web.

The study of microfossils of planktonic organisms in sediment cores has yielded much information on past temperature changes in the ocean and the atmosphere. These changes are associated with changing atmospheric carbon dioxide. Both these past warmings and the present warming in the North Pacific were only a few degrees above the long-term mean. But they were persistent over extensive areas. This had very great effects on large-scale ocean systems, especially ecosystems.

These lessons from the past twenty years have greatly increased our basic understanding of the global ocean. Such understanding will need improvement if we are to predict the consequences of future global changes.

RESEARCH PRIORITIES

- A global array of ocean monitoring systems is needed to measure climate change and variability in the oceans on a long-term basis, measuring physical, chemical, and biological variables.
- Studies are required of how the structure and functioning of ocean ecosystems change through time and space, to relate these to climate change and variability and to human activity.
- Studies of the present-day ecology and ecophysiology of plankton indicator species are needed to improve interpretation of the past record provided by those whose fossils are found in sediments.
- Large-scale air and sea studies of mixing and upwelling of nutrient-rich water are needed to interpret and predict likely side effects of proposals to engineer (through iron or other fertilization) enhanced primary production to mitigate carbon dioxide increase in the atmosphere.
- Studies are needed to give sharper, more certain knowledge of the biodiversity and species structure of cardinal marine ecosystems, their boundaries, and their geographic patterns.
- Research is needed to improve methods of scaling down from the coarse scale of global ocean models to predict regional and local effects on the scales important for biological processes and fisheries.
- Research is needed to improve methods of scaling up from observations at the local scale to improve estimates of primary and secondary production at the global scale.

ENDNOTES FOR FURTHER READING

Aebischer, N. J., J. C. Coulson, and S. M. Colebrook. 1990. Parallel long-term trends across four trophic levels and weather. *Nature* 347:735–55.

Broecker, W. S. 1997. Thermohaline circulation, the Achilles heel of our climate system: Will man-made CO_2 upset the current balance? *Science* 278:1582–88.

McPhaden, M. J., A. J. Busalacchi, R. Chevvey, J. R. Donqury, K. S. Gage, D. Halpern, Ming Ji, P. Julian, G. Meyers, G. T. Mitchumm, P. Niiler, J. Pacaut, R. W. Reynolds, N. Smith, and K. Takauchi. 1998. The tropical ocean-global atmosphere observing system: A decade of progress. *Journal of Geophysical Research* 103(12):14169–240.

Roemmich, D., and J. A. McGowan. 1995. Climatic warming and the decline of zooplankton in the California current. *Science* 267:1324–26.

Sarmiento, J. L. 1993. Ocean carbon cycle. *Chemical and Engineering News* 31:30–44.

PART I
ISSUES

Chapter 3

The Coastal Zone: An Ecosystem Under Pressure

*Han Lindeboom**

The world's coastal zones are long narrow features of mainland, islands, and seas, generally forming the outer boundary of the coastal domain (from 200 m above sea level to 200 m below). About 60 percent of the people in the world live in this relatively small but highly productive, highly valued, dynamic, and sensitive area. Coastal zones, which in our definition include the entire continental shelf, occupy about 18 percent of the surface of the globe, supplying about 90 percent of global fish catch and accounting for some 25 percent of global primary productivity. At the same time, they are among the most endangered areas. Pollution, eutrophication, changing sediment load, urbanization, land reclamation, overfishing, mining, and tourism continuously threaten the future of coastal ecosystems. The major challenge facing us today is managing the human use of this area, so that future generations can also enjoy the fantastic visual, cultural, and edible products that it provides.

Another major challenge is the constant change of coastal systems. Changing wave and current regimes, climate, morphological processes and fluxes from land, atmosphere, and oceans create a very high natural

---

*With indispensable contributions from Ulrich Saint-Paul, Chris Crossland, Martin Hemminga, Kees Camphuysen, Bill Burnett, Ong Jin Eong, Wooi-Khoon Gong, Roel Riegman, Henrik Enevoldsen, Wim van Raaphorst, Carlos Duarte, Julie Hall, Jan Helge Fosså, Steve Smith, Jan Boon, Hartwig Kremer, Neil Adger, Jozef Pacyna, Henk de Kruik, Frank van der Meulen, Harald Rosenthal, Vladimir Ryabinin, and Eduardo Marone A.O.

variability in coastal systems. This variability is still imperfectly understood. And now, with their increasing technological capabilities, humans are further influencing this already highly variable ecosystem. Although most impacts are still regional, the scale of development along the world's coasts is increasing to such an extent that it has become a global issue. Sustainable use and protection of this very vulnerable but crucial part of "System Earth" are now high on international agendas. Meanwhile, international instruments, like the United Nations Convention on the Law of the Sea (UNCLOS), Agenda 21, the Biosphere Reserve Program, and the Ramsar Convention on Wetlands play an important role in regulating developments.

From the point of view of ocean science, many advances have been made in recent decades in our understanding of the processes that affect the continental shelves and their boundaries with the land in the coastal zone. In this chapter we focus on the nearshore part of the coastal zone, because many of the interactions of humans with the sea take place in this environment. This is where sustainable development and science interact most intensely. Fisheries, of course, can extend seaward across the entire continental shelf within the coastal zone. However, they are dealt with in another chapter (chapter 5). Because the physics of coastal seas have been described in comprehensive detail recently in two volumes of *The Sea* (Brink and Robinson 1998; Robinson and Brink 1998), this aspect of the science is not treated in detail in this chapter. Thus, we largely ignore here significant processes such as coastal upwelling, fronts, and exchanges between the continental shelves and the open ocean that may have an impact on the open ocean fisheries of the continental shelf. Readers interested in these topics can find them addressed by Mann (2000) and Mann and Lazier (1996).

Despite our rapidly increasing knowledge about coastal ecosystems, crucial questions on the causes of variability and the effects of human impacts remain unanswered. And although the perception of politicians and managers of our coasts is shifting from a mainly short-term economical approach toward a long-term economical and ecological perspective, the consequences of this shift—changes of management practice—are often ignored or difficult to sell.

In this chapter, we try to identify to what extent our increasing knowledge and understanding of coastal seas and adjacent land areas has helped to improve management of the coastal zone. We also present an inventory of the expected major issues, both managerial and scientific, for the coming twenty years. Of course, land-based processes, activities, and modifications farther inland have enormous direct and indirect impact. But this chapter emphasizes the aquatic rather than the terrestrial issues and scientific capabilities. For the same reason, with the exception of the Caspian Sea, we will only deal with saline waters with open connections to the oceans. Although we recognize the importance and problems of saline lakes, these are not addressed in this chapter.

CHANGE AND VARIATION IN THE COASTAL ZONE

(Natural) Variability

Coastal marine ecosystems are not in steady state, but exhibit continuous changes in production and species composition. The question we have to answer is to what extent these changes are due to natural variation or to the impact of human activities. Our awareness and scientific understanding of this variability has increased during the past decades. For example, long-term data sets on phytoplankton, zooplankton, macrobenthos, fish, and birds have now been collected in the Wadden Sea and North Sea. Until recently, these data sets were mainly used to demonstrate the effects on the ecosystem of human (mis)use. However, when the various data sets are combined, a striking picture emerges. Certain changes are sudden rather than gradual, as one would have expected from a progressively increasing human impact.

The algal biomass in the western Wadden Sea doubled between 1976 and 1978, followed by the macrobenthos in 1980 (figure 3.1). Similarly, the breeding success of eider ducks suddenly increased by several orders of magnitude. In the North Sea, sudden changes in the phytoplankton and zooplankton species composition were reported. And in the northeastern Pacific, a major shift in the marine ecosystem was recorded in the second half of the seventies, pointing to the possibility that there was a large-scale cause for the observed phenomena. Other examples include the recent collapse of the cod stocks in the eastern Atlantic and of certain North Sea fish stocks at the end of the previous century.

In many data sets there are obvious cyclic changes. For example, three-year maxima have been observed in the abundance of *Noctiluca scintillans*, while a six-year cycle has been found in shrimp catches around the North Sea. A similar cycle has been observed for the recruitment success of shellfish. And, when Gray and Christie (1983) analyzed plankton data from the North Atlantic, they found evidence for 3–4, 6–7, and 10–11 year cycles, whereas benthic data suggested 6–7, 10–11, and 25–30 year cycles. Another well-known cycle is the El Niño–Southern Oscillation (ENSO) that results

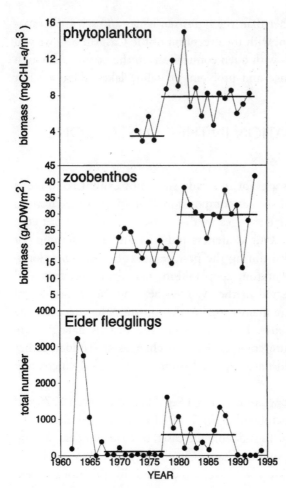

Figure 3.1. Variations in the biomass of phytoplankton and zooplankton and in the numbers of eider duck fledglings in the western Wadden Sea, The Netherlands. Reproduced from Bergman and Lindeboom 1999, with kind permission from Kluwer Academic Publishers.

in the nearly complete failure of fisheries off Peru and Northern Chile and many other worldwide ecological deviations every four to seven years.

Despite an increasing number of examples from many areas around the world, experts disagree on the nature—and even the existence—of cyclic behavior in coastal seas. Are these cycles really the result of complex physical-biological interactions, or are they merely a statistical artifact of data sets? Cyclical phenomena have been well documented in freshwater systems and in tree rings on land. In marine systems, there is growing evidence for cyclic behavior in sediments (see Pike and Kemp 1997), corals, and shellfish growth. An increasing number of authors are suggesting links with solar activity.

But the picture is still not clear, challenging the existence of predictable cycles. Very long data sets show alternation between periods with clear cycles and periods where there is no obvious pattern. Bergman and Lindeboom (1999) concluded that many long data series indicate interannual and decadal variability, with other less predictable phenomena occurring, such as rapid changes, gradual changes (e.g., in the direction of trends), changes in monthly or seasonal variability, changes in species dominance, and cyclic variation.

Other authors have suggested links with short-term or large-scale weather patterns, wind, winter and/or summer temperatures, or rainfall. A shift in storm frequencies or wind directions might cause changes in sediment water exchange or mixing. In temperate regions, the occurrence of cold winters strongly influences the species composition of intertidal benthic communities. Possible causes of these phenomena also include changes in water or nutrient fluxes from the landward or seaward side and in internal processes in the marine ecosystem.

We therefore need more long-term data sets, in combination with the results of experimental laboratory and field studies, if we are to decide whether we are looking at explainable phenomena with clear cause-and-effect relationships or at a more or less chaotic and unpredictable behavior of the marine ecosystem. These data are vital in order to separate human-induced changes—which can therefore be managed, at least in principle—from natural variations, about which little can be done. The data would also help to show up a delayed response of the marine environment that could otherwise mislead status assessment studies. But, at present, limited funding and lack of long-term commitments of governments often hamper the continuous collection of data.

Coastal Habitats

MANGROVES

Mangroves are salt-tolerant forest ecosystems bound to the intertidal shallow-water areas, lagoons, and river estuaries of the tropics and subtropics. They play a particularly significant role as ecosystems populated by animal species that are often of economic importance. They function as nursery areas for numerous fish and crustacean species, while providing natural protection against the surf, currents, and tides. They also supply wood and are thought to serve as an important carbon dioxide sink. But, apart from natural disasters, mangroves are exposed to numerous threats from human activity. These include unsustainable logging, clearance to make way for fish and shrimp farming, industrial and domestic pollution, dredging, and land reclamation (box 3.1). An estimated 80 percent of previously existing mangroves have already been degraded. Even though this damage is still continuing in several places around the world, it has received little public attention. Mangroves do not yet have a powerful lobby to defend them.

Box 3.1. Mangroves and Aquaculture: Malaysia Case Study
Jin E. Ong and Wooi K. Gong

Mangroves are a dominant coastal ecosystem in Malaysia, constituting about 2 percent of the total land area. They come under the jurisdiction of the forest departments of the various states, and some have been successfully managed on a sustainable yield basis since the early 1900s (Ong 1982). The 40,000-hectare Matang mangrove forest on the west coast of Peninsular Malaysia is arguably one of the best exploitatively managed mangrove forests in the world. Yet, some 20 to 25 percent of all the mangroves have been lost or degraded in the past two decades or so. The textile industry, which converts wood chips to rayon, has degraded a few hundred thousand hectares of mangroves in the Malaysian states of Sabah and Sarawak. More visible and widespread is the conversion of mangrove to aquaculture ponds (mainly for rearing of the tiger prawn, *Penaeus monodon*). Aquaculture is thus seen as the biggest threat to mangroves.

Mangrove forests are chosen for the culture of tiger prawns because of the brackish water in which they grow. But, above all, mangroves are both abundant and cheap. In Malaysia, almost all mangroves belong to the various state governments. They can be "appropriated" for other uses, often through the issue of temporary occupancy licenses (for a nominal charge). In the past, officials did not understand the real value of mangroves. But this is no longer the case. Federal government policy is to preserve all mangroves except when there is a strategic need. But many states simply ignore the directives. Land is a state matter and the federal government policy has no legal standing.

The culture of tiger prawns is potentially lucrative, especially if the mangrove areas can be obtained for almost nothing. But the risks are high, especially if the first few crops fail. Technically, mangrove areas are not the most suitable areas for prawn ponds. Almost all mangrove soil is potentially acid sulphate, requiring expensive liming. Mangroves are also natural nurseries and feed areas for prawns. So the use of mangroves for prawn culture is tantamount to robbing Peter (the capture fisheries) to pay Paul (the culture fisheries). And there is the added penalty of destroying a scarce coastal ecosystem, while losing the other functions that the mangrove ecosystem performs.

> Aquaculturalists should not only avoid using mangrove coastal areas, but they also need to reduce or treat water pollution from pond effluent while looking at hygiene and disease control.
>
> One scientific question remains to be answered: how much does the mangrove ecosystem contribute to the adjacent coastal ecosystem in terms of the outwelling of carbon and nutrients? We have already spent considerable effort addressing this problem (e.g., Nixon 1980), but a satisfactory answer is needed for the sustainable use of mangroves.
>
> The challenge in the coming decades will be the ability to integrate natural science with the socioeconomic elements.

Although there are a great many publications on mangroves, very few embody the integrated coastal zone management concepts that are necessary for sustainable stewardship. To fill this gap, a combination of different modeling approaches is needed. These should combine the respective benefits of global system analyses (e.g., a trophic model of biomass flows), detailed process studies (e.g., a "forest model" simulation package), and joint, ecological and socioeconomical risk assessment. In other words, a transdisciplinary approach is essential, allowing for dynamic interplay between theory and empirical study. The German/Brazilian integrated project on Mangrove Dynamics and Management (MADAM) is a good example, illustrating the scientific basis of this concept (see also box 11.4).

CORAL REEFS

Coral reefs are complex ecosystems based around physical structures built by coral and algae that are themselves dependent on light. Globally, coral reefs are changing as a result of widespread impacts of both human use and "natural" influences of global climate dynamics. Human impacts are both direct and indirect. Direct impacts include overfishing and cyanide/dynamite fishing, habitat destruction by tourist divers and recreational fishers, limestone mining, land reclamation, and pollution. Indirect impacts include global warming as well as coastal and catchment developments that cause high turbidity and sedimentation. The majority of the world's coral reefs are already degraded. Many have been devastated. But most human-induced impacts can be managed to ensure sustainable, economic benefits. The missing ingredients are usually education, political will, and action.

In the face of natural change, reef ecosystems are quite resilient, having

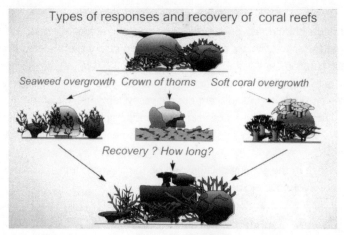

Figure 3.2. Possible damage to and restoration of coral reefs. Supplied by Terry Done.

Box 3.2. Great Barrier Reef Case Study: Wise Use from Partnerships with Science

Chris Crossland

BACKGROUND

The Great Barrier Reef (GBR) covers about 350,000 km^2 stretching more than 2000 km along the northeastern coast of Australia. A Marine Park (1975), it is contained within the Great Barrier Reef World Heritage Area (listed in 1981; see Harvey 1999). There is a clear goal for management, supported by a strong political (and community) will.

KEY USE AND ISSUES

Conservation of the natural heritage is a key factor in the GBR, which is managed as a large multiple-use area, with human activity generally subject to permitted use in different zones. The direct economic annual value from human use is about $3 billion. The main human activities include tourism and recreation, fisheries (recreational, commercial trawl, and line), shipping (transit and ports), aboriginal traditional hunting, and research. Oil drilling and mining are prohibited. Water quality and the effects of changes in adjacent land use (e.g., urbanization, agricultural development, and activities) are central issues.

The resolution of conflicts over use is an important part of management. The approach is generally one of education rather than regulation. Community consultation with participatory management underpins the stewardship actions for the reef. In 1994, more than sixty interest and user groups (policy, management, industry, community, and research) agreed on a common vision and twenty-five-year strategic plan for management direction and outcomes. A procedure for public reporting on the environmental state of the GBR has been introduced.

SCIENCE

Scientific research and information are vital aspects of management of the GBR. While the "precautionary principle" is applied, scientific research innovation and information remain the major management demands.

As the importance of the GBR for the community has grown, so has marine science capacity. There have been major developments at James Cook University and the Australian Institute of Marine Science, in particular, attracting a high level of recognition and application elsewhere. The links between reef research and management have not always been close. But, in the past decade, a deliberate and reasonably successful alliance has been established. In 1993, the Cooperative Research Center for the GBRWHA (CRC Reef) involving reef users (more than 1100), reef managers (national and state government), and research agencies was funded under a national program. This has provided a unique platform for the alliance (www.gbrmpa.gov.au/~crcreef/new/). A vigorous program communicates research findings in a readily understood form to the users and community, actively involving the public media.

Scientists have realized that there is a demand from users outside the academic world. Meanwhile, managers have recognized the uncertainties associated with research. Communication, trust, partnership, and recognition of roles, along with improved institutional dimensions, remain vital.

One of the key scientific questions is to separate the effects of natural variability from human-related activities that are able to be managed. This involves both fundamental science and socioeconomic concerns. Environmental management is about managing people and their actions, so a high emphasis must continue to be placed on improving socioeconomic research, data, and information.

adapted and survived major global changes in climate throughout geological time. For example, 16,000 to 22,000 years ago, sea level was some 100 m lower than today. The sea level rose so fast about 14,000 years ago that corals were unable to keep up. This created major differences in reef structure across the world's ocean. Science is just beginning to understand these changes as well as the processes and limits involved. Apparently natural phenomena, such as widespread crown-of-thorns starfish outbreaks and coral bleaching, are being studied actively. Human impacts can exacerbate their effects but it is still not certain to what extent humans are actually responsible for them. Global warming appears to be one cause for the increases in coral bleaching observed today.

The fundamental structure of reefs depends on corals and algae using carbon dioxide and accumulating calcium carbonate (calcification). Recent findings show that rising atmospheric CO_2 results in increased total inorganic carbon in the surface ocean, an increase in aqueous CO_2, and a significant decrease in carbonate ion concentration, in turn affecting the carbonate saturation state. The strong dependence of both organism and community calcification rates on saturation state implies that calcification rates of reefs and organisms will diminish further with elevation of atmospheric CO_2, and this will be reflected in shifts in community composition and metabolism (Kleypas et al. 1999).

From preindustrial times (around A.D. 1800), calcification has decreased by an estimated 10 percent and could decrease a further 20 percent by A.D. 2100 if the atmospheric CO_2 concentration doubles. The Red Sea, west central Pacific, the Caribbean, and high latitude reefs such as Bermuda are likely to experience the greatest changes. The ramifications of a forecast such as 30 percent decrease in reef calcification after 300 years of industrialization need urgent study, both at a whole reef system scale and at the level of organism physiology and reproduction.

SEAGRASS MEADOWS AND SALT MARSHES

Seagrass meadows occur in all marine waters, except polar waters, fulfilling several important roles. They serve as a habitat for animal species, increasing local biodiversity and acting as nursery areas for the recruitment of important fisheries species. They play a role in sediment retention and the mitigation of coastal erosion. They are important sites for the production and retention of organic carbon and associated biogenic nutrients. Finally, they are sensitive indicators of water quality.

Over the past decades the fauna of seagrass meadows has been a major research topic. It appears that seagrass meadows generally harbor a more diverse and abundant fauna than bare areas, often serving as a nursery habitat for many commercially important species. Meanwhile, molecular biology now provides researchers with the necessary tools to study genetic diversity in seagrass meadows. But the results are not yet clear. Some studies indicate a very low level of diversity, whereas other studies suggest a much more diverse pattern in beds of these clonal plants.

Over the past twenty years dozens of cases of large-scale, human-induced decline in seagrass meadows have been described. Several studies have investigated the causes of decline (including ecophysiological studies), giving considerable insight into the major threats.

The main research challenges for sustainability include:

- the prediction and identification of the causes of large-scale degradation of seagrass populations worldwide;
- the capacity to predict recovery times of seagrass meadows once the perturbations responsible for their decline disappear or are attenuated;
- a quantitative analysis of the roles of seagrass meadows, which so far have often been anecdotal, excluding their inclusion in integrated coastal management efforts;
- the reliable determination of the habitat requirements of seagrasses, so that clear habitat quality targets can be incorporated in integrated coastal management plans.

Salt marshes are another important coastal ecosystem. Also highly productive, they are characterized by relatively high below-ground biomass. For example, in Spartina marshes, below-ground plant parts represent two to four times more biomass than that above ground. In both seagrass and salt-marsh systems macroalgae, benthic microalgae, phytoplankton and attached algae contribute to total production. In areas with high nutrient inputs, epiphytes might hamper the growth of seagrasses and marsh plants. Temperature, salinity, light, currents, waves, water transparency, and sediment loads are major factors that determine the growth and species composition of the meadows.

Major threats are eutrophication, land reclamation, increased turbidity, and climate change. In Europe, salt marshes have been reported as systems with a net import of organic matter. But in other areas, a net export has been found that is proportionate to the net primary production. More research is needed into the factors that limit marsh distribution, with the emphasis on the effects of extreme events. We also need to know more about the role these marshes play in the life cycle of other organisms, not just those with a commercial value.

ROCKY SHORES AND KELP BEDS

Apart from harbor and industrial constructions, rocky shores themselves are not under real threat. But the ecosystems they support are often in great danger, mostly as a result of pollution and overharvesting of algae and macrofauna. Seaweeds are common inhabitants of rocky shores, attaching themselves by means of their specially adapted root systems. Some seaweeds, such as kelps, thrive as forests in the cool clear waters of higher latitudes and in subtropical areas. A blooming industry has grown up, harvesting these systems for food, fertilizer, and a variety of cosmetic products. Giant kelp beds can rival coral reefs for their physical beauty, productivity, and ability to attract numerous other organisms. They deserve the highest protective status. Scientifically relevant questions include genetic variability in the different populations and the thresholds for permanent damage.

ARCTIC SHORES

The Arctic shores are characterized by large permafrost regions and long periods of ice and snow cover. The projected climate warming will probably affect the polar areas most, with a reduction in the polar ice sheets. This, together with warming of the air and water, will affect the climate of the coastal zone. This will lead to melting and/or weakening of the permafrost, which, along with the fast ice belt, plays an important role in polar coast protection. This, in turn, may accelerate the destruction of the polar coast, with a consequent recession of the coastline. This process is already being observed in some areas of the Russian Arctic. Some parts of the Yamal Peninsula can withdraw at a rate of up to 6 to 8 m a year. Acceleration of coastal retreat will have a wide range of ecological consequences. There is a need for focused research in order to predict and, if necessary, mitigate the consequences.

SANDY BEACHES

Sandy beaches are important coastal areas for a number of animal species and to the economies of several countries, for example, through tourism. The great mobility of sand makes beaches and tidal flats highly dynamic systems, with tides, wind-induced waves, and storms all playing a role in shaping them. Under normal circumstances, beach shape and the changing processes find a long-term equilibrium. However, sea-level rise, subsidence, a decrease in nourishment, changing wind regimes, and activities such as sand mining and protective constructions can cause a major departure from this equilibrium. The importance of beaches for tourism often renders protective measures such as breakwaters and dikes less desirable. Expensive alternatives, like artificial beach nourishment, may become daily practice.

In several parts of the world barrier islands are part of the sandy coastal system. If these islands are allowed to stay in equilibrium with the transport

Box 3.3. The Dutch Coast: A Case Study
Henk de Kruik and Frank van der Meulen

After a long history of fighting the sea, the Dutch tamed their dynamic deltas and coastal sands with huge fixed structures. But now they realize that this solution is both expensive and at odds with the dynamics of the coast and its intrinsic natural values. So, the coastal management challenge is to find the right balance between socioeconomic development and nature: building with nature. This leads to six main management aims that should reinforce the basic qualities of the coast without lowering the safety factor. Through dialogue, opportunities should be sought for high-quality, innovative, and sustainable solutions for the longer term that are supported by the government and a broad group of the population.

ROBUST DUNES

Narrow and low dunes are the weakest spots in the coastal defense system. These are reinforced by nourishing and enlarging the sand buffer. Broad dunes offer a more flexible means of coastal defense. They also present less resistance to wind activity and allow the dunes to be mobile, enhancing their natural ecological value. A more flexible coast is better adapted to the consequences of accelerated sea level rise.

WATER MANAGEMENT

The low-lying hinterlands behind the dunes are kept dry by a complex system of water management. This is a vulnerable situation. The costs of water management and flood prevention are increasing. Sea level is rising while land is subsiding. Water management has to be adapted to these processes. It will be necessary to study where intensive and costly water management is really needed and where a less strict regime can be applied.

IMPROVING THE QUALITY OF WETLANDS

There are two real, remaining estuaries in the Netherlands: the Westerscheldt in the south and the Eems-Dollard in the north. The other former estuaries have been dammed off from the sea, resulting in ecological deterioration of the coastal wetlands, affecting water quality, oxygen and nutrients, foraging, and nursery functions. Since estuaries belong to the world's most productive ecosystems, the potential for restoration has to be studied. Where and when transitions between saltwater and freshwater are restored depends on the other functions of the area in question.

SAND FOR THE WADDEN SEA

It will be increasingly difficult to maintain sand and mudflats in the Wadden Sea. Sea level is rising, and sources of sediments have been cut off by dikes and polders. The

(continues)

> **Box 3.3.** Continued
>
> sand balance needs to be restored in order to maintain the ecological quality of the Wadden Sea as a wetland and habitat for seals.
>
> **COASTAL QUALITY**
>
> Variety is the hallmark of the coast, accounting for the range of coastal resorts, from remote spots with few facilities and small-scale, traditional family resorts, to up-market seaside resorts with an international reputation. The present trend toward large-scale tourism is threatening the quality of coastal resorts.
>
> **CARE FOR THE SEA**
>
> The unique physical and ecological characteristics of the North Sea make it a priceless resource. Uses of the North Sea are already multiple, including fisheries, oil and gas exploitation, shipping, harbors, sand extraction, cables, and pipelines. Meanwhile, there is a push to reclaim land from the sea to extend the available space for development. A truly integrated approach is urgently needed.

mechanisms, they change shape or position. But, when dikes and other structures force these islands to maintain their position, continuous artificial sand nourishment may become a prerequisite for their continued existence. This is the case with the Frisian Barrier Islands along the north coast of the Netherlands. A better understanding of the sand transport processes is essential for future management of these sandy systems.

Beaches and mud flats play an important role in the lives of turtles and seals. Disturbance of their egg-laying and incubation activities is threatening sea turtles with extinction. Despite significant conservation efforts, this threat is still very real, and more protected areas are needed. But until we have a better idea of all the requirements of these organisms, these efforts might prove futile.

ESTUARIES

Estuaries are semi-enclosed basins permanently connected to the sea, whose waters are diluted by freshwater drainage, often from rivers. Estuaries maintain exceptionally high levels of biological productivity and play important ecological roles (Clark 1996). These include "exporting" nutrients and organic materials to outside waters through tidal circulation, providing habitats for a number of commercially or recreationally valuable fish species, and serving the needs of migratory nearshore and oceanic species that require nursery areas for breeding and/or sanctuary for their young. A key feature of estuaries is their variable salinity.

Major threats to this environment include damming of rivers, pollution, urbanization, harbor and industry development, construction of flood protection devices, mariculture, and recreation.

COASTAL HAZARDS

Coastal natural disasters (e.g., flooding, earthquakes, storm surges, sea-level rise) or human-related hazards (e.g., oil spills, other pollution sources) cut across all economic sectors. Their effects can hit tourism, fishing, port operations, public works, transportation, housing, and industry. Along many densely populated coastlines, population growth and unmanaged development projects, including residential and industrial urban development, are increasing the risks of natural disasters.

Tropical cyclones (hurricanes) mainly form over the western parts of warm oceans (at least 26°C) where there are no cold currents. Apart from the effects of wind and rain, storm surges and waves have a major impact and are responsible for major loss of life, particularly in low-lying densely populated coastal areas such as Bangladesh or China.

Tsunamis remain a serious threat to people living in the coastal zone, particularly in the Pacific Ocean. Originating from subsea tectonic movements, the danger of tsunami also exists in other parts of the world, like the Caribbean and even in the Mediterranean region. The tsunami warning system in the Pacific Ocean needs to be properly maintained. It is also highly desirable to increase the lead warning times, while minimizing the frequency of false alarms. Advances in earthquake prediction could have spin-off for tsunami prediction.

A more widespread coastal hazard is associated with sea-level rise, with the resulting increased coastal erosion and threat to people and property (box 3.4). The Intergovernmental Panel on Climate Change (IPCC) has largely driven international approaches to coastal vulnerability assessment through its Coastal Zone Management Subgroup. IPCC sees coastal zone management planning as a prerequisite for sea-level rise mitigation policies and recommended that all coastal nations should implement integrated coastal zone management as soon as possible. The latest IPCC "best

> **Box 3.4.** The Caspian Sea Problem
> *Vladimir E. Ryabinin*
>
> The Caspian Sea presents a unique problem of coastal zone inundation, related to an unexpected and unparalleled (in its rate) mean sea level rise. The actual rise in twenty years reached more than 2.5 m—at least an order of magnitude faster than the projected sea level rise of the world's oceans.
>
> Urgent response measures were needed to cope with the resulting inundation of vast territories, increased pollution, and more catastrophic storm surges. Research carried out by Russia's Hydrometcentre predicted an end to the rise in mean sea level. The causes of the rise were attributed primarily to changes in the atmospheric circulation and corresponding variations of precipitation.
>
> The terrible consequences of this large-scale environmental disaster did, however, provide a valuable scenario in the development of coastal management decision support systems.

estimate" prediction based on doubling of CO_2 is that sea level will rise by 49 cm in the next 100 years (IPCC 2001).

In 1998, IPCC highlighted the vulnerability of small island states to climate change, but stressed the need to consider other factors that contribute to their overall vulnerability (IPCC 1998). Hence, there is a need for a more integrated approach.

Because natural hazards are so important for the dynamics of global change in coastal zones, the physical and social impacts and elements must also be taken into account. Extreme events can bring about the greatest periodic changes in natural systems. These then act as threshold changes within ecosystems, moving resilient systems to new equilibrium through nonmarginal changes in state.

A long-term temporal framework is needed when addressing the physical vulnerability of coastlines with respect to these hazards in order to separate "natural variation" from impacts due to human activities. This kind of assessment can take various forms, depending on the hazard being analyzed. Superimposed on this is the need to evaluate the contribution of human activities and settlement to changes in thresholds of driving processes on land and in the sea (table 3.1).

Table 3.1. The interaction of physical and social vulnerability to coastal hazards in a pressure-state-impacts-response framework. Note that the pressures are driven by global change processes or by development associated with land use change. Based on Adger 1996.

Pressure	State	Impact	Response
Typhoons	BASELINE PHYSICAL VULNERABILITY		
River flooding	Return period, topography, infrastructure	Flooding impacts, infrastructure damage, large scale sediment flows, threshold ecosystem effects	Policy response, zoning and planning, water resource and coastal infrastructure altering morphology
Tsunamis and earthquakes	BASELINE SOCIAL VULNERABILITY		
	Poverty, dependency on risky resources, distribution of resources, institutional adaptation, insurance market	Livelihood disruption, risk perception	Insurance claims, informal collective action, migration and other coping strategies

INCREASING HUMAN PRESENCE AND PRESSURES

Demographics

Already some 60 percent of the world's human population live close to the coast. And this proportion is expected to increase, along with growing urbanization, industrialization, and transportation, putting even greater pressure on the living and nonliving resources of the coastal ocean (table 3.2). Some of this migration toward the coast is temporary, albeit significant (Cook 1996). For example, the Mediterranean coastal zone, which has a population of about 130 million, swells to 230 million for most of the summer, increasing transportation and pollution problems.

There is a need for systemic studies of the ecosystems associated with large coastal urban agglomerations. Growth in the so-called megacities adds to a tendency of people to concentrate in the coastal zone anyway. Clearly, this extends the range of impacts on the marine environment beyond

Table 3.2. Large Coastal Cities with the Highest Animal Growth Rate

No.	City	Population (millions)	Annual Growth Rate (%)
1	Dhaka	5.88	6.2
2	Lagos	7.74	5.8
3	Karachi	7.96	4.7
4	Jakarta	9.29	4.4
5	Bombay	12.22	4.2
6	Istanbul	6.51	4.0

traditional sewage and waste, adding things like increased risk of disasters, excessive noise levels, and thermal pollution.

The tremendous population increase also puts a heavy burden on coastal zone management. The obvious global demand for proper guidelines to cope with these increasing pressures presents the science community with a major challenge, namely to supply scientific information on possible solutions, and on the predicted effects of the different measures.

River Runoff and Load

Human activities have greatly altered the flow of freshwater and the substances it contains to the coastal zone. In some arid regions, such as the Nile (Egypt) and Colorado River (USA/Mexico), where the availability of freshwater on land is a major factor limiting human activities, discharge has diminished to 10 percent or less of natural flow. In other regions the issue is the management of water, with the seasonal pattern of discharge having been greatly modified. Water loss and a change in seasonal discharge can have major impacts on coastal ecosystems. Human activities have also altered the patterns of sediment discharge (table 3.3). Human land use (especially agriculture) has increased erosion, with increased sediment delivery, while water reservoirs can have a contrary effect, by increasing the quantity of sediments that are trapped. Thus, some regions experience artificially elevated sediment discharge, and others experience a marked reduction. To an ecosystem acclimatized to a particular level of sediment load, change in either direction can be detrimental. For example, severe erosion

Table 3.3. Comparison of the Drainage Areas, the Annual Sediment Load, and Yield per Square Kilometer for Various Tropical Rivers

River	Area (10^6 km^2)	Load (10^6 ton/year-1)	Yield (ton km^2 year-1)
Yangtze	1.9	480	252
Mississippi	3.3	210	120
Ganges/Brahmaputra	1.48	2180	1670
Mekong	0.79	170	215
Fly	0.076	116	1500
Cimanuk	0.0036	15.7	6350

Adapted from Wolanski et al. 1996.

without sediment replacement may occur in systems (such as the Colorado River delta) that are poised to receive high sediment loads. In contrast, ecosystems such as coral reefs are generally acclimatized to low sediment discharge. Large amounts of sediment can bury or otherwise damage reefs. Meanwhile, an increase in the turbidity of coastal waters also affects the photosynthetic activity of macroalgae and coral reefs, reducing the range of their depth.

The full list of materials discharged by humans, via rivers to the coastal zone is very long, including nutrients, pesticides, herbicides, heavy metals, industrial chemicals, and so forth. Human activities have generally led to an increase in the discharges. Some countries have done better than others in effectively regulating and controlling them.

Groundwater Discharge into the Coastal Zone

Although not as obvious as river discharge, continental groundwater also discharges directly into the sea. Like surface water, groundwater flows down gradient, directly into the ocean wherever a coastal aquifer is connected to the sea. Artesian aquifers can also extend for considerable distances from shore, underneath the continental shelf. In some cases, these deeper aquifers may have fractures or other breaches in the overlying confining layers, allowing groundwater to flow into the sea.

Within the past decade scientists have begun to recognize that, at least in some cases, submarine groundwater discharge (SGD) may be both volumetrically and chemically important. Estimates of global SGD vary widely;

some are as high as 10 percent of the river flow, but most are considerably lower.

So far we have learned that direct groundwater flow into the ocean is characterized by high degrees of temporal and spatial variability, is often a composite of fresh groundwater and recirculated seawater, and occurs to some degree almost everywhere we look. Although there is no clear pattern as yet, we can safely conclude that SGD is an important pathway for nutrients and other chemicals into some coastal areas.

The marine community must join with hydrologists to address the SGD issues. Up to now, the compartmentalization of disciplines has impeded progress. Specifically, hydrologists, ecologists, and coastal oceanographers should work together to achieve a number of goals:

- Improve measurement techniques including natural geochemical tracers, thermal probes, automated seepage meters.
- Perform intercalibration experiments to compare all measurement and assessment techniques available. Such experiments should be multifaceted and should be repeated in several environments.
- Develop generic models that are useful for a particular "type" of shoreline. This way we can better extrapolate the more detailed estimates around the shorelines of the world with greater reliability.

Eutrophication

Eutrophication is the increase of nutritional resources to a particular water body and includes the supply of mineral nutrients (N, P, Si, trace elements) as well as organic carbon (Richardson and Jørgensen, 1996) (figure 3.3 and box 3.5). In the recent past, eutrophication of coastal waters has been most pronounced in industrialized countries (NAS 2000). But it will undoubtedly become increasingly important in the developing countries of Asia, Africa, and Latin America in the near future (Nixon 1995). This anthropogenically enhanced eutrophication is particularly a reflection of the growing use of artificial fertilizers as well as discharges of sewage from growing coastal populations.

For management purposes, the focus should be placed on fundamental research leading to quantification of the relations between nutrient load, retention time, and ecosystem response. This will make it possible to establish efficient water quality measures and cost-benefit analyses, with the ultimate goal of sustainable exploitation of coastal areas. These measures are likely to vary from region to region.

Figure 3.3. Foam from decaying algae and protozoa, mainly *Phaeocystis* and *Noctiluca*, on the beach, thought to be the result of eutrophication. Supplied by Roel Riegman.

Box 3.5. Nitrogen, Phosphates, and the Eutrophication of Coastal Waters
Han Lindeboom

Nitrogen is of paramount importance in causing and controlling eutrophication in coastal waters, and problems are likely to increase as the use of fertilizers increases (NAS 2000). Nitrogen is widely applied in artificial fertilizers to stimulate crop growth on farms. But when too much fertilizer is applied, the excess will be washed away by runoff when it rains. The global production of nitrogen is keeping step with global population growth, mainly because of increased application of nitrogenous fertilizer and the burning of fossil fuel. It is now predicted that, by 2050, the global export of anthropogenic inorganic nitrogen to the coastal zone by rivers will be more than double the 1990 levels. Anthropogenic inorganic nitrogen reaching the ocean through the atmosphere has also increased significantly and already exceeds the natural nitrogen supply in the North Atlantic Ocean basin (e.g., see Howarth 1996). This increased supply of nitrogen is expected to affect biogeochemical cycles on a global scale. At present, the interactive effects on the carbon cycle of anthropogenically induced changes in the nitrogen cycle (including the dynamics of greenhouse gases within both cycles) are poorly understood.

Meanwhile, a turning point seems to have been reached with respect to phosphate discharges. Technically, it is cheaper and easier to reduce the phosphorus discharge in

(continues)

Box 3.5. Continued

coastal areas, since this element originates mainly from point sources like sewage plants. Through political consensus, an increase in sewage treatment in wastewater plants has reduced the anthropogenic phosphate discharges of various European countries by more than 50 percent.

The effects of increasing the nutrient supply vary according to the degree of eutrophication. At moderate levels, these can range from increased growth of phytoplankton, benthos, and fish to changed species composition. When eutrophication is severe, the effects include blooms of nuisance-causing or toxic algae to mass growth of certain species and mortality of others. Where the algae consume all the oxygen, this leads to anoxic conditions and mass mortality. More severe conditions may be expected in enclosed bodies of water, where the flushing time (i.e., resupply of oxygen) is low. But there can be widespread development of anoxic bottom waters in the open ocean under certain conditions. This is the case, for example, over much of the continental shelf off the mouth of the Mississippi River, where eutrophication appears to be related to the widespread use of fertilizers in the North American hinterland up to several thousand kilometers away (Goolsby 2000; Rabalais et al. 1999).

Phenomena induced by algae, such as oxygen depletion of the water column and consequent mortality of animals, can be prevented by a general reduction of nutrient discharges. At present, numerical models on ecosystem dynamics are reliable enough to estimate dose-effect relationships. However, the residence time of the nutrients and the role of sediments as temporary sources or sinks on local and regional scales remain important topics for further research.

Changes in N:P:Si ratios may also cause a shift in species composition. Consequently, increased supplies of nitrogen are likely to alter pelagic and benthic communities. Local environmental conditions have a major influence on which species are stimulated by eutrophication. In some instances, nitrogen supply may stimulate the production of harmful algal species. Although there is general consensus that there is a global increase in harmful algal blooms, it is not yet clear whether there is a direct and simple link between these blooms and nutrient supply.

Pollution in Relation to Water Quality

Until the early 1990s, organic-pollution attention focused on the occurrence of classical organochlorine compounds such as PCBs, DDTs, the cyclodienes, and hexachlorobenzene (HCB). This was partly due to anticipated problems, but also to the limitation of analytical facilities in many laboratories to gas chromatography with electron capture detection.

In the marine environment, the acute toxicity of compounds is only relevant after accidental spills. The only important exceptions to this rule are the bird casualties following operational spills of oil and surfactants from shipping. From a scientific point of view, it is more important to focus on chronic exposure to relatively low concentrations and their effects on reproduction, immunology, and carcinogenicity.

New classes of compounds should be identified on the basis of their predicted environmental concentration/no (observed) effect concentration (PEC/N(O)EC) ratios, bioaccumulative properties, persistence to environmental degradation processes, and known concentrations in the marine environment or their production figures.

Meanwhile, compounds used in alternative antifouling treatments are of special interest, since they are designed to display their toxic effects in the marine environment. Moreover, their use is expected to increase following an expected International Maritime Organisation (IMO) ban on the application of tributyltin-based antifoulings by 2003.

Another problem is the tendency of certain pollutants to concentrate somewhere in the ecosystem. For example, PCBs are concentrated in the colder parts of the world oceans and from there in the fatty tissues of marine mammals and seabirds. In this way, even very small amounts of pollutants discharged into our coastal waters may have a devastating effect in unexpected and often very remote places. Apart from being aware that new types of pollutants are created all the time, we need to be vigilant regarding this kind of problem. The group of substances that disrupt endocrine functions in aquatic animals is of great concern.

Basic research is required to strengthen the scientific foundation of risk assessment, for example, baseline studies on endocrine dysfunction across classes of animals. This will reduce the uncertainty associated with species extrapolations.

Environmental chemical and ecotoxicological research should expand its scope of interest to a number of compounds affecting the health of ecosystems during chronic exposure at low concentrations. Biomarkers of biological effects could be useful tools to identify such hazardous effects. It will be difficult to disentangle effects on populations from the high pressure of some other anthropogenic activities, such as fishing. New developments in analytical chemistry will often be necessary to achieve these goals.

FOOD PRODUCTION

Aquaculture

Aquaculture has grown substantially over the past three decades. According to Food and Agricultural Organization statistics, overall production figures were around 7.5 million tons in the early 1970s and have reached around 37 million tons by the beginning of this millennium. Shrimp farming and other marine aquaculture affect coastal systems, whether on land (shrimp ponds), in wetlands (tambaks), or the sea (salmon cages or artificial reefs).

While most finfish production is in freshwater, a growing proportion of food, including seaweed, shellfish, and crustaceans, originates from coastal brackish and marine farming systems. And while finfish production usually involves fish species low in the food chain, a growing proportion comes from high-quality products from species with higher trophic levels.

As the industry has grown, so have the resulting environmental problems. Major advances have been made in the past decade to cope with many of these environmental problems. These include improved site selection criteria and husbandry techniques (e.g., better stress management and improved disease control). With the development of vaccines for some key diseases it is now possible to avoid the use of antimicrobials in modern farming systems to a great extent. Finally, environmental assessment methodologies have been developed that include predictive models of benthic impacts of finfish cage farms and carrying capacity in coastal shellfish farming. These take the assimilative capacity (finfish) and the productivity (shellfish) of a coastal area (bay or inlet) into account.

A major problem remains the "energy efficiency" of aquaculture. Protein consumption is often much higher than the amount of protein produced, even when the added value is taken into account. And production of "fourth trophic level fish" may not be very sustainable. But aquaculture will continue to grow, with overall production figures for finfish, invertebrates, and marine algae predicted to account for about half of the world fishery production by the year 2020.

In order to achieve this overall production, research needs will be dictated by growing competition for resources from other users of the coastal zone, especially the availability of suitable sites, nutrients (new feeds), energy, and space. With the growth of the industry new environmental concerns will arise which have not yet been addressed and are in need of urgent attention. These include:

- better assessment methodologies on the impacts of aquaculture on regional and global scales (from both an environmental and economic viewpoint).
- improved management and system efficiencies in several branches of the aquaculture industry (e.g., replacement of fish meal in fish feeds for carnivore species, better understanding of the interaction between system design and the behavior of cultured species [stress reduction], better product quality control, etc.).
- multidisciplinary studies to improve understanding of the impact of other water resource users on aquaculture performance in coastal areas. Apart from tourism, there is also the effect of ballast water released from shipping, with the unintentional transmission of exotic species and/or disease agents. These can have devastating effects on coastal culture systems.
- interactions between aquaculture and other stakeholders around inland backwaters and estuarine systems and marine coastal waters (e.g., multiple resource uses, co-culture and integrated farming systems, ranching, and interaction between aquaculture and fisheries).

If aquaculture expansion is not properly controlled, there may be severe long-term consequences. Many of the environmental implications are well known (Clark 1996): preemption of critical fishery habitat (e.g., mangroves); pollution of lagoon waters; excessive exploitation of natural stocks of larvae and juveniles; imports of inferior, sick, or noncompatible seedstock; and introduction of exotic species, parasites, and diseases.

Other research priorities need to address problems associated with new species (product diversification), new vaccines, new methods of disease control, and the socioeconomic aspects. Integrated farming systems in marine coastal habitats also need to be studied, where waste from one resource user becomes a new resource for others. Finally, aquaculture may serve as a model to demonstrate the usefulness of expert systems to support integrated coastal zone management policies.

Fisheries

Fishing is often an important economic activity in the coastal zone. But the potentially high benefits come with equally high risks of ecological deterioration through overfishing, killing of unwanted bycatch, and habitat destruction. These risks are described in detail in chapter 5. It is important to integrate fisheries into coastal zone management, with the aim of ensuring sustainable yields, while protecting vulnerable areas and species. More research is needed into the long-term impacts of fisheries and their effect on populations of target and nontarget species—indeed, on the entire ecosystem of the coastal zone.

RECREATION AND TOURISM

Beaches, Swimming, and Recreational Boating

Although beaches may be important areas for tourism, the increase in population pushes many areas to their sustainable limits. There are clear feedback mechanisms in beach tourism: nice beaches attract people. But when the number of tourists is too high, a beach loses its attractiveness. And while good beaches are worth billions of tourist dollars, degraded beaches are worth little (Clark 1996). Without controls on the scale and type of activities, tourism can have negative impacts, including a reduction in biodiversity, depletion of resources, and human health problems. One management measure is to set a maximum limit on tourist numbers. But, in practice, economic pressures often mean that the restrictions are relaxed once these limits are reached. Scientific methods are needed to calculate and define clear maximum limits.

Recreational boating also increases with affluence. In some countries, harbors and marinas built for pleasure craft may disturb more of the coastal zone than commercial and industrial use. Their environmental impacts depend on site location, design, construction methods, and "housekeeping." Careful site planning can help avoid or minimize many of the impacts of marinas.

Ecotourism

In recent years a new kind of tourism has evolved around opportunities to observe wildlife in its natural environment. The presence of seabirds and marine mammals, particularly cetaceans, in coastal areas offers excellent opportunities for this so-called ecotourism (figure 3.4). Spectacular seabird colonies and seal rookeries are becoming increasingly popular tourist destinations. Similarly, tourist agencies organize whale-watching trips, giving advice on where and how to observe whales from headlands and coastal promontories (Taylor 1988). But ecotourism has also been creating some concern (Coultier 1984; Woehler et al. 1994). As a result, codes of ethics and best practice guidelines for ecotourism have been published, with most of the major tourist organizations formally agreeing to follow them.

Figure 3.4. Too close? A tourist too near a walrus. Supplied by Kees Camphuysen.

Tourists should be given the opportunity to enjoy and learn from visits to seabird colonies, seal rookeries, beaches with egg-laying turtles, coral reefs, and whale-watching trips, etc. But it is in the long-term commercial interests of both tour operators and managers of nature reserves to develop and implement long-term management plans to prevent deterioration of these ecosystems. Ecotourism should strengthen conservation efforts and enhance the natural integrity of places visited, not the reverse.

SUSTAINABLE DEVELOPMENT OF THE COASTAL ZONE

Coastal Protection and Engineering

Severe beach erosion is a problem shared by many countries (Clark 1996). Structural solutions to beach erosion and protection of shoreline property from hazards of sea storms may be expensive and are often temporary or counterproductive. Groins, seawalls, breakwaters, dikes, and other popular protection structures often have complex and unanticipated secondary effects. These can result in major downstream erosion and, quite often, total loss of beach. For example, on the Senegal coast, a dike increased erosion so much that two roads recorded on maps of the early twentieth century have disappeared today.

The Dutch case study (see box 3.3) gives some new perspectives for future management.

MAMMALS AND BIRDS

The practice of nature conservation has developed during the twentieth century, particularly in industrialized countries. As a result, the (over)exploitation of seabirds and marine mammals, even of species that were generally considered as pests (e.g., cormorants, gulls, and seals), largely came to a halt. Most commercial whaling operations ceased in the 1960s and 1970s, and the implementation of adequate measures to protect coastal wildlife, such as seabirds, was successful. The populations of many seabirds have increased markedly, and the first signs of a recovery of whale populations are becoming apparent. Some authors have attributed the dramatic increase in seabird populations to a drop in the rates of persecution during the twentieth century (Camphuysen and Garthe 1999). But a careful examination of patterns and trends in many species often shows that this explanation is not sufficient.

Another explanation is the availability of new sources of food as a byproduct of human activities (e.g., fishery waste and garbage). We do not yet know enough about the feeding ecology of seabirds to predict what would happen if this additional source of food was reduced or removed. The effect of fishing on populations of predator species (e.g., seabirds) is still difficult to predict. We know quite a lot about the numbers, distribution, and energetic requirements of seabirds and marine mammals, but much less about their foraging strategies, aggregative and functional responses, and factors influencing prey availability. Any change in fishery policy could therefore have unexpected and perhaps undesirable side effects.

HARMFUL ALGAE

Proliferation of microalgae in marine or brackish waters can contaminate seafood with toxins, killing large numbers of fish and altering ecosystems in ways that humans perceive as harmful. The scientific community refers to these events with a generic term, harmful algal bloom (HAB). Nevertheless, because a wide range of organisms is involved and some species have toxic effects at low cell densities, not all HABs are "algal" and not all occur as "blooms." Referring to modern systematics, "algae" and "phytoplankton" are colloquial terms, not well defined or natural groupings. The only common feature of HAB organisms is that they belong to the kingdom of protists. A broad classification of HABs distinguishes two groups of organisms: the toxin producers, which can contaminate seafood or kill fish, and the high-biomass producers, which can cause anoxia and indiscriminate kills of marine life after reaching dense concentrations. Some HABs have characteristics of both.

Although HABs occurred long before human activities began to transform coastal ecosystems, a survey of affected regions and of economic losses and human poisonings throughout the world demonstrates that there has been a dramatic increase in their impact of over the last few decades. The HAB problem is now both widespread and serious. And the harmful effects of HABs extend well beyond direct economic losses and impact on human health. When HABs contaminate or destroy coastal resources, the livelihoods and food supply of local residents are threatened.

The HAB problem raises two kinds of questions: one practical, the other scientific. What can be done about HABs in a practical sense? What information is needed to manage affected marine resources efficiently, protect the health of the ecosystem and the human population, and contribute to policy decisions on coastal zone issues, such as waste or sewage disposal, aquaculture development, or dredging? Are human activities making the HAB problem worse? If so, what steps should be taken to minimize further impacts? Clearly, there is an increasing demand to develop effective responses to the threat of HABs through management and mitigation. This requires scientific investigation into the population dynamics of HAB species.

The major research questions include:

- What are the unique adaptations of HAB species that determine when and where they occur and the extent to which they produce harmful effects?
- How do HAB species and their community interactions respond to environmental forcing?
- What are the effects of human activities (e.g., eutrophication, transfer of organisms)?
- What are the effects of interannual and decadal climate variability (e.g., El Niño or the North Atlantic Oscillation) on the occurrence of HABs?
- What are the different toxins produced, and what is the effect of these toxins on marine species and/or humans?

INVASIVE SPECIES

Human activity has resulted in nonindigenous invasive species being introduced to marine ecosystems for centuries. A significant number of these species have now become established, particularly in estuarine port areas. Their impact ranges from beneficial to extremely destructive and costly (see box 7.3 on ballast water). It is often difficult to identify nonindigenous species, because the indigenous flora and fauna is frequently not well recorded. But there is some well-documented evidence. For example, in the United States, between 60 and 212 nonindigenous species have been found in each of the estuaries studied (Ruiz et al. 1997). In the Mediterranean, more than 240 nonindigenous species have been identified, with 75 percent attributed to shipping moving through the Suez Canal.

Some nonindigenous species have significant adverse effects on the ecosystem. For example, the introduction of the Asian clam into San Francisco Bay has resulted in clam densities of up to 10,000 m^2. This has resulted in the displacement of benthic organisms and a decrease in the planktonic communities in the bay (Cloern 1996; Nichols et al. 1990). The indirect effects on the ecosystem include impacts on the food web structure, nutrient cycling, and sedimentation rates. The introduction of the American comb jelly into the Black and Azov Seas contributed to the collapse of the fisheries (Harbison and Volovick 1994). The introduction of some nonindigenous species can also have implications for human health. For example, the increasing frequency of red tides that threaten both shellfish aquaculture and human health are partly due to the transfer of dinoflagellates and dinoflagellate cysts in ballast water (Hallegraeff 1993; Hallegraeff and Bolch 1991). There is also recent evidence that cholera has been transported via ballast water in South and Central America (Ruiz et al. 1997).

There are a number of ways in which humans mediate the transfer of nonindigenous invasive species into a new ecosystem. These include transport on vessel hulls, intentional and unintentional release of aquaculture, bait and biomanagement species, movement via connecting waterways, and movement via ship ballast water. The most likely method of introduction of nonindigenous species today is via ship ballast water. An estimated 10 billion tons of ballast water are transported around the world annually (Stewart 1991). Recent surveys in Australia, Canada, Germany, the United Kingdom, and the United States show that about 500 different species are known to have been transported by ballast water.

We are currently unable to predict when and where a nonindigenous species will become established. Eradication of an introduced species is also very expensive or impossible. So it is very important to prevent introduction in the first place. Some countries have recently introduced both voluntary and mandatory mid-ocean ballast water exchange requirements. If properly enforced, this practice should lead to a decrease in the number of potential introductions into coastal waters while methods of ballast water treatment such as heating, filtering, and chemical treatment are investigated.

INSTRUMENTATION

Special Instruments and Equipment Needs

Chapter 9 reviews the major groups of instruments used to monitor changes of various chemical, physical, meteorological, hydrological, and biological parameters. The following groups of technology are of special interest for research in the coastal zone:

- Shore stations, including tide gauges and coastal radar
- Moored and drifting buoys
- Biological sensors
- Data management technology, including data assimilation
- Acoustic tomography and acoustic thermometry
- Acoustic listening arrays
- Satellite technology
- Airborne remote sensing
- Ship-borne instrument packages

Special equipment is needed in order to produce a reliable, accurate, long-term data stream for detecting environmental change in the coastal zone. This data stream should be obtained with the absolute minimum of personnel needed to go to sea. Moored buoys serving as platforms for various types of sensor packages and as transmitter stations have to be established at crucial points along the coast and in front of estuaries.

The development of the Global Ocean Observing System (GOOS) and the emerging availability of real-time data from the open ocean make it both feasible and desirable to develop a set of models of the coastal zone, which would benefit from the use of open ocean boundary conditions. In doing so, fully coupled models of all relevant physical processes (circulation, sea ice, wind waves, tides, storm surges, mixing, pollution transport, etc.) should become routine tools. In addition, coupled hydrodynamic-biological-geochemical models of a set of useful spatial-temporal scales should be developed by the year 2020. Assimilation of observations into the models should become a routine procedure as well.

COASTAL MANAGEMENT TOOLS

Coastal management as a means of sustainable co-evolution of environmental and socioeconomic systems mainly relies on the underpinning multidisciplinary science. Scientific programs, such as the International Geosphere-Biosphere Global Change Research Program (IGBP) core projects, Joint Global Ocean Flux Study (JGOFS), Global Ocean Ecosystem Dynamics program (GLOBEC), or Land-Ocean Interactions in the Coastal Zone (LOICZ), continue to provide not only data but information (i.e., data processed against various questions) and tools for application in modeling and monitoring. However, the key feature needed here is to take an issue-driven approach that enables appropriate scaling. Special attention needs to be paid to reconciling the various societal, economic, administrative, and (inter)national scales of certain coastal change regimes with the natural—that is, environmental or habitat—scales. Examples are the LOICZ typology approach, the Driver, Pressures, State Impact and Response approach (DPSIR), and the Sim Coast (see McGlade 1999) approach.

Two instrumental levels are needed for the successful implementation of coastal zone management, namely legal instruments on the one hand and capacity building (both human and institutional) on the other. Scialabba (1998) mentions the following legal principles:

- The precautionary principle
- The principle of preventive action
- The "polluter pays" principle
- The principle to avoid transboundary environmental damage
- The principle of rational and equitable use of natural resources
- The principle of public involvement

They all need careful consideration and employment in horizontal (e.g., cross-sector) and vertical coastal management efforts (i.e., policymaking from regional through national to smaller administrative scales). These underpin the need to charge science with a serious brokering function in participatory processes. They also reiterate the frequently mentioned need for capacity building. This has to include institutions, scientific and nonscientific bodies, and local people.

The key instrument to be employed here is most likely to be science communication, addressing the various levels of understanding and information needs through a common language. This will help to generate joint ownership in issues and actions by providing mentor links between the different coastal stakeholders and the science community. It will also facilitate the process of joint development and use of market and nonmarket instruments, while creating incentives to regulate the exploitation of systems toward sustainability criteria if this is needed.

CONCLUSIONS AND RECOMMENDATIONS

Major Management Issues

- A major issue in future coastal management is the perception of the coastal zone by the public, politicians, managers, and scientists. Environmental management is not managing "nature"; it is about managing people and their actions. Therefore, socioeconomic research should become an integral part of coastal sciences.
- A serious problem is the invisibility of most of the underwater parts of marine ecosystems, making it very difficult to convince politicians that something is wrong and should be changed. Science should support management by developing clear visions about the entire ecosystem, its natural variability, the boundary conditions that sustain the system, and the managerial possibilities. It is vitally important to inform and educate politicians and the public about what is known and where the uncertainties lie.
- Clear definitions of sustainable use and goals for nature conservation are needed. Marine protected areas (MPAs), also in open seas, are an essential part of nature conservation. And when allowing human uses that have expected environmental effects, a precautionary principle should be applied. Science has the duty to identify these effects and their long-term consequences.
- One of the keys to successful management of sustainable exploitation is controlling the exploiters in such a way as to maximize yield and to minimize impacts. An accessible global library of case studies with success stories and failures needs to be established. It is important to include in these case study descriptions, the social, cultural, political, and economic framework under which they were implemented. A unified DPSIR may facilitate the exchange of crucial information.
- For Integrated Coastal Zone Management (ICZM) the following major objectives have been identified (after Clark 1996): maintain a high quality coastal environment, protect species diversity, identify and conserve critical habitats, enhance critical ecological processes, control pollution, identify lands for development, protect against natural hazards, restore damaged ecosystems, encourage participation, and provide planning and development guidance.
- Increasing threats to the coastal ecosystem include harmful algal blooms, overfishing and damage by fishing gear, destruction by tourist activities, shore construction, weather hazards, eutrophication, and chronic exposure to toxins, including endocrine disrupters and invasive species. Scientifically sound countermeasures need to be developed and advocated.
- The conservation of biological diversity is an urgent coastal matter. Thousands of species and subspecies of wild plants are threatened with extinction. The most serious threat is habitat destruction. Further protection of special or crit-

ical littoral habitats, including mangrove forests, coral reefs, kelp beds, seagrass meadows, salt marshes, brackish lagoons, beaches, and mudflats, should be high on the international agenda. At the same time, the introduction of non-indigenous invasive species is a significant threat to the integrity of marine ecosystems. A high priority should also be given to restoring critical coastal habitats and avoiding unwanted introductions of species.
- The complexity of coastal ecosystems and the interactions between all their components require that each coastal ecosystem be managed as a system. The challenge for science is to define and understand the natural and socioeconomic boundaries and limits of these systems.

Scientific Priorities

- A major priority for research in many ecosystems should be the description and measurement of their natural variability. This is needed to set (variable) exploitation limits and to provide the framework for understanding human-induced changes. Collection of long-term data series and proper monitoring programs are prerequisites.
- The parameters selected for monitoring should provide insight into the variability of both abiotic driving forces and the key biotic components. Development in time and space needs to be recorded, while remembering that marine organisms do not stay behind human-drawn boundaries. There is a need for scientifically sound and cost-effective data collection, storage, and analyses programs, which are linked and coordinated via programs like GOOS.
- Because there have only been moderate changes in climate over the past 150 years, the observational and instrumental record is much too short to detect the whole range of natural variability of the systems involved. Proxies should be developed to generate long-term series of environmental change data, using sediment cores raised from coastal zones and the shelves, as well as coral reef drillings.
- The possible direct effect of increasing global CO_2 concentrations and increased ultraviolet flux on calcification of coral reefs needs urgent attention.
- Submarine groundwater discharges are often overlooked. There is a need to improve measurement techniques, intercalibration experiments, and generic model development.
- The size, distribution, and genetic composition or variability of populations with high environmental and/or ecological value needs further research. Changes in biodiversity, their causes and possible consequences, should be addressed.
- For management and restoration of vulnerable areas like mangroves, coral

reefs, salt marshes, seagrass meadows, kelp beds, and estuaries, there is a need for in-depth knowledge of natural processes as well as of the relevant institutional, cultural, economic, social, and political frameworks, within a transdisciplinary concept. Suitable models should be developed and used to analyze the short- and long-term function of the ecosystem to forecast changes and to answer management-related questions, preferably in effective decision support systems. The development of dynamic models, taking account of the (natural) variability of the system and risk assessment should be encouraged. A global coastal typology approach may facilitate better access to all relevant information.
- A better understanding of sand transport processes and the further development of calibrated models is a prerequisite for future management of sandy beaches. The expected rise in sea level, climate changes, and subsidence put this issue high on the scientific agenda.
- Improved early warning systems need to be developed on the undesirable effects of eutrophication, toxic substances, fisheries, and invasive species.
- Politicians and managers expect clear, substantiated advice from scientists. At the same time, our understanding of the complex marine ecosystem is limited. In several cases, the credibility of scientists has been questioned, either because of a lack of useful advice or a lack of scientific support. An active communication and education program must be incorporated with scientific involvement in the coastal zone.
- There is a need to improve research and monitoring capabilities, along with continuous management advice in many parts of the world. Assistance and partnership with advanced laboratories are essential for developing countries on their (long) way toward sustainable management of their coastal systems and resources (see also chapter 11 on capacity building).
- There is increasing tension between applied and fundamental research. In many projects there is a request for applicable results. But direct applicability and academic output seems to be in contradiction. Perhaps a concept of "usable" science can be a way to bridge applied and basic science. To avoid disappointment, clear definitions of objectives, expectations, and products should be agreed upon before starting coastal research projects.

ENDNOTES FOR FURTHER READING

Alongi, D. M. 1998. *Coastal Ecosystem Processes*. London: CRC Press.

Bergman, M. J. N., and H. J. Lindeboom. 1999. Natural variability and the effects of fisheries in the North Sea: Towards an integrated fisheries and ecosystem management. In J. S. Gray et al. (eds.), *Biogeochemical Cycling and Sediment Ecology*, 173–84. Amsterdam: Kluwer Academic Publishers.

Brink, K. H., and A. R. Robinson, eds. 1998. *The Global Coastal Ocean: Processes and Methods. The Sea, Volume 10.* Chichester: Wiley.

Cicin-Sain, B., and R. W. Knecht. 1998. *Integrated Coastal Zone Management: Concepts and Practices.* Washington, D.C.: Island Press.

Mann, K. H. 2000. *Ecology of Coastal Waters, with Implications for Management.* 2nd ed. Malden, Mass.: Blackwell Science.

Chapter 4

Climate Change and the Ocean

Gerald R. North and Robert A. Duce

The oceans are an integral part of the global climate system. Other parts include the atmosphere, the land surface, the biosphere, and the cryosphere (sea- and land-based ice). Of these components, the atmosphere changes most rapidly, bringing the weather patterns that pass overhead on a daily basis. But averaged over months the slow changes in the ocean's distribution of surface temperatures systematically affect weather patterns over distant parts of the world. The oceans not only affect our weather over periods of a few months, but they also have an intricate two-way relationship with the atmosphere as well as with the other components that affect climate in the long term.

At present we do not fully understand all of the interactions and exchanges between the climate system components, making the relationship between oceanography and the climate system a very active field of research. The observational and modeling difficulties associated with this research are such that oceanography and its application to global climate will continue to be of great interest through much of the twenty-first century.

In this chapter we discuss some of the important climate-related issues in oceanography. First, we present a short description of the climate system and then clarify the distinction between "natural variability" and "forced climate change." We then list a few of the important areas of current research that illustrate the great variety of ways in which the ocean participates in climate change through interactions and exchanges of heat, momentum, and chemical species. For instance, sea ice areas shrink as the climate warms, but these changes in themselves can amplify climate change through a feedback mechanism. Three other important mechanisms are discussed along similar lines:

Box 4.1. The World Ocean Circulation Experiment and the Ocean's Role in Climate

W. John Gould

The oceans play a key role in maintaining Earth's climate. They cover 70 percent of its surface, contain 96 percent of its water, and have physical properties that profoundly influence their behavior. Only in the past twenty years have we been able to explore this influence.

In 1968, the SeaSat satellite showed that surface currents could be monitored by measuring sea surface slopes and that scatterometers could measure surface winds. In the 1990s, a new generation of ocean-observing satellites exploited these techniques. To complement these satellite measurements, the World Ocean Circulation Experiment (WOCE) used ship-based observations and computer models to quantify the oceans' role in climate (figure BX4.1).

WOCE was a huge undertaking (costing between $0.5 and $1.0 billion). It involved thirty countries over twenty years, including intensive observations between 1990 and 1998.

WOCE collected a global set of temperature, salinity, and chemical tracer data (including measurements of CO_2) of unprecedented scope and accuracy. It revealed that over tens of years temperatures and salinities throughout the ocean change by significant amounts. These changes record past atmosphere and ocean interactions and provide clues as to how future climate will influence the oceans.

The satellite altimeters mapped the position and strength of currents both seasonally and from year to year. They also showed the seasonal monitored heating and cooling as the seawater expanded and contracted.

Tracers such as chlorofluorocarbons (CFCs) and the tritium produced by bomb tests

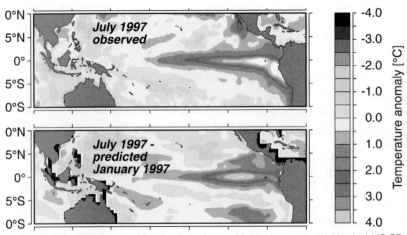

Figure BX4.1. Positions of full-depth hydrographic sections occupied in the WOCE one-time survey (1990–98). Courtesy of WOCE International Project Office.

enabled the measurement of the rate at which water sinking due to cooling near the poles moves through the abyssal ocean. This is essential in understanding how the climate will react to increased levels of CO_2. The mountainous mid-ocean ridges have been found to be important sites for mixing this cold deep water back to the surface.

Ocean models are now much better than they were before WOCE at representing both the small-scale eddies in the ocean and the oceans' transports of heat and freshwater. This improvement has come from using observations to identify defects in the models.

WOCE developed a new instrument that could measure currents below the surface (down to 2 km) and also measure and transmit profiles of temperature and salinity via satellite. The technique is now the basis of a global observing system that will be implemented starting in 2001.

The benefits of sustained ocean observations and the power of models were clearly demonstrated by the prediction of the onset and evolution of the 1997–98 El Niño event. Much remains to be done to assimilate satellite and in situ data into models for use in climate research and prediction. A new research program on Climate Variability and Predictability (CLIVAR) will continue the research that led to the El Niño (ENSO) prediction and that from WOCE. The development of an operational Global Ocean Observing System (GOOS) will enable future changes in the ocean to be speedily detected and their impacts assessed. The scientific challenge remains to understand fully the oceans' role in the ocean-atmosphere system and to assess the extent to which climate is predictable on time scales ranging from seasons to centuries.

More information can be found at the following sites:
http://www.soc.soton.ac.uk/OTHERS/woceipo/ipo.html
http://www.soc.soton.ac.uk/OTHERS/ICPO

sea level, the thermohaline circulation, and the carbon cycle. In a third short section we briefly mention some problems in the important field of paleoceanography, the study of past ocean circulation. Box 4.1 describes the World Ocean Circulation Experiment and the ocean's role in climate, and box 4.2 homes in on the role of circulation in the South Atlantic in controlling climate at the regional scale.

NATURAL VARIABILITY OF WEATHER, CLIMATE, AND THE OCEAN

It is virtually impossible to predict atmospheric motions beyond a few weeks. Instabilities in these motions are such that small errors (for example, due to gaps in the observed initial state specification) grow rapidly and

Box 4.2. The South Atlantic and Climate
Ilana Wainer and Edmo Campos

Variations of the South Atlantic Ocean circulation patterns and sea surface temperature (SST) can occur over time scales ranging from subseasonal to seasonal and interannual. These variations are thought to be strongly influenced by interactions between the opposing flows of the Brazil Current and the Malvinas Current, which in turn are affected by the basin-scale wind field and other atmospheric features. During the past few decades, climatic variations have had an important economic and social impact on the region (Garzoli et al. 1996). Drought periods have produced changes in cattle population, drained the water supplies of large cities, and caused shortages of hydroelectric power. Furthermore, a westward shift in rainfall that occurred during the 1970s has been related to a significant expansion of farming in the south of Argentina, Uruguay, and southern Brazil (Garzoli et al. 1996). Though the mechanisms behind these climatic fluctuations remain unclear, the few existing observations point to relations between climate variability and the large SST changes observed in the open ocean.

The most dramatic contrasts in SST of the entire South Atlantic occurs at its western boundary when the warm and salty waters of the southward flowing Brazil Current meet the colder fresh waters of the northward flowing Malvinas Current (see figure BX4.2 in the color section), where temperature gradients are as high as 1°C per 100 m (Campos et al. 2001). The confluence zone between these two currents migrates up and down the continental margin at seasonal, interannual, and possibly longer time scales, which in turn impacts on the atmosphere, with likely effects on cyclogenesis and regional rainfall distribution. Indeed, the links between SST anomalies in the southwestern Atlantic Ocean and rainfall anomalies in the entire region are more pronounced during October–December and April–July. The fact that the years that show positive peaks in these links do not necessarily coincide with those obtained for the Pacific Ocean indicates that the SST anomalies in the Atlantic may also contribute to rainfall anomalies over Uruguay and southern Brazil.

There is significant evidence for the importance of monitoring South Atlantic SST anomalies for regional climate and weather prediction in southern South America. But much work is still needed to understand the details of how these SST anomalies (sometimes short-lived) affect rainfall regimes, weather patterns, and how (or if) these, in turn, will modify the SST distribution.

propagate from small scales to planetary scales. Beyond just a few weeks, weather statistics (i.e., climatology) provide the best forecasts. These motion instabilities are inherent properties of the nonlinear form of the governing equations of moving fluids. The same instabilities lead to the phenomenon known as chaos—highly irregular motions even in the presence of fixed external conditions. For example, no stirring is needed to generate these irregular motions or eddies. The upshot is that we expect weather to fluctuate even though the external conditions are fixed. Large anomalies are expected with certain recurrence frequencies, although precise timing is not predictable (Lorenz 1968, 1969).

If the atmosphere is taken as our system and the ocean is considered as external, providing only a lower boundary condition, we can imagine a climatology (weather statistics) for the distribution of fixed sea surface temperature (SST). An atmospheric general circulation model (AGCM) might generate such an imaginary climatology. If the SST pattern is altered, the statistics of the atmospheric variables and patterns will be affected even far away from the changes because of large-scale wave structures in the atmosphere. In this limited framework we can consider that the shifts of the SST field force a climate change on the atmospheric circulation system. The irregular El Niño cycle (often denoted ENSO, for El Niño–Southern Oscillation) is just such a phenomenon and is discussed next. When we take the ocean-atmosphere system as a single system, we have an ocean-atmosphere climatology. Considered in this form, the El Niño cycle is just part of the natural variability of the coupled system. In addition to the El Niño example, natural variability of the ocean-atmosphere system occurs on all time scales from thousands of years to a few days.

Atmospheric Climate and Sea Surface Temperature

The most studied connection between atmospheric circulation patterns and SST is the El Niño cycle. The normal state is one in which the waters off the coast of Peru are cold because of upwelling along the eastern boundary of the Pacific. These upwelling waters bring nutrients from below, favoring a variety of marine life—and the local fishing industry. But from time to time warm water poor in nutrients replaces the cold upwelling waters. Since a smaller version of this phenomenon occurs near Christmas every year, the local inhabitants of South America have dubbed this elimination of the cold upwelling *El Niño* (the boy child, or infant Jesus).

Through improved observing systems and computer modeling over a period of twenty years, the mystery of El Niño is gradually being solved. It appears that the phenomenon is a cooperative oscillation between the ocean and atmosphere dynamic systems. In their normal state, the easterly trade winds in this region tend to push surface waters into the western part of the Pacific, raising surface height by roughly a meter compared to that on the eastern boundary of the basin. As this accumulation occurs in the west, upwelling of cold, nutrient-rich waters occurs naturally and prolifically in the east. Eventually, a threshold is crossed and the buildup in the west collapses, bringing a slow wave-like procession of raised surface height (a few tens of centimeters) across the Pacific from west to east. This wave train works its way across in a matter of months, depressing the thermocline and eventually covering the cold upwelling water along the Eastern margin, leading to the demise of fishing. Where upwelling continues, the thermocline is sufficiently deep that only warm nutrient-poor water reaches the surface. This picture is a very simplified version of what really happens in this very complicated dynamic system. The same pattern rarely happens twice, and the magnitudes of the events as well as their precise space and time distributions are quite variable from one event to another (Schopf and Suarez 1990).

As an El Niño event proceeds, the huge atmospheric convection centers and associated heavy rainfall shift from their normal location in the East Indies out into the middle of the Pacific Ocean, near the date line. The normally perennially wet regions such as Borneo, Indonesia, and even Northern Australia begin to experience drought. The convective cell known as the Walker Circulation shifts dramatically to the east. The shift of convection and precipitation from the East Indies into the central tropical Pacific has huge energetic implications for the circulation of the global atmosphere. The precipitation now occurring in the mid-Pacific means that latent heat is released in the middle troposphere along the equator near the date line, rather than much farther to the west. The formation of cloud and raindrops releases huge quantities of latent heat several kilometers above the surface. This in turn excites long, stationary waves along great circular paths that pass over North America, guiding the storm tracks, especially in northern winter. The patterns of the storm tracks make a dramatic shift as the El Niño proceeds through its cycle. Storm track predictability in North America out to several months is often possible with the unfolding of an ENSO event.

The Climate Prediction Center of the U.S. National Oceanic and Atmospheric Administration (NOAA) makes routine weather and climate outlooks for the United States. During the El Niño that began in late 1997 and lasted into 1998, the center was able to take the SST and run computer forecasts using long-range weather prediction models. By running an ensemble of such forecasts with slightly different (but observationally indistinguishable) initial conditions they were able to make very good climatological forecasts for monthly average precipitation and temperature over the United States. The results were spectacular, considering that nothing of the sort had ever been done before. This was a triumph for the weather models. It appears then, that there is considerable atmospheric predictability over North America during an intense El Niño or its opposite, La Niña, especially in winter.

The goal is to predict the onset of an ENSO or a La Niña six months in advance. A number of research centers are working on this problem, with a combination of statistical and coupled ocean-atmosphere dynamic (numerical) models. The results are mixed so far. Most of the models failed to predict the intensity of the 1997–1998 El Niño event, but models are improving. And there is every reason to believe that this program can be carried out successfully during the next decade (Cane 1992). Figure 4.1 (in the color section) shows the prediction of the geographical occurrence of warm surface waters.

The El Niño cycle is a natural phenomenon of the coupled atmosphere-ocean system. Banding from cores taken from coral reefs off the coast of Australia and other similar evidence indicate that it has been going on for hundreds and probably thousands of years. But does the El Niño cycle change as the overall climate changes? There is some indication that the frequency and intensity of the El Niño cycle do change with phenomena like global warming. But the evidence is hardly compelling at the moment.

Perturbing the System by External Forcing

In addition to the natural variability of the climate system (figure 4.2), there are perturbations that can induce climate change. Four examples of these are discussed here.

The coupled system can be perturbed by, for example, a *volcanic eruption*, which we will consider here as an external perturbation. A more global view might take it to be part of the system, but that would be inconvenient for our present purposes. The dust veil from a volcanic cloud can sometimes reach the stratosphere and spread out during a few months to cover the entire globe. This will subsequently block some of the sunlight that would otherwise have entered the lower atmosphere and thus heated the system. Volcanic dust veils reaching the stratosphere can last for up to several years, as was shown by the spectacular eruption of Mount Pinatubo in the Philippines in 1991 (Krueger et al. 1995; Bluth et al. 1992).

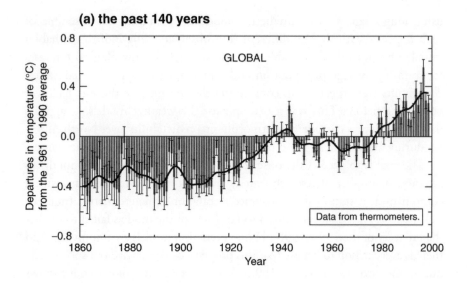
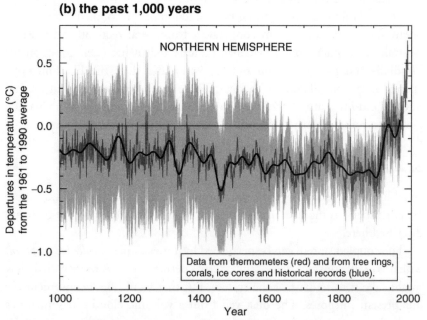

Figure 4.2. Variations of the Earth's surface temperature. Top: Over the past 140 years. Bottom: Over the last millennium. From IPCC 2001. Courtesy of IPCC; reproduced with permission.

Another external forcing that might perturb the coupled system is the *variability of solar luminosity* or brightness. It is now well established from precise satellite observations that the solar constant varies by by about 0.1 percent with an eleven-year period, in phase with the familiar sunspot cycle (Willson and Hudson 1988; Pap 1997). Some solar physicists argue that there are longer periods of variation, implying that the sun might have been brightening over the past few centuries (Lean 1991; Hoyt and Schatten 1993).

The most prominent external forcing that has been suggested recently is that of the *greenhouse effect.* This forcing agent is primarily anthropogenic, deriving from the burning of fossil fuels, clearing of tropical forests, and various industrial processes. Direct measurements suggest that the concentration of atmospheric carbon dioxide has been increasing at about 0.5 percent per year for the past half century, and indirect evidence suggests that this rate has been nearly constant for several centuries. Other greenhouse gases such as methane and certain chlorofluorocarbons have also been increasing, with suggested links to human activities. Climate simulation models expect a response in the globally averaged surface temperature of 1.5 to 4.5°C over the next century. When all the greenhouse gases are combined, the effect is roughly equivalent to a 1 percent increase in carbon dioxide per year, extrapolating to a doubling of the effective concentration over the next 70 to 100 years (Houghton et al. 1994). The greenhouse effect is the most likely candidate to explain the warming of the planet over the past century (see figure 4.2 in the color section). For up-to-date summaries of changes in the large-scale temperature fields over the past century, see www.ncdc.noaa.gov and www.giss.nasa.gov.

The final external forcing to be mentioned here is that of the *anthropogenic aerosols,* which are produced primarily from the burning of fossil fuels. These aerosols are generally restricted to the troposphere and have a lifetime of only a few weeks at most before being cleansed by the precipitation process. They are nevertheless replenished at a sufficient rate to form a veil of generally sub-micrometer-sized particles over much of the northern hemisphere and some of the southern hemisphere (mainly downwind from highly industrialized populations). The so-called direct aerosol effect consists of the direct reflection of sunlight off these suspended particles back to space—a cooling effect on the planet and to some extent a potential agent against the greenhouse effect. The indirect effect, which is much less well understood, is the potential that such aerosols might have to form clouds. The effects of clouds on the heat balance pose a difficult problem for modelers, since, on the one hand they reflect sunlight back to space and yet also absorb infrared radiation from below, rather like the glass of a greenhouse, two almost opposite effects. Furthermore, cloud processes operate on spatial scales that are much too small for present simulation models. The confusing introduction of anthropogenic aerosols into models therefore poses a problem that is insurmountable at this time. A major limitation in the aerosol problem is the lack of sufficient observational data to incorporate into the models (Houghton et al. 1994).

In summary, changes in atmospheric variables such as temperature and pressure can either be part of the natural variability of the atmospheric dynamic system or they can be forced by external (e.g., anthropogenic) perturbations that alter weather statistics. Of course, it is not always obvious which cause is behind a given anomaly. Researchers are putting considerable effort into distinguishing between the origins of apparent changes, discussed next.

ATTRIBUTION OF CAUSES OF CLIMATE CHANGE

Given that there are different types of climate variability, natural and forced, can we attribute changes in the record to specific causes? Several research groups have addressed this issue over the past two decades. The approach taken by most of the groups is equivalent to multiple regression or least squares fitting in statistics. One assumes that there are several known signals or "fingerprints" in the system, along with the natural variability (Barnett et al. 1999; North and Wu 2001). The signals are the patterns in space-time that are the separate responses to the greenhouse gases, anthropogenic aerosols, volcanic dust veils, and solar variability. The object is to estimate the strengths of the fingerprints from the measured records of surface temperature. Using the methods of regression analysis, it is also possible to assess the statistical significance of the individual fingerprint contributions. The shapes of the fingerprints used in the exercise must come from model simulations. In addition, natural variability statistics have to come from computer models.

An important question is what variables and measurement-site-configurations to use in the detection and attribution exercise. Certain conditions must be met, including the availability of a long record of reliable observational data, the representativeness of the data collected at the sites chosen, the ability of models to simulate the variables (e.g., surface temperature) at the points chosen, and the relevance of the sites and variables for the problem at hand. The leading candidate is near-surface temperature, for which we have more than a century of data at tens of stations spread over the globe (North and Stevens 1998). In the past few decades we also have some upper air measurements that can be added to the mix (Santer et al. 1996).

Currently, groups working on the detection and attribution problem agree almost unanimously that both the greenhouse gas and volcanic fingerprint strengths are statistically significant (i.e., the margin of error for the strength is much smaller than its estimated size). There is less unanimity regarding the aerosol and solar signals, which are much weaker in the record. These investigators would conclude that the greenhouse signal has emerged significantly from background climatic variability and that it can be declared "detected." The volcanic signal strength is also significant and can be used as a consistency check on the anthropogenic climate signal detection. Aerosol and solar signals remain problematic (North and Wu 2001; Tett et al. 1999).

Changes in Extreme Events

Atmospheric climate is a statistical summary of "weather" variables. What happens to the extreme excursions of weather as climate changes? To a first order of approximation we can imagine that the means shift but the variances do not. As a simple example, consider the bell-shaped curve of frequencies describing the statistics of global annual surface temperature. Its mean is about 14°C and its standard deviation is about 0.18°C. This means that 97.5 percent of the annual means of the global temperature for individual years fall be below 14.36°C. Annual average temperatures above this would be considered rare (a few times per century). But if a climate shift of only 1°C occurs, the average itself would lie 0.64°C (more than three standard deviations) above this value!

It is perhaps more meaningful to the average person to consider this problem at the local scale. Roughly speaking, the standard deviation for monthly temperature averages is about eight times that of the global average (of course, it varies with season and location). This means the standard deviation is about 1.44°C. Without a climate shift, a standard deviation warm anomaly of 2.88°C will occur about 2.5 percent of the time. We might call this a local "heat wave"—a once-in-forty-year occurrence. A climate warming (shift of the mean) of 1°C means that a fluctuation of just 1.88°C would bring about what the old-timers would call a heat wave. But this anomaly after global warming will occur about once in every ten years. A 2°C warming leads to the old-timers' heat wave once in every 3.5 years. So we see that even a small shift in the mean can lead to the frequent occurrence of what we would call locally rare heat waves (for some model simulations, see www.gfdl.noaa.gov and www.giss.nasa.gov).

There is a body of theory and model evidence that extreme events will occur more frequently from a dynamic point of view. For example, it has been argued that the occurrence of extreme rainfall events in the United States has increased in frequency and intensity over the past century (Karl et al. 1995; see also www.ncdc.noaa.gov). An elementary but appealing argument is that if the globe warms, the amount of water vapor in an air column will increase because of the strong dependence of the saturation vapor pressure upon temperature. So, the argument goes, if there is more airborne water there will be an increase in the fluxes within the hydrological or water cycle. Atmospheric simulation model results are consistent with this conjecture (Manabe and Stouffer 1994). Some model results also suggest an increase in the frequency of tropical storms or El Niño events (Manabe and Stouffer 1994). Another phenomenon that could increase in frequency of occurrence or strength is the tropical storm. Again, model evidence is not yet conclusive on this problem. A caveat worth mentioning is the danger of drawing conclusions about extreme events with too little data or with poorly performing models—recalling that numerical models do not simulate the individual components of the hydrological cycle with much skill.

CLIMATE AND OCEAN: INTERACTIONS AND CONSEQUENCES

In this section we discuss five examples of phenomena in the ocean system that are likely to interact strongly with the atmospheric component of the climate system. In most cases the interaction is two-way; that is, each component significantly interacts with the other during a climate shift.

Delay of Forced Climate Change by Ocean Mixing

When an external perturbation is imposed on the climate system, such as increasing the concentration of greenhouse gases, the surface temperatures will begin to increase. The rate of the increase will depend on the effective heat capacity of the system. If the planet were all land, we might expect the response to be rather fast (a few months) since the thermal conductivity of solid material is very small and most of the heating will be confined to the first few meters of the surface. On the other hand, if the planet were all ocean, we might expect a much longer response time. For example, if only the upper boundary layer of the ocean (a layer well mixed by turbulence due to the winds above, usually between 50 and 100 m thick) were participating in the heating, we might expect a delay of the warming of this mass by a few years. But this is only a small part of a complicated story of how the ocean mixes heat into its interior. If the entire column of ocean is well mixed and participates equally in the warming, the delay would be thousands of years. The answer obviously lies somewhere in between.

The oceans can mix heat vertically in several ways. One example is the idealized thermohaline circulation discussed below (see also chapter 2). In this case, denser waters in specific areas of the polar regions sink to the bottom and then spread in a very systematic way back toward the equator along the ocean floor. The sinking waters are fed by the large-scale surface circulation of the oceans. There are regions where intermediate depth waters are supplied from above. One such mechanism involves exposure of equal density surfaces along the eastern boundaries of ocean basins. These equal density surfaces slope downward as they extend westward across the basin to depths of several kilometers. Water warmed at the surface can slide along these surfaces relatively unimpeded. Coupled ocean-atmosphere models do not have this problem very well constrained at the present. In fact, it appears that the different models produce quite different results with respect to the vertical mixing of heat during forced climate change. This is one reason that climate models with quite different sensitivities to external forcing seem to agree on the response to forcing over the last century. This area should be fertile for observational and theoretical studies over the next few decades.

Sea Ice

Sea ice is the oceanic counterpart of clouds in the atmosphere. And, indeed, there are a number of parallels between sea ice and cloud properties, processes, and their role in climate. However, whereas clouds are virtually everywhere, sea ice exists only in polar regions. Sea ice forms and melts seasonally, its global area oscillating between 20 and 30 million square kilometers (Gloersen et al. 1992). On an annual mean basis, perennial sea ice in the Central Arctic and around the coast of Antarctica covers about 7 percent of the ocean surface.

Sea ice is an important component of the climate system because it significantly modifies the heat and salt/freshwater exchanges between the atmosphere and the ocean. First, sea ice drastically increases the reflectivity of the ocean surface, particularly if the ice is covered by snow, as is normally the case from autumn to spring. Thus, an ice-covered sea absorbs significantly less sunlight energy than an ice-free sea. Second, sea ice acts as a nearly impenetrable barrier to the turbulent transfer of heat and moisture between the ocean and the atmosphere. This is, however, complicated by the fact that long cracks and holes (called leads and polynyas, respectively) form in the ice, allowing partial communication of heat and moisture between air and water. Third, due to the rejection of salt into the water column during freezing as well as the influx of freshwater during ice melting, the cyclic appearance and disappearance of sea ice modifies the vertical distribution of both water salinity and density and, hence, the oceanic circulation.

Sea ice exists as a result of a rather vulnerable balance of heat and mass fluxes at the ocean surface (e.g., Martinson and Iannuzzi 1998). Consequently, sea ice covers are expected to be very responsive to climatic changes and, therefore, to serve as early indicators of global warming. Modeling studies indeed reveal that anthropogenic global warming is likely to become amplified over regions covered by sea ice (Murphy and Mitchell 1995; Rind et al. 1995; Gordon and O'Farrell 1997). The model developed at the Geophysical Fluid Dynamics Laboratory (NOAA) suggests that with a quadrupling of CO_2, the Northern Hemisphere sea ice would completely disappear by early spring every year (see www.gfdl.noaa.gov). Recent observations show that in fact important changes are taking place in the Arctic. From 1978 to 1998, the areal extent of winter sea ice in the Arctic has been decreasing at a rate of 2 percent per decade, and the multiyear sea ice has been decreasing at a rate of 7 percent (Cavalieri et al. 1997; Johannessen et al. 1999). Data obtained during the SHEBA (Surface Heat Budget of the Arctic) experiment shows that multiyear ice in the Central Arctic has thinned and that the upper ocean has become warmer and fresher, presumably due to ice melting (McPhee et al. 1998). It is yet unclear whether these changes are part of the natural variability of the polar climate or are related to an ongoing anthropogenic climatic change.

Computer simulation of sea ice is problematic because most sea ice processes operate at scales smaller than the current generation of climate models can resolve. Climate models have therefore so far incorporated rather crude representations of sea ice. Some of them simply prescribe the location of the ice margin as coinciding with the freezing line of surface air temperature (Hyde et al. 1990). The majority, however, now contain sea ice thermodynamics, with a more or less sophisticated parameterization of top and bottom heat fluxes, but few of them include realistic sea ice dynamics (Schmidt and Hansen 1999). In contrast, current sea ice models are capable of resolving time and space variations in

- snow accumulation and melt-pond formation on top of sea ice;
- ice thickness distribution;
- ice internal heat content;
- generation and energy budget of leads and polynyas; and
- ice-ice interactions, including ice fracture (e.g., Ebert and Curry 1993; Flato 1995; Fichefet and Morales Maqueda 1997; Afshan and Curry 1997; Hibler 1997).

Recent simulations with the climate models at the Geophysical Fluid Dynamics Laboratory and at the Hadley Centre in England have been able to simulate the decrease of Northern Hemisphere sea ice to a satisfying degree (Vinnikov et al. 1999). It is hoped that the next generation of climate models will incorporate most of these sea ice processes, and, in combination with observational data, help gain a deeper understanding of sea ice–climate interactions.

Sea Level

The reservoirs of water on the planet consist of the world oceans and seas; freshwater lakes; underground waters, including soil moisture, rivers, and mountain glaciers; and, most relevant to our problem, the continental ice sheets covering Antarctica and Greenland. The main question to be answered is, What happens to this ice under various future global warming scenarios? We already have evidence that mountain glaciers are receding and that this may be contributing to a sea level rise of about 0.5 mm per year (Meier 1984). In addition, the thermal expansion of seawater may be causing a rise of similar proportions (e.g., as seen in coupled ocean-atmosphere model simulations by Mitchell et al. 1995). All of this is complicated by the subsidence of some heavily populated coastal areas such as Houston, Texas (North et al. 1995). Overall, the observed sea level rise over the past 100 years is estimated to be about 18 cm, with a range of uncertainty from 10 to 25 cm (table 4.1) (Warrick et al. 1995). When estimates of the various causes thought to be operating are totaled, we can only account for about 8 cm, with an even wider range of uncertainty (table 4.1). This discrepancy and the

Table 4.1. Estimated Contributions to Sea Level Rise over the Past 100 Years

Component Contributions	Low (cm)	Middle (cm)	High (cm)
Thermal expansion	2	4	7
Glaciers/small ice caps	2	3.5	5
Greenland ice sheet	−4	0	4
Antarctic ice sheet	−14	0	14
Surf and ground water storage	−5	0.5	7
Total	−19	8	37
Observed	10	18	25

From Warrick et al. 1995.

associated ranges of uncertainty give an idea of our poor understanding of this very difficult problem (for some modeling simulations of the thermal expansion effect, see www.gfdl.noaa.gov).

The major concern is the continental ice sheets. These are in polar regions, where the complex physics of the processes involved renders the answers far from simple to obtain. Adding to this complication is the difficulty of access to these regions and the long time scales over which the processes act. The two combine to give us very little data to work with. We can build numerical models, but these are hard to check or tune because of the lack of good data and the unfamiliar physics of ice under great pressure. Aside from our poor understanding of glacial dynamics below the surface, a major problem is the snow/ice budget over the ice sheets. If the planet is slightly warmer in winter more snow might be deposited, with a buildup of glacial ice. Melt-back occurs mostly in summer. If summers are much warmer over the ice sheets, there can be a considerable loss of land-based ice. The annual competition between these sources and sinks leads to the difference between two large magnitudes, and the hazards of modeling such a process are obvious. Our ability to model these processes is clearly limited at present, but the bulk of modeling evidence suggests that the danger of large-scale or catastrophic melting is not large for many decades, possibly more than 100 years (van der Veen 1992).

The best estimates based upon our current (rather poor) understanding of the models and observations is that the Greenland ice sheet volume might be shrinking, but only slightly, while the Antarctic may be growing slightly

(Warrick et al. 1995). Current best estimates based mainly on models suggest that over the next century the sea level will rise about 30 to 100 cm, with an average estimate slightly less than 50 cm (Warrick et al. 1995).

Thermohaline Circulation

The part of the ocean circulation driven by density differences (caused by temperature and salinity differences) of seawater is called the thermohaline circulation (see figure 2.1). In contrast to the wind-driven ocean currents, this form of circulation is not confined to near-surface waters, but engages the full depth of the ocean in a large-scale slow overturning motion (Gordon 1986; Broecker 1991; Schmitz 1995). Water sinks down in a few specific regions in high latitudes where it reaches its highest density as a result of cooling: in the Greenland Sea, the Labrador Sea, and around the coast of Antarctica (Warren and Wunsch 1981). This process is called deep water formation. No deep water forms in the North Pacific, because salinity there is too low. From the sites of deep water formation, the cold dense waters spread across the deep ocean.

The large heat transport of thermohaline circulation, rivaling that of the atmosphere, makes it important for climate. The North Atlantic region in particular benefits from this heat transport. The thermohaline circulation there transports about 1 petawatt (PW) of heat poleward across 24°N (Roemmich and Wunsch 1985), warming the climate of northwestern Europe by 5 to 10°C. However, data from ice cores and deep sea sediments show that this circulation pattern has not been stable in the past. Throughout the last Ice Age, several sudden flips in the circulation occurred, causing major rapid climate swings in the region (the so-called Dansgaard-Oeschger events). Some cold climate episodes started with a temperature drop over Greenland of 5°C over a few decades or even less, but lasting for centuries. We have more to say about the past behavior of the ocean in the section on paleoceanography below.

This nonlinear flip-flop behavior of the thermohaline ocean circulation can be understood with the help of computer model studies (see review in Weaver et al. 1993; Rahmstorf et al. 1996). These studies have identified two basic positive feedback processes that underlie the nonlinear behavior. The first, called advective feedback, is caused by the large-scale northward transport of salt by the Atlantic currents, which in turn enhances the thermohaline circulation by increasing density in the northern latitudes. The second, called convective feedback, is caused by the fact that oceanic convection creates conditions favorable for further convection. These (interconnected) feedbacks make convection and the large-scale thermohaline circulation self-sustaining within certain limits, with well-defined thresholds where they shut down.

When the respective thresholds are exceeded, two responses are possible: a complete shutdown of the Atlantic thermohaline circulation, or a regional shutdown of convection. Both these responses have been found in global warming scenario simulations. Manabe and Stouffer (1993) simulated a complete shutdown for a quadrupling of atmospheric CO_2, and Rahmstorf and Ganopolski (1999) used a transient peak in CO_2 content. The time scale of the complete shutdown is several centuries in both cases. Wood et al. (1999) simulated a regional shutdown of convection in the Labrador Sea (while the second major Atlantic convection site in the Greenland Sea continued to operate); see also the discussion by Rahmstorf (1999). For this process, the time scale can be very rapid, as little as a few years, with the change occurring in the first decades of the twenty-first century in this particular scenario.

The possible impacts of such circulation changes have not yet been studied systematically. A complete shutdown of the Atlantic thermohaline circulation would represent a major change in the heat budget of the Northern Hemisphere. Consequently, a shutdown would lead to major cooling of this region relative to the rest of the globe. The impacts of a regional shutdown of convection would be much smaller but probably still serious. For the surface climate of Europe, the loss of the Greenland Sea convection would probably have a much stronger effect than that of the Labrador Sea convection, as the northward extent of the warm North Atlantic Current depends mainly on the former. In either case, the consequences of circulation changes for marine ecosystems and fisheries could be severe.

Neither the probability or timing of a major ocean circulation change, nor its impacts, can at present be predicted with confidence, but such an event presents a plausible, nonnegligible risk that requires further study.

Carbon Cycle

The ocean is an enormous reservoir for carbon. Estimates suggest that ≈38,000 gigatons of carbon (GtC) are stored in the deep layers of the ocean, compared to ≈1000 GtC in the surface waters, only ≈750 GtC in the atmosphere, and ≈2200 GtC on land in vegetation, soils, and detritus. (GtC = thousand million tons of carbon; the symbol ≈ here means "in rough approximation.") The fluxes of carbon (in the form of carbon dioxide) are of the order of 60 GtC per year out of the terrestrial biosphere, and ≈61 GtC per year enters. By way of contrast, ≈90 GtC per year leaves the ocean surface, going into the atmosphere, and ≈92 GtC per year returns. Human-derived sources through fossil fuel combustion, cement production, and changing land use may cause a net flux into the atmosphere of ≈6.6 GtC per year (Schimel et al. 1994).

Much of the annual flux of carbon into and out of the oceans is due to biotic material in the surface waters. Also, the solubility of carbon dioxide in the ocean is an inverse function of temperature. Both contribute to a large seasonal cycle of CO_2 in the air. Because of the difficulties in modeling the small-scale features of these processes, we do not have reliable simulation tools. On the other hand, we can use oceanic general circulation models to estimate the patterns of ocean temperature and attempt to bridge to the biological and chemical processes that occur near the surface. Some future scenario modeling exercises have been conducted over the past few years, but the results are at best tentative. Such modeling exercises do suggest that the residence time for carbon in the atmosphere is of the order of 200 years, a result that appears to be robust (e.g., Sarmiento 1992).

At present we cannot completely account for the global budget of carbon with sufficient certainty to pin down the reasons for changes in the trends of CO_2 in the atmosphere. For example, in the late nineties the trend upward slowed from about 0.5 percent per year to only about 0.3 percent per year (Hansen et al. 1998). This change is surely not due to the curtailing of fossil fuel consumption, but rather to shifts in other parts of the system, for example, the terrestrial biosphere. Recent studies of the Tropical Pacific suggest that large changes in carbon flux occur during the El Niño cycle. Gaining a better understanding of the carbon cycle is one of the most pressing parts of the global warming problem.

It has been suggested that in a warmer climate phytoplankton might bloom earlier because a warmer ocean would provide a more stable, shallow stratified surface water layer. Organisms that graze on these primary producers might develop at the regular time of year because the length of daylight determines their natural cycles. This mismatch might significantly disrupt some marine ecosystems and change the pattern, timing, and amount of the exchange of carbon dioxide as well as that of dimethyl sulfide gas produced by plankton. For instance, the mismatch could lead to a higher fraction of organic carbon being recycled by bacteria and photooxidation, which in turn could result in a greater fraction of photosynthetically fixed carbon returning as carbon dioxide to the atmosphere. In contrast to the effect of increasing the flux of CO_2 to the atmosphere, an increase in the emission of dimethyl sulfide may enhance formation of atmospheric sulfate aerosol particles and cloud condensation nuclei. These atmospheric aerosols can screen out sunlight and therefore have a role in offsetting the greenhouse effect (Charlson et al. 1987).

Another potentially important marine carbon issue is the calcification of coral reefs. Increased sequestering of atmospheric CO_2 in the ocean would result in a lowering of the oceanic carbonate (CO_3^{2-}) concentration. Since the calcification of coral reefs depends on the saturation of the carbonate mineral aragonite, it has been suggested that increasing atmospheric CO_2 could result in a decrease in the aragonite saturation state in the ocean by 30 percent and biogenic aragonite precipitation by 14 to 30 percent (Kleypas et al. 1999). This could result in a significant decrease in the reef building process.

THE VALUE OF PALEOCEANOGRAPHY

Any human-induced changes in climate and ocean circulation will be superimposed on a background of natural climatic variations. Hence, to understand future climatic and oceanic changes, it is necessary to document the space and time scales of natural climate variability. Useful as instrumental studies may be, they are limited by the available data record, which generally extends back less than 150 years (see figure 4.2). This duration is too short to extract the full range of variability likely to be present. Nevertheless, data from multiple sources suggest that the rate, duration, and amount of warming has been much greater in the twentieth century than in any of the previous nine centuries (see figure 4.2).

The North Atlantic Drift carries warm water northward. This water is cooled almost to freezing point before sinking to the abyss. This process releases an enormous amount of heat into the atmosphere and is responsible for the mild climate of Western Europe. The study of marine sediment cores (see chapter 2) has shown that the strength and pattern of the thermohaline circulation, which drives the oceanic heat transport, has changed significantly during the last climatic cycle (the past 150,000 years). These changes occurred not only between glacial and interglacial periods, but also within the last glaciation and within interglacial periods as well. Heat transport was significantly smaller during the last glacial maximum, 20,000 years ago. One of the most fruitful interactions between the modern and paleo scientific communities has been the study of multiple states of the thermohaline circulation under various forcing factors. This characteristic feature of coupled ocean-atmosphere models would have been left aside without the abundant evidence provided by paleoceanographers.

Tropical ocean-atmosphere systems orchestrate climate variability worldwide over interannual to decadal time scales, and the tropical ocean is the primary source of energy and water vapor to the global atmosphere. Paleoclimatic records from corals and sediments have been used to extend the observational baseline of tropical variability and document the sensitivity of these systems to past changes in forcings, such as volcanism, solar activity, or greenhouse gas emissions. Such reconstructions offer new opportunities to gain insight into the natural variability of tropical systems and validate numerical model simulations of regional and global climate variability. In addition, records from Pacific corals spanning the past four centuries have revealed simultaneous shifts in periodicity among annual, interannual, and multidecadal modes of Pacific variability, implying a coupling across time scales, which is still to be understood.

The recent past as revealed by instrumental observations (see figure 4.2) is not representative of climate over longer periods. This is highlighted by paleoclimatic evidence that the climate system has repeatedly switched, in a matter of years to decades, between significantly different climatic modes during the last glacial period (see chapter 2). Many of these changes were apparently episodic pulses of century/millennial duration, but also included abrupt (i.e., of annual to decadal-scale duration) changes in climate state and ocean circulation. More recently, there has been the growing realization that abrupt climatic events also characterized the present Holocene interglacial period (past ten thousand years), both in the ocean and over low to mid-latitude continents, where the hydrological cycle can shift from heavy rain to long-term drought.

The significance of past abrupt climatic changes is heightened by the fact that they cannot be studied using data collected directly from instrumentation deployed by scientists over the past hundred years or so. Careful work is needed to map out the space and time patterns of change associated with past abrupt events, to determine their causes, and to determine if they are predictable. It is quite plausible that abrupt changes in the Holocene could be of the type that might occur in the future. Equally important is the manner in which the climatic system responds to abrupt change. It is possible that trace-gas-induced warming could manifest itself as a geologically abrupt event, and studies of past abrupt events are necessary for developing and evaluating predictive models that can simulate abrupt change of the coupled ocean-atmosphere system.

CONCLUSIONS AND RECOMMENDATIONS

The oceans participate in climatic change through feedback mechanisms such as the sea-ice albedo effect and delays due to heat storage below the surface and its subsequent slow release to the air. The oceans may also exhibit internal chaotic behavior on long time scales that is not driven by any external mechanism. Major drivers of climate change are thought to be anthropogenic carbon dioxide and aerosols, both released by humans into the air. The oceans are a major reservoir for carbon, and large exchanges between the oceans and the atmosphere occur every year. Alterations in these fluxes could buffer or enhance climate change. The thermohaline circulation carries heat poleward in the North Atlantic, and its strength may be sensitive to the climate at the surface. When global climate warms, the ocean volume expands and land-based ice melts, providing freshwater to the oceans. These two processes lead to a rise in sea level that could threaten vulnerable coastal communities. There is conjecture that a climate change could increase the frequency and/or intensity of precipitation events, including tropical storms. In addition, changes in the runoff of freshwater into estuaries and seas can change the composition of these waters locally and lead to changes in ecology and consequently, fisheries.

A major concern is our ignorance of the many processes at work in the ocean-atmosphere system. Hand in hand with our growing ability to perform computer simulations of the ocean-climate system is the need to gather data from the ocean's surface to within its floor. There is a great need to address these problems so that our understanding will advance to a point that well-founded science forms the basis for public and private sector policy. The consequences of changes in climate and the related changes in the ocean are important to all the countries and economies of the world. Oceanic research is expensive and requires international cooperation. We believe progress is being made, but at a pace too slow to be useful in private and public policy decisions that need to be made soon. We urge a renewed effort to increase our knowledge of the oceans and their relation to the climate system.

RESEARCH PRIORITIES

The many unknowns in the interaction between the atmospheric and oceanic components of the climate system merit a strong response from the world's agencies sponsoring research.

- We need a better understanding of the subduction of heat into the oceanic interior, since this is a major unknown in the delay of global warming or other climatic changes.
- The oceans represent a huge reservoir of carbon; the fluxes of this element, usually in the form of carbon dioxide, are large both into and out of the oceans. We need better understanding of this important climate feedback mechanism.
- Considerable improvement is needed in understanding the role of sea ice in the climate system, and in its treatment in climate models. More observations are needed in the field, including improved use of satellite data.
- Coupled models of the oceans and the atmosphere need refinement, calling for dedicated man-hours and larger, faster computers.
- More global, systematic in situ measurements of the oceans need to be conducted to better understand its structure and composition, as the basis for further improving models of its behavior.
- The ancient ocean needs more scrutiny through the study of its history as represented in seabed sediments, which can tell us about climate change through time.
- More observations of sea level are needed as the basis for a better understanding of why it is exhibiting its current behavior (thermal expansion, melting of land-based ice, etc.).
- Better observations of precipitation (and evaporation) over the oceans are important because the salinity of the surface waters helps to drive the oceanic circulation.
- There needs to be a better understanding of the general circulation of the oceans to clarify such concepts as the thermohaline circulation and its possible shifts in positions and strength through time.
- We need a better understanding of the basic mechanisms involved in the interaction between the atmosphere and the oceans at their interface as regards the fundamental exchanges: momentum, chemical species transfer, heat, and moisture.

ENDNOTES FOR FURTHER READING

Crowley, T. J., and G. R. North. 1991. *Paleoclimatology*. New York: Oxford University Press.

Harvey, L. D. Danny. 2000. *Global Warming: The Hard Science.* London: Prentice Hall.
IPCC. 2001. Summary for policymakers. In *Climate Change 2001: The Scientific Basis,* 1. Geneva: Intergovernmental Panel on Climate Change.
Philander, G. G. H. 1990. *El Niño, La Niña, and the Southern Oscillation.* New York: Academic Press.
Trenberth, K. (ed.). 1992. *Climate System Modeling.* Cambridge: Cambridge University Press.

Chapter 5

Fisheries and Fisheries Science in Their Search for Sustainability

Gotthilf Hempel and Daniel Pauly

WHAT ARE FISHERIES AND FISHERIES SCIENCE ABOUT?

Fisheries is the exploitation of the living resources of the sea. In a broad sense, fisheries encompass not only the catching of finfish and invertebrates such as shrimps and squids, but also the collection of cockles and other bivalves and of seaweed, and the (largely past) hunting of marine mammals. The farming of fish and aquatic invertebrates is also often, if confusingly, considered by many to be part of fisheries, as well. Fisheries employ craft ranging from outriggers to factory trawlers and nets ranging from beach seines to open-ocean driftnets dozens of kilometers long. Fisheries products may be consumed by subsistence fishers and their families or sold on a strongly integrated global market. For millennia, fished organisms and/or their shells have been used as raw material for jewelry or other nonfood products. More importantly, one-third of the world landings are diverted from direct human consumption to feed pigs and other farmed animals, with an increasing fraction of global fish meal and oil production being used to feed salmon and other farmed fish and aquatic invertebrates, particularly shrimp.

Fisheries date back to the Stone Age, but still represent one of the key uses of the world ocean. Globally, annual marine landings peaked at about 80 million metric tons in the late 1980s (figure 5.1). About 20 to 30 million metric tons of fish caught as bycatch, and subsequently discarded, may be added to this, resulting in a global annual catch (= landings + discards) of more than 100 million tons in the waning years of the late twentieth century.

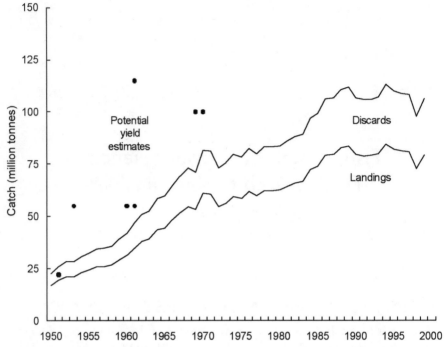

Figure 5.1. Marine fisheries statistics from 1950 to 1997, contrasting nominal landings (from www.fao.org) with an estimate of discarded bycatch for the early 1990s (from Alverson et al. 1994), scaled to the landings from other periods. Dots represent some earlier published estimates of potential yield for the global ocean. Note that nominal landings in the 1990s are probably overestimated by about 10 million tons, due to excessive reports from East Asia; also note that global catch composition is rapidly changing, with small pelagics and invertebrates increasingly replacing larger bottom fishes, whose absolute catches are declining. This results in the trends illustrated in Figure 5.3.

Marine fisheries provide about 20 percent of the world protein supply. For large parts of the world population, particularly in East and Southeast Asia, fish in the broad sense is the most important source of animal protein; 200 million heads of cattle would be required to substitute for this (Alverson et al. 1994; Hubold 1999). From the early 1950s to the late 1980s, capture fisheries landings and aquaculture production grew faster than the global human population, leading to a substantial increase in fish supply per person. This trend reversed in the 1990s, and it is only the large reported increase in global aquaculture production that has prevented a rapid decline of fish supply per person.

During the past fifty years, world fisheries have seen drastic changes due to technological developments and the emergence of the new Law of the Sea, the collapse of the Eastern Bloc distant water fisheries, and the globalization of much of the fishing industry and of the markets for marine products. All those changes had substantial effects on the living resources themselves, through various ecological and economical feedback mechanisms. Of all human impacts on marine ecosystems, fisheries are the most important, particularly in shelf seas.

Most of the world's landings are taken from about 200 fish stocks, more than half of them fully used or decreasing because of overfishing. The stocks of much sought after bottom fishes, like cod, groupers, hake, and sole, are in a particularly deplorable state. As a consequence, the fisheries have turned to species that were formerly less valuable and have expanded to deeper and more distant fishing grounds.

It was concern about the stagnation and decrease in wild-fish landings—in spite of increasing fishing effort—that more than 100 years ago gave birth to fisheries science, whose task was to provide advice for a better management of the fisheries, based on scientific insights into the dynamics of the fish stocks as parts of marine ecosystems. More ominously, other aspects of fisheries science were devoted to making fishing operations more effective, notably by improving location and catching methods.

Our chapter deals largely with the problems faced in the attempt to enhance the sustainability of fisheries and their associated ecosystems at the high level of exploitation prevailing nowadays.

ORIGINS

The hunter-gatherer ancestors of modern humans would usually stay at a given spot as long as its fauna and flora provided enough food. When the "patch" in question was too disturbed to remain productive, they then moved on. With regard to fisheries, we often still act as "patch disturbers."

The community of professional fisheries scientists that had emerged in Western Europe by the end of the nineteenth century was well aware of the impact of heavy fishing on the abundance and productivity of fish populations. The International Council for the Exploration of the Sea (ICES), founded in 1902, immediately created an Overfishing Committee and concentrated on obtaining practical results on the question of overfishing. The ICES scientists were particularly interested in the effects of fishing on the flatfishes in the southern part of the North Sea. This was the first area and species complex to be fully exposed to the impact of steam trawling (Went 1972; Cushing 1988).

SUSTAINABILITY AS A GOAL OF FISHERIES SCIENCE

Overfishing made fisheries science one of the first natural history disciplines to be fully structured around the concept of sustainability. Major conceptual advances in fisheries science include the first-catch curves and yield per recruit analyses (Baranov 1918), the first functional definition of overfishing (Russell 1931), of "maximum yield" (Bückmann 1938) and of "surplus production" (Graham 1943), all of which may be seen as attempts to render the concept of sustainability operational. Such interpretations then logically led to the concept of maximum sustainable yield (MSY) (Schaefer 1954; Beverton and Holt 1957), a concept which, for several decades, was at the very heart of fisheries science (figure 5.2). Given basic economic considerations, this leads to maximum economic yield, an even worthier goal.

From the very beginning there has been a gap between the theories and advice of fisheries scientists and the praxis of fishing industry (Mace 1997). This has largely limited the profession to one that diagnoses and prognoses, but cannot solve problems on its own. Fishery regulations and management have been, so far, the responsibility of governments and industry. The results have been stock declines, and a sustainability that appears very difficult, if not impossible to achieve (Ludwig et al. 1963), even for well-studied, staple fish such as cod in the North Atlantic, where fisheries research has emerged as scientific discipline in its own right, with a high concentration of prestigious scientific fisheries research institutions.

We are faced with a worldwide surplus of high technology fishing vessels. Excess fishing mortality in single-species fisheries implies—at least in theory—that fishing becomes unprofitable when low biomass levels are reached. This should result in a reduction of fishing effort; thus, in theory, fishing effort should be self-regulating. But in practice this is not the case, partly because of the massive subsidies that governments provide to fisheries (see figure 5.2) and partly because of differential fishing costs within heterogeneous fleets. Rothschild (personal communication) suggests that a lack of property rights might make a significant contribution to excess capital in fisheries. This is illustrated in box 5.1, which deals with an unsteady market-driven fishery on blue whiting and hake on the Patagonian shelf.

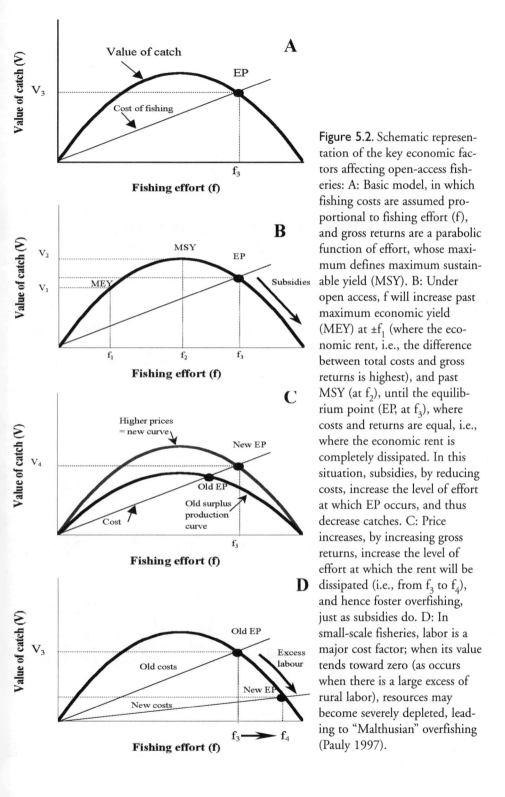

Figure 5.2. Schematic representation of the key economic factors affecting open-access fisheries: A: Basic model, in which fishing costs are assumed proportional to fishing effort (f), and gross returns are a parabolic function of effort, whose maximum defines maximum sustainable yield (MSY). B: Under open access, f will increase past maximum economic yield (MEY) at ±f_1 (where the economic rent, i.e., the difference between total costs and gross returns is highest), and past MSY (at f_2), until the equilibrium point (EP, at f_3), where costs and returns are equal, i.e., where the economic rent is completely dissipated. In this situation, subsidies, by reducing costs, increase the level of effort at which EP occurs, and thus decrease catches. C: Price increases, by increasing gross returns, increase the level of effort at which the rent will be dissipated (i.e., from f_3 to f_4), and hence foster overfishing, just as subsidies do. D: In small-scale fisheries, labor is a major cost factor; when its value tends toward zero (as occurs when there is a large excess of rural labor), resources may become severely depleted, leading to "Malthusian" overfishing (Pauly 1997).

Box 5.1. The Role of Science in an Unsteady Market-Driven Fishery: The Patagonian Case

Ramiro Sánchez

It was not until the second half of this century that commercial fishing and fishery science began to move ahead in Argentina. During the 1960s and 1970s the government took a series of measures that aimed at promoting the expansion of the offshore fleet, the hake fishery, and the development of Patagonia. There was a need to carry out systematic research, with established programs covering the main resources inhabiting the largest continental shelf in the Southern Hemisphere, which had been latent for decades. This need was met first through international technical cooperation programs, noticeably a Food and Agricultural Organization project (1966–1971), and later by the creation of Instituto Nacional de Investigación y Desarrollo Pesquero (INIDEP) (1977) and the incorporation of three research vessels.

These initiatives, however, were not enough to ensure the sustained growth of the fishery. For nearly thirty years, political and economic instability affected both industry and science in different ways (figure BX5.1). The crisis in the early 1980s also revealed some structural weakness in the organization of the industry and the first reduction in the biomass of the hake stock—a notion that did not go beyond the scientific milieu.

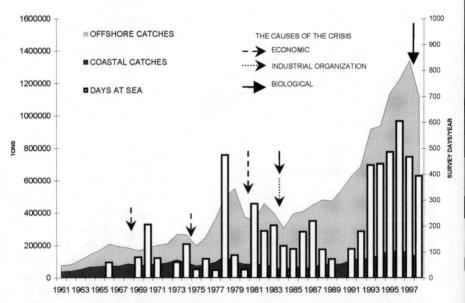

Figure BX5.1. The evolution of fishery research (vertical bars = survey days) in Argentina with fleet development and fishing crises as a background. Sources: Argentine official statistics and Fishery Economy Project, INIDEP/University of Mar del Plata.

The Malvinas (Falklands) War in 1982 drew the attention of the international community toward the southern extreme of the continental shelf off Argentina. After the war, the fishing scenario in the region changed dramatically. The activity of the international fleet grew constantly throughout the decade. In spite of acknowledged information gaps, Argentinean scientists expressed their concern for the high levels of exploitation of squid, hake, and blue whiting (INIDEP 1986). The downfall of the local economy caused a gradual decline in the support given to marine science. There were political and economic changes and a new legal framework aimed at reinvigorating the industry and encouraging foreign trade and investments. These were the basis for a threefold rise in the catches of the offshore fleet from 1988 to 1997. Several factors contributed to this uncontrolled rise, including the development of a managed squid fishery with a chartering regime, the incorporation of large factory vessels, and the unwanted outcome of a fishing agreement with the European Union. A historic peak in catch (1.34 million tons) was reached in 1997. A large fraction (80 percent) of this was accounted for by only three species: squid, hake, and southern blue whiting.

The Argentine fishery is, at present, undergoing the worst crisis of its comparatively short history. It is also unique in that it is largely based on the biological collapse of several of its main fish resources. Blue whiting and hake show clear signs of uncontrolled exploitation (figure BX5.2). The high catch levels internationally suggested for the blue whiting during the 1980s (Csirke 1987) could clearly not be maintained. No other resource has been so greatly affected by the present expansion of the local fishery as hake. Fall of the spawning biomass, recruitment failure, reduction in the size and age of the population structure, and other symptoms are well documented (Pérez et al. 1999). Technical advice is definitely not lacking. The allocation of sea survey time to this resource is unprecedented. But aside from science, the recovery of the stock will necessarily imply intensified control. Nursery grounds, although well identified, are not yet completely closed to the fishery. Selective gear to reduce the unwanted catch of juvenile hake and the bycatch in the shrimp fishery have been designed, though not yet enforced or effectively applied.

A new legal tool was drawn up in 1998 as a means to regulate fishing capacity and effort. The basis for the assignment of quotas and the species to be included in the new regime were matters of considerable debate. The system has not yet been implemented.

The present crisis brought with it a drastic change in the relationship between fishery scientists, administrators, the industry, and the press. The avenues of communication between the different participants are quite open, and we have the feeling that scientific results matter, even if they are not immediately enacted. A new generation of scientists was forced to move from the backstage to the footlights. In spite of being molded by the old—and valuable—tradition, they have so far succeeded in meeting the challenge. Research, albeit proactive, must transcend the constraints of operational management. The academic world has yet to react accordingly. The future of the Argentine fishery will require a new type of scientist.

(continues)

Box 5.1. Continued

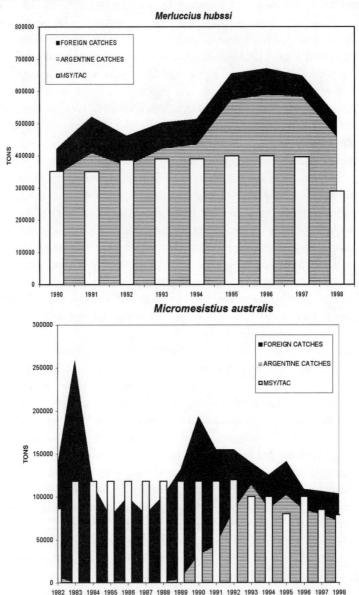

Figure BX5.2. Exploitation beyond scientific advice and regulations by Argentine and foreign vessels. Bars indicate maximum sustainable yield/total allowable catch (MSY/TAC) values for hake and blue whiting. Based on INIDEP Hake Assessment and Austral Fisheries Projects.

ENVIRONMENTAL INFLUENCES ON FISH STOCKS

Fisheries exploiting, for example, coral reefs are very different from those exploiting temperate soft bottom shelf seas, deep continental slopes, eastern boundary upwelling regions, or the open ocean realm. Each of these systems, therefore, calls for an individual management strategy.

In the various parts of the world ocean and its shelf regions, temperature and current regimes differ in the pattern and extent of their variations. Fish stocks are susceptible to those changes, which impact their recruitment, but also their distribution and spawning migrations. Much of the controversy among fisheries biologists about the relative importance of "environment" versus "fishing" as causes of stock variations results from the differences in the susceptibility of various stocks to changes in the environment. Strategies for sustainable exploitation of fish stocks will vary with the longevity and growth of the targeted species, but also with the natural variations of their environment. Thus, fishing is not the only reason why fish stocks fluctuate. Sediment cores in upwelling regions demonstrate periodic shifts in the abundance of anchovies and sardines alternating back in time long before fishing commenced. Long-term observations have further suggested that ocean climate and fisheries might interact to cause multidecadal fluctuations in stocks.

But, factors others than ocean climate are also involved. The various fish species that are targeted by fisheries are parts of ecosystems, which also must persist in a certain state for the fisheries to be sustainable. This requirement was not widely acknowledged until recently. It seemed hard enough to deal with the assessment of single species populations, let alone whole ecosystems.

THE MULTISPECIES PROBLEM

In multispecies fisheries, the self-regulatory process hypothesized above does not occur even in principle, because a fishery that continues to exploit a much-depleted species (e.g., of large fish in a shrimp trawl fishery) catches fish that it has not targeted. Hence, shrimp fisheries contribute substantially to the discarding problem mentioned above (see figure 5.1 and Alverson et al. 1994). The implications of removing such large amounts of fish and invertebrates from the marine ecosystems, then dumping them—mostly dead—back in the water, are still not understood.

Bottom trawl fisheries are able to control the overall numbers of fish that are killed, but not the relative mortality of the different species in a multi-species fishery. In a commercial fishery not using highly selective gear, some species will always be subjected to excessive, nonsustainable mortality, leading populations to drop to very low levels. But each species reacts differently to changes in stock size and to changes in the environment, making predictions even more difficult.

Biological interactions cannot be ignored, especially feeding interactions. First of all, it is not possible to exploit simultaneously a species of predatory fish as well as its prey species and to expect both species to produce their stock-specific MSY. Catching the prey fishes reduces the food base for the predators and hence their productivity and potential yield. Many predators feed on a variety of fish and invertebrates simultaneously, and hence predator depletion can lead to unwanted prey becoming abundant. On the other hand, exploiting forage (prey) species can reduce stocks of predator fish. This may seem trivial, but there are very few governance arrangements for predator fisheries that explicitly reduce fishing on the prey species, for example, by compensating those who want to target the latter. One example where such a policy is practiced is Iceland, where attempts are made to boost the cod (predator) population by limiting the shrimp and capelin (prey) fisheries.

FROM SPECIES CHANGES TO FISHING DOWN MARINE FOOD WEBS

Changes in species dominance in ecosystems, and hence in fishery catches, have been known to occur for decades. Indeed, not all of these changes are due to fishing (Daan 1980; Skud 1982). But analyses of a large number of cases of species changes in different regions, at least over the past two decades, show a disturbing, common pattern. Almost everywhere, large, long-lived fish high in the food web tend to decline, both in the catches and in the ecosystem. They are then replaced by species of small, fast-growing fish with lower trophic levels (Pauly et al. 1998.) (figure 5.3). This process of "fishing down food webs" would be acceptable—at least to some—if the decline of mean trophic levels was accompanied by a corresponding increase in catch size. But, while such an increase often occurs at the outset of fishing down, it is usually short-lived. Moreover, average fish sizes in the catch decrease, whereas the proportion of invertebrates in the catch increases. This might be welcome if they are highly prized shrimps, but not if they are starfish.

CONFLICTS BETWEEN USERS

Coastal habitats differ widely, ranging from sandy and muddy shores and estuaries to hard rock, mangroves, and coral reefs. Each habitat has its spe-

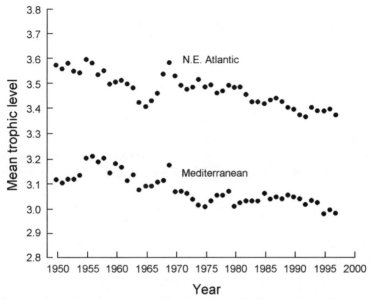

Figure 5.3. Trends of mean trophic level of fisheries catches in the Northeastern Atlantic and the Mediterranean, documenting that fisheries increasingly rely on fish from the lower parts of marine food webs. Similar trends are observed in other parts of the world (Pauly et al. 1998). Note that most commercial fish have trophic levels between 2.5 and 4.5, implying that the trends in question cannot continue for long without major disruption of marine food webs. Based on data in FishBase; see Froese and Pauly 2000 and www.fishbase.org.

cific fauna, interacting with that of neighboring habitats and with the fish stocks of the open sea. The natural resources include not only many kinds of fish, but also mollusks, shrimps, and other invertebrates as well as algae.

Nearshore fisheries are more ancient and more diverse than open-water fisheries. Fishing methods and practices vary widely, and most of the nearshore catch consists of a multitude of species. Landings are spread over many small places, often without passing through a centralized sales system. As a result, statistics of efforts, catches, and landings for nearshore fisheries are often poor. Still, small-scale, nearshore fisheries make a substantial contribution to world fisheries in terms of total landings and value. In addition, these fisheries are the main source of both protein and income for large parts of the coastal populations, particularly in tropical and subtropical zones.

Naturally, the inherent wealth and diversity of coastal waters attract not only fishers and fish farmers, but also a variety of other uses, such as tourism, transport, and the extraction of nonliving resources (e.g., sand, coral stone, oil, and gas). Coastal waters are also abused as sinks for pollutants and nutrients.

With such a diversity of stakeholders, some conflict is inevitable, with the fishers often on the losing side. Fishing may appear to be incompatible with, for example, some industrial uses like pipelines, wind parks, and traffic lanes. But conflicts also occur within the fisheries sector itself, for example, trawlers versus longliners, recreational versus commercial fishing, artisanal fishers versus large-scale operators, or inshore capture fisheries versus mariculture. Conflict resolution thus must include representatives of these stakeholders as well as conservation biologists, and nongovernmental organizations (NGOs) representing the interests of local communities or advocating nature conservation.

Knowledge of the key features of the coastal and oceanic food webs is essential for the biologists involved here, along with information on the demographic, social, and economic features of the people who depend exclusively or partly on fishing. A new generation of integrated approaches will have to combine ecological and fisheries knowledge with information on the socioeconomic framework of fishing. Those approaches, even in their simplest form, must also include a co-management component, allowing for the interaction of fishing communities, industry, NGOs and conservationist groups, and government managers.

Fisheries management uses a number of regulatory tools, such as catch and effort limits, minimum landing size, gear restrictions, closed seasons, and, recently, marine protected areas. But we need to know more about their effects on the exploited and unexploited parts of the resource, the economics of the fisheries, and their impacts on biodiversity and exploited ecosystems.

Fisheries, like most other human activities at sea, alter the marine environment on various scales of space and time. Recently, Jackson et al. (2001) identified overfishing as the most important human impact on a variety of shelf and coastal ecosystems, with past overfishing of large marine animals blamed not only for collapses of the targeted stocks, but also for drastic changes in the overall structure of the underlying ecosystems. In nearshore waters, semiclosed seas and estuaries, eutrophication may play a similar role by causing algal blooms and oxygen depletion. Other anthropogenic impacts include the introduction of alien species as predators, competitors, and pathogens. In the long run, our impact on the climate change will, as well, add to the effects of those impacts on marine ecosystems.

Apart from top-down effects of fisheries on marine food webs resulting from removal of predators, marine ecosystems are often impacted directly, for example, by bottom trawls. The fishing grounds of the North Sea and the northern North Atlantic shelves are ploughed again and again, many times per year and for many decades: the rougher the sea floor, the heavier the gear.

In their search for new resources, the patch disturbers reach now for pristine areas 1000 m down the continental slopes. There, the unique communities of the cold-water coral *Lophelia* become endangered even before they have been scientifically explored (see box 5.2). The rock-hopper gear of heavy bottom trawls bulldozes the coral banks hitherto exploited only by nondestructive longlines and gill nets.

Modern fisheries science has to provide the scientific knowledge needed for sustainability, that is, for obtaining a continuous supply of fish and other seafood. On the one hand, fisheries science should analyze the ecological impacts of fisheries and mitigate these with regard to biodiversity and ecosystem integrity. On the other hand, fisheries science will have to contribute its bit to the peaceful coexistence of fisheries with other marine stakeholders. In order to meet all those demands, fisheries science must become a multidis-

---

Box 5.2. Deep-Water Coral Ecosystems
Johan H. Fosså

Several species of deep-water corals form reefs. *Lophelia pertusa* is the most widespread and best known. It is almost cosmopolitan, but occurs most commonly in the Northeast Atlantic. It thrives in oceanic water, usually between 6 and 12°C, at depths of between 60 m to several thousand meters and on hard substrates of several kinds: on continental slopes, seamounts, ridges, and banks and in fjords (e.g., Rogers 1999). Deep-water banks of these corals can be up to 13 km long and 350 m wide.

A highly diverse fauna is associated with these banks, where the complex architecture of the skeletons provides a variety of different microhabitats. Along the Norwegian coast alone, a total of 614 associated species have been documented. Almost nothing is known about ecological relationships within the *Lophelia* reef community. Longline fishing shows that catches of *Sebastes marinus*, *Molva molva*, and *Brosme brosme* are significantly higher in coral areas than on the surrounding bottoms.

Bottom trawling and oil exploration seem to be the two most probable threats to the reefs. *Lophelia* reef areas have traditionally been rich fishing grounds for longline and gill-net fisheries. During the past decade, bottom trawlers with rock-hopper gear have extended their fishing grounds into the coral reefs. Trawlers often crush the corals to clear the area before fishing starts. Reports from fishers suggest that large areas have been cleared and that catches are significantly lowered in cleared areas. There is growing evidence that the destruction of corals is one of the most severe forms of habitat destruction in Norwegian waters (Fosså et al., in press). Extensive

(continues)

Box 5.2. Continued

damage to reefs has also been reported from Iceland, the Faeroe Islands, and the western side of Great Britain. Investigations on southern Tasmanian seamounts have revealed extensive damage to deep-water reefs formed by *Solenosmilia variabilis*, caused by a trawl fishery for orange roughy and oreos.

Because *Lophelia* grows slowly, the recovery rate of damaged reefs is also likely to be extremely slow, which makes the damage of these habitats a serious environmental problem.

To gain information about the presence and status of deep-water coral reefs, the most promising approach seems to be mapping the seabed by multibeam echo sounder, combined with deep-towed side-scan sonar and information from fishers. Provisional maps can then be validated using remotely operated vehicles (ROVs) equipped with video cameras to determine the exact location of the reefs in order to advise fishers where to set nets or longlines to avoid losing gear or damaging the corals. Enforceable regulations to protect the corals are urgently needed. In Norway, fishers, scientists, and environmental organizations persuaded the government to create regulations to protect the reefs. As a result, trawling is now banned in the Sula area, where coral is abundant. It is also forbidden to damage corals deliberately. All bottom trawling can be forbidden in new areas.

To improve understanding of these corals and the ecosystems of the coral reefs, fundamental research is needed on the biology of *Lophelia* (where the larva is completely unknown) and on other reef-building deep-water corals. Research is further required into the possibilities for restoring damaged reefs. We need to know more about the potential importance of the reefs as nursery areas, feeding chambers, and refuges for commercially important fish.

ciplinary, ecologically, and socioeconomically oriented science, while maintaining the skill of its practitioners in the areas of single-species and multi-species modeling.

FISHERIES IN ECOSYSTEMS

Fish populations are increasingly understood as elements of marine ecosystems. Over their life spans, fish interact with other animals while progressing from one trophic level to the next during the development from the larval stage to the adult fish. Fisheries interfere with the interactions this implies, by removing fish (whether targeted species or bycatch) from the ecosystems in which they act as both predators and prey. Fisheries also destroy habitats (box 5.3) and disturb fish and especially marine mammals acoustically. Many modern fisheries ecologists see their primary role in analyzing and mitigating these effects to make fisheries tolerable for the ecosystem in the long run.

The ecosystem approach to fisheries management has been adopted for the entire Southern Ocean by CCAMLR (Commission for the Conservation of Antarctic Marine Living Resources). The exploitation of seals and whales, followed by the catch of finfish and of (relatively small quantities of) krill, led to very successful scientific studies on the life history of the exploited species, their interactions, and their dependence upon the oceanographic regime and sea-ice extent, and consequently upon primary and secondary production and the varying composition of plankton, benthos, and sea-ice fauna. For example, the waters around South Georgia and their potential for

Box 5.3. The North Sea Herring: A Case Study for Single Species Fisheries Management

Christopher Zimmermann

The North Sea herring is believed to have had a spawning stock biomass (SSB) of more than 2 million tons before large-scale fishing started in the 1950s. Heavy exploitation led to a collapse of the stock in the late 1970s (figure BX5.3). The closure of the (directed) herring fishery for four years, accompanied by years of good recruitment, led to a recovery of the North Sea herring stock. After the fishery reopened in 1981, it adopted a simple total-allowable-catch (TAC) approach applicable to adult fish for human consumption to avoid detrimental stock depletion in the future. However, the stock showed another rapid decline in the mid-1990s, mainly due to two factors: TACs were significantly exceeded and, even more important, greater numbers of juvenile herring were caught as bycatch in the increasing industrial fishery for fish meal and oil production, targeted on sprat.

The North Sea herring therefore became the first stock to be managed according to the precautionary approach in European Union waters. Now an emergency plan restricts catches on adults and juveniles to allow for fast rebuilding of the stock as soon as the SSB falls below the agreed reference points. Equally important, a new control system has been implemented to limit the bycatch of juveniles in the industrial fishery effectively.

(continues)

Box 5.3. Continued

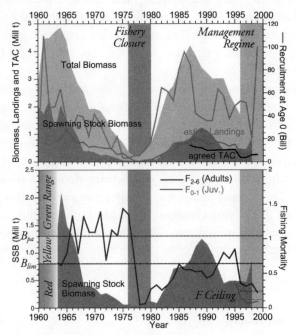

Figure BX5.3. North Sea herring (autumn spawners). Top: Total biomass, spawning stock biomass (SSB), recruitment at age 0, landings, and agreed total allowable catch (TAC). Bottom: SSB and fishing mortalities (F; adult and juvenile) and management reference points. The phases of the fishery closure and the management regime are indicated. See text for more details and explanation of terms.

The management plan is based on biomass reference points, derived from stock recruitment relationships, as follows:

- Red range below the limit biomass reference point B_{lim}, set to 800,000 tons—the emergency plan to reduce fishing mortality (F) on adults and juveniles automatically comes into force.
- Yellow range between B_{lim} and the upper reference point, B_{pa}, set to 1.3 million tons SSB.
- Green range above a SSB of 1.3 million tons—the fishing mortality is allowed to reach F = 0.25 for adults and F = 0.12 for juveniles. All bycatch is considered for the bycatch quota. The industrial fishery for fish meal is closed as soon as the ceiling is reached.

Soon after the management regime was implemented in 1996, there was a significant recorded reduction in fish mortality. After a few years of good recruitment, the stock rapidly rebuilt itself. Since then, fishing mortalities have been limited to $F_{2-6} = 0.2$ and $F_{juv} = 0.1$. As a result, the SSB was estimated to exceed B_{lim} at the beginning of 1998, so it can be expected that status quo fishing in 1999 and 2000 will lead to an SSB of 0.9 million tons in 2000.

Scientific advice and biomass estimates on the North Sea herring stock are provided by the International Council for the Exploration of the Sea (ICES 1999a, and 1999b). A variety of different input data build the basis for the assessment: commercial landings (usually official figures) and sampling data as well as indices from (currently) four fishery-independent surveys conducted yearly. Both are provided by the National Institutes; only the surveys are internationally coordinated.

North Sea herring recruitment did not seem to be strongly dependent on environmental or climatic fluctuations. The currently conducted single-species assessment and management benefits from the fact that there is little species mixing in the managed herring fishery. However, there are some severe problems with the management. The impossibility of differentiating stocks or stock components within the North Sea stock complex may lead to the collapse of small populations, for example, the Downs herring, even under the current management regime. Furthermore, sampling of the commercial catches is often not adequate. This could be solved by coordination of the national sampling schemes (e.g., by a common European fisheries institute). Misreporting as well as exceeding TACs may lead to an overshoot of TACs by up to 30 percent. This introduces a significant (and avoidable) uncertainty into the assessment, which is finally to the detriment of the fishermen.

North Sea herring management is now used as an example of the way European fish stocks can be managed (ICES 1999a, 1999b). However, there are serious limitations in transferring the single-species herring regime to multispecies fisheries, like most demersal fisheries. In order for these to be managed intelligently, there needs to be progress in multispecies modeling in the coming years. This will require a significant increase in our knowledge of ecosystem mechanisms, which can only be achieved by a substantial increase in targeted research effort.

harvesting at different trophic levels have been the subjects of long-term studies for over seventy years since the series of Discovery Expeditions, which, incidentally, was financed by the revenues of the whaling industry. We now understand the waters off South Georgia as a cold, highly productive ecosystem, surrounded by the less productive open Southern Ocean (Atkinson et al. 2001). In some, but not all years, krill play a prime role both as consumers of phytoplankton and as the staple food for large colonies of seals and penguins. Krill, however, does not reproduce off South Georgia but is advected into the area from the south.

Long-term, comprehensive studies on the upwelling systems off California and southwestern Africa are further examples of programs that were initiated—and are still driven—by concern about the decline of fisheries. These studies grew into holistic approaches for the analysis of the functioning and interannual and decadal variability of ecosystems, with the aims of predicting fluctuations in catch and of managing the fisheries in an ecologically and economically sound way.

Similar studies have demonstrated the importance of total stock abundance and age composition on the geographical distribution of cod, Atlanto-Scandian herring and capelin, sardines, and anchovies (see, e.g., Bakun 1996).

SUSTAINABLE FISHING? REASONS FOR CAUTIOUS OPTIMISM

We have questioned the utility of much fisheries research with regard to sustainable exploitation of the living resources of the sea. The tools of input control (fleet size, engine power, mesh regulations, restricted areas, closed seasons, etc.) and output control (total allowable catch, landed catches, minimum fish sizes, catch quotas, etc.) have proven inadequate and inefficient in many cases. They did not prevent the "tragedy of the commons" syndrome in fisheries, where the resources are inevitably overexploited. An ever-increasing complexity of regulations and restrictions fosters neither common sense nor altruistic behavior. Instead, it creates a climate of tension, leading to resistance and confrontation between the fishing industry, public authorities, and scientists.

Nowadays, fisheries science increasingly accepts the notion of maintaining both fish populations and marine ecosystems as public assets. The ultimate goals are continuous high yields on the one hand, and protection of the marine environment and its biodiversity on the other. This calls for multidisciplinary research programs, supported by scientific capacity building, particularly in developing countries, where there is an urgent need for the management of rapidly dwindling marine resources.

In recent years there has been evidence of positive trends in the management of marine living resources and ecosystems. In some countries and fishing industries, there is a rapidly increasing public awareness of the need for precautionary measures in fisheries. Also, there are a number of cases where scientific advice has succeeded in stabilizing single species fisheries (see box 5.3).

Improvements in resource modeling should help identify critical factors and thus enhance the credibility of scientific advice by warning of impending disasters and helping to track the collapse and rebuilding of stocks. User-friendly and graphic-oriented simulation models could help managers visualize the effects of a planned management decision. These models would provide the nonexpert with clear illustrations of the effects of varying fishing efforts. This might improve the dialogue among scientists, administrators, the fishing industry, and conservationists.

One of the reasons for cautious optimism regarding greater sustainability in fisheries is the rapidly growing realization, among fisheries scientists, that marine protected areas (MPAs; also known as marine reserves) may help address some of the overfishing and sustainability issues discussed above. This is particularly the case in tropical and subtropical seas, but not exclusively so. It took a great deal of research, notably in the Philippines, New Zealand, and off the southeast coast of the United States, for a consensus to emerge (see Roberts et al. 1995; NRC 1999) that MPAs "enhance fisheries, reduce conflicts and protect resources" (Bohnsack 1993).

Small-scale fisheries tend to generate little discarded bycatch, because they frequently use highly selective, passive gear. The closeness of their ports to the fishing grounds also means they are more energy efficient. This proximity also gives small-scale fisheries a stake in the state of local resources, a feature often lacking in distant-water fleets. However, ocean resources cannot be exploited by small-scale fisheries alone. Large-scale fishing operations supply the global market for human consumption and animal feeds, and some companies are players on a global scale. Fisheries science must therefore continue to address the concerns of those fisheries.

In areas where small- to medium-scale vessels can adequately exploit the resource (as was the case for cod around Newfoundland), allocation of exclusive access to coastal resources would go a long way to putting fisheries, at least in developed countries, on a path toward sustainability. This is with the proviso that management strictly follows the best possible scientific advice and resists the pressure of various stakeholders in the community. In developing countries, population growth and lack of land work against sustainability in small-scale fisheries (Pauly 1997).

Research to monitor small-scale fisheries needs close collaboration with the fishing communities themselves. This would probably have beneficial effects on both the scientists and the science, making it both human centered and ecosystem based. The increasingly important voice of the conservation community also needs to be heard, because it often expresses a broader-based desire for the oceans and their resources to remain "healthy," to be a legacy for future generations. The expectation of society at large is that the resource will continue to be valuable in the long term.

Fisheries management has traditionally worked top-down. Government regulations are formulated on the basis of stock assessments, and the industry often strives to get around or subvert the regulations. Recently, the concept of co-management has modified the regulation/implementation part of this top-down model. Co-management means a form of shared responsibility between the government and the industry, which is increasingly involved in monitoring and tactical management. In some areas, one or two major multinational firms own all the fishing fleets and dominate the market. These firms claim to be abandoning a short-term "pillage and run" mode of operation and exhibiting more interest in sustainable management of the resources and in long-term strategic ecosystem-oriented approaches. This claim, however, must be taken with a grain of salt, because the fishing industry tends to go for high discount rates.

Some new approaches to fisheries management are structured around partial or even complete privatization of fisheries resources, that is, a move toward private ownership, sometimes within a system of co-management by stakeholders from the public and the fishing sectors. Except in a few cases where complete privatization can be achieved without major social conflict, consultative approaches will have to be developed to manage fisheries in which fishers and the various sectors of the industry, public authorities, scientists, conservation organizations, etc., all take part in decision making and management of the fishery. Such integrated groups will have to

- define the socioeconomic framework for the fishery (e.g., a balanced relation between fleet sectors, limited ownership, regional coastal development, etc.);
- define ecosystem-based management objectives (e.g., minimum stock sizes of commercial species, acceptable impacts on nontargeted species and on biodiversity);
- decide on (and pay for) all measures deemed necessary for input or output control;
- decide on scientific stock assessment procedures, gear development, technical measures, etc., to obtain the desired economic results;
- punish infringements, e.g., by withdrawing fishing licenses from individuals or producer organizations; and

- mitigate conflicts between fisheries, other users of the sea, conservationists, and the general public.

There is no global formula for managing the living resources of the sea. The variety of the targeted species and the complexity of the natural systems, on the one hand, and conflicting societal, economic, and political interests, on the other, call for different kinds of scientific advice. And not all marine animals should be considered as "resources" to be exploited by fisheries, even if, like whales and turtles, they are edible.

FISHERIES RESEARCH BEYOND 2000

The great economic and social importance of fisheries has been a strong driving force for marine research in general, including basic science. Fishery research and its time-series observations have made major contributions both to the development of population ecology and to monitoring changes in the marine environment and its food webs. Nowadays, fisheries scientists are particularly confronted with the tasks of understanding recruitment and multispecies interactions in relation to changing environmental conditions, as well as to fisheries and their management. Furthermore, fishery scientists see the living marine resources and any fish as part of a complex system and understand fisheries as an element of coastal and marine management.

Technological developments over the past decade in locating, catching, and processing fish have rendered fishing operations and fisheries science more effective. New instrumentation makes it possible to monitor—and hence predict—stocks and their environment with greater precision and to maintain surveillance of sea areas and fishing fleets. Differential GPS (global positioning system) allows fishing vessels to return to productive fishing areas, to trawl around obstacles, to recover lost gear, etc. The safety benefits of accurate position fixing are obvious, while the built-in monitoring systems in the vessel provide full external control of a multitude of processes and maneuvers. Modern sensors and communication systems make fishing vessels suitable platforms for many kinds of routine observations. The U.N. Food and Agricultural Organization is currently pushing for the use of these technologies, including "black boxes" with built-in GPS for satellite monitoring of fishing vessels. Both are important tools for the control and surveillance of fishing operations. Improvements in satellite imagery in terms of spectral bands and resolution in space and time have turned tuna fishers into fishery oceanographers. The demands on fisheries science and technology as seen by the large-scale fishing industry are summarized in box 5.4.

In recent years, the market for fish and fisheries commodities became increasingly globalized, with over 75 percent of the world catch sold and consumed in countries other than those in whose exclusive economic zones it was caught. Fish from Lake Victoria partly replaced North Sea plaice in German food stores; shrimps from Thailand, and Peruvian fishmeal, are in demand everywhere. European distant-water fleets operate off western and eastern Africa, and fleets from East Asia operate worldwide, catching tuna, squid, and other high-price species. Much of the large-scale fisheries and fish

---

Box 5.4. Demands on Fishery Science and Technology: A Fishing Perspective
Richard Ball

OBJECTIVES

Commercial fishing and fishers have a common interest in effective resource management, research, and logically consistent policies. Much of the responsibility for poor fisheries management lies at the political level and at the level of institutional management. Answers lie in the development of logically consistent fisheries management decision strategies (operational management plans), and the rigid application of cost-benefit analyses. There is insufficient use of market mechanisms to encourage and require responsible behavior.

PRIMARY OBJECTIVES OF COMMERCIAL FISHERIES: TO MAXIMIZE THE NET PRESENT VALUE OF CATCHING STRATEGIES—WITHIN CONSTRAINTS

- Commercial fishers have common interests in markets, sustainability, and stability.
- The best protection for a resource is that it is owned by those who have an interest in its good management and have assurance of their long-term participation.
- The fisher is the last residue of the hunting culture, where the benefits of the rights of property and legal certitude have had only limited application.

TECHNOLOGICAL IMPROVEMENTS IN PAST DECADE

Improved *position fixing* leads to:

- identifying previous viable areas;

- trawling accurately around hazards, recovering lost gear, and safety benefits;
- building better fishing database;
- vessel monitoring systems.

Satellite imagery of temperature and chlorophyll can identify thermoclines, gyres, fronts, and upwelling and can measure biological activity.

Improvement in *resource modeling* helps:

- identify critical factors;
- credibility;
- move toward parameterized decision making; and
- multispecies management, including interspecies, impacts, and environmental collateral damage control.

Comparative analysis of *alternative fishing strategies:*

- Technological advances should allow commercial platforms to partly substitute for expensive survey vessels and research ships.

REALISTIC OBJECTIVES OF THE NEXT TWENTY YEARS

- Move to bioeconomic analysis
- Cost-benefit disciplines
- Congruence of objectives with Marine Stewardship Council
- Better understanding of macro events like El Niño
- Improved management by integrating technologies above
- Effective application of electronic networks to deliver information and for dialogue with interest groups
- Application of farming and stock enhancement technologies for the development of genetically modified species and more active intervention in resource recovery
- Cooperation between marine science and commercial fishing, which have common interests (cost effectiveness and profit) and common threats: social concerns about resource management.

trade are in the hands of a few big firms, though globalization also affects many small-scale local fisheries.

How should marine science be tailored to the requirements of fisheries to help them meet the growing demands of a global market as well as the needs of local populations? Government-sponsored fisheries science has, until recently, largely concentrated on providing expertise for fishing industries exploiting large, often mixed populations of pelagic and demersal fish in the open shelf seas. Very different advice is required to manage inshore fisheries, which intensively exploit coastal fish populations, and fisheries science is not yet very advanced in this area. Scientific and cultural adjustments to the prevailing practice of fisheries science are required (Finlayson 1994). We have to improve our understanding of several factors, such as the impact of local fishing practices and other marine activities (e.g., tourism, shipping, pollution) as well as shifts in species interactions, global climate change, and climate-driven teleconnections. Of course, attempts to improve the status of our fisheries should not wait until "everything is known." But in the long run, sustainable exploitation has to be based on broad scientific knowledge for ecosystem management.

Fishing mortality and recruitment are the most important factors determining the size, age, structure, and hence productivity of heavily fished stocks. In addition, to avoid overfishing in the traditional way by regulating fishing mortality, we must quantify the relative impacts of top-down fishing and of bottom-up physical and biological oceanographic factors, both of which affect abundance, size composition, and productivity of fish stocks largely by controlling recruitment. Mann (2000) argues that the most important principle of management in the foreseeable future is to match the level of fishing effort to the environmental fluctuations. Stocks should not be fished hard when they are experiencing adverse environmental conditions. Long-lived fish in cool-temperate waters (e.g., cod) have evolved to survive decadal-scale environmental fluctuations by living long enough to have successful recruitment once in a while after many years of unfavorable conditions. Heavy fishing removes the large, old, and most fecund fish that are no longer there to spawn when a "good year" recurs. The combination of heavy fishing and decadal-scale environmental variation was presumably the cause of many stock collapses, particularly in temperate and cool conditions (Longhurst 1998). Managers must develop ways to allow survival of older fish to restore these fisheries. This principle may also apply to the shorter-lived sardine and anchovy fisheries of upwelling systems, on a shorter time scale (e.g., Mullin 1993). In years of good recruitment, heavy fishing of anchovy will do little harm, but in years of poor recruitment, fishing pressure should be relaxed to allow survival of short-lived fish to spawn again.

The future of fishery research will be in the right blend of studies in population dynamics and fisheries economics on one hand and in fisheries ecology and ecosystem research on the other hand. Studies in the feeding and migratory behavior of fish in relation to anthropogenic and natural changes in the environment will help to direct fishing actions, in order to make them more economical and less destructive to the bycatch and marine habitats and their biodiversity.

Ideally, scientific advice to fisheries management has to be based on broad ecological knowledge of single-stock dynamics, including the "classic" parameters of growth, recruitment, and mortality but also of population genetics as well as of ecosystem structure and functioning. We need to improve the monitoring of fisheries, that is, statistics of catches, discards, and landings. The geographical information systems (GIS) have great potential here, given their capacity to combine data on the geographical and oceanographic features of a given area with its resources and fishing operations, while considering variations in space and time.

The first task for fisheries science vis-à-vis stakeholders from both the fishing industry and the public sector is population assessment (see box 5.4). This means providing managers with information on the state of the stock of fish and other organisms and the ways that stock biomass and composition fluctuates with changes in environment and fisheries. In addition, multispecies interactions must be investigated on both an experimental and a strategic basis. A next step is to develop precautionary approaches for the formulation of management objectives that consider the sustainable use and preservation of the ecosystems and their resources. These approaches must assess the impact of several factors, such as fisheries, pollution, and climate, on the populations and systems. And, for this, a baseline of long data series is needed.

This opens a wide field of academic research on genetic population structure, recruitment processes, sources of mortality, climate effects, and so forth. Similarly, the specific socioeconomic constraints on the fisheries sector must be investigated within interdisciplinary groups aiming at a theoretical basis for approaches that limit the ecosystem impacts of fishing.

The overall goals of sustainable use of the living resources of the ocean are widely accepted. But they are difficult to implement, because they conflict with the immediate demands of coastal populations and with global market forces. The basic needs to protect the biodiversity and other marine bounties and beauties conflict with the tendency to maximize ocean exploitation.

Ecosystem-oriented research, combined with socioeconomic studies, should provide blueprints for the further development of both fisheries and nature conservation. New ways of management have to be found, based on new scientific approaches.

Both old and new questions continue to arise and have to be addressed. There is the issue of anthropogenic versus natural causes for population declines and stock collapses, for example, or small-scale local fisheries versus large fleets and industries operating globally, and the new protective measures, including marine reserves versus the traditional system of catch quotas and minimum length and mesh regulations.

Fisheries science for nearshore areas needs to be a mix of applied marine ecology and socioeconomics. Its main objectives must be to advise local user groups and national administrations on more sustainable management approaches, taking into account the common goal of maximizing national income as well as the social and economic constraints of local people.

Mariculture plays an increasing role, although fish and shrimp cultures consume more fish biomass (small pelagics and other fishes turned into fish meal and oil) than they produce in the form of shrimps, salmon, and other table fishes. Many believe in the continued rapid development of mariculture, provided that measures can be developed to protect fish and shrimp farms against diseases and to minimize the negative effects of such farms on both the environment and on wild fish populations, including those that serve as the basis for fish feed.

Here again, fishery research and management can no longer be seen as a matter of simply applying ecological recipes, but must involve issues of economics and equity at all levels, from the global market to the fishing village.

CONCLUSIONS AND RECOMMENDATIONS

The Marine Stewardship Council, devoted to ecolabeling of fisheries products, has identified three principles of fisheries sustainability:

1. Exploitation should not result in overfishing. In case of overfishing or exhaustion, the stocks have to be rebuilt.
2. The fishery should retain structure, productivity, function, and diversity of the ecosystem, including its habitats and nontargeted species.
3. The fishery should be governed by effective management systems.

Each of those principles requires a set of research activities:

- Detailed life history studies, stock assessment and predictive modeling for precautionary management under the conditions of varying recruitment and of multispecies interactions.
- Ecosystem research at different scales of space and time to understand the interactions of the targeted species with the system as a whole and to identify the collateral effects of fishing operations.

- Socioeconomic research to optimize fisheries within the ecological limits of fisheries and nature conservation.

RESEARCH PRIORITIES

- Defining the concept of "ecosystem-based management" in operational terms.
- Devising indicators of ecosystem status (or "health") and their implications for management of single-species populations.
- Identifying generic rules for optimal size and location of marine reserves as a major ecosystem-based fisheries management tool.
- Devising management regimes that provide fishers with incentives for conservative modes of resource exploitation.
- Developing fishing technologies with small ecological footprints (especially low habitat and other ecosystem impacts, and low energy consumption).
- Understanding recruitment processes in relation to environment and the interactions between top-down and bottom-up control in early life history.
- Understanding shifts in population genetics caused by fishing.

ACKNOWLEDGMENTS

The authors wish to acknowledge a multitude of comments and suggestions by participants of the Potsdam workshop and in particular Richard Ball, Jacqueline McGlade, John McGowan, Gerd Hubold, Brian Rothschild, Ramiro Sánchez, John Steele, and Christopher Zimmermann.

ENDNOTES FOR FURTHER READING

Jackson, J. B. C., M. X. Kirby, W. H. Berger, K. A. Bjorndal, L. W. Rotsford, B. J. Bourque, R. Cooke, J. A. Estes, T. P. Hughes, S. Kidwell, C. B. Lange, H. S. Lenihan, J. M. Pandolfi, C. H. Peterson, R. S. Steneck, M. J. Tegner, and R. R. Warner. 2001. Historical overfishing and the recent collapse of coastal ecosystems. *Science* 293:629–38.

Longhurst, A. 1998. Cod: Perhaps if we all stood back a bit? *Fisheries Research* 38:101–108.

Mann, K. H. 2000. *Ecology of Coastal Waters, with Implications for Management.* 2nd ed. Malden, Mass.: Blackwell Science.

Pauly, D., V. Christensen, J. Dalsgaard, R. Froese, and F. C. Torres Jr. 1998. Fishing down marine food webs. *Science* 279:860–63.

Watson, R., and D. Pauly. 2001. Systematic distortions in world fisheries catch trends. *Nature* 414:534–536.

Chapter 6

Ocean Studies for Offshore Industry

Colin P. Summerhayes and Karin Lochte

In this chapter we focus on recent advances and future needs in marine science and technology in support of offshore industries. We cover oil and gas, touch on leisure and environment, include the dumping of waste of various kinds, and address the construction, cable-laying, marine mining, and biotechnology sectors, mentioning environmental protection issues connected with these industries. We exclude shipping, defense, fisheries, and instrumentation, which are dealt with in other chapters.

The major issue is the growing demand for energy, stimulated by population growth, which is driving up the production of energy from offshore development—as well as increasing the volume of hydrocarbons transported by sea—and causing a search for alternative sources of energy from the sea. These increases have to be evaluated against their costs in environmental impact, especially on coasts. Fossil fuel will meet the bulk of the demand for energy in the near term, but oil reserves will begin to run out, making oil more expensive. Much of the petroleum in shallow-water oil and gas reservoirs has now been found and is being exploited. The trends are now moving toward three main alternatives:

- Cheaper and more complete development of mature oil and gas fields.
- Developing lower quality oil and gas fields in conventional water depths.
- Exploring for and developing new fields in deep water and in harsh environments, like the Arctic.

Deep-water gas hydrates are not yet commercially exploitable, so solar, wind, wave, tidal, and ocean thermal energy conversion (OTEC) will gradually provide more of the energy required (Kaku 1998). To keep fossil fuel costs down—and in response to mounting pressure from environment lobbies—the motor industry is developing cars powered partly or wholly by fuel cells. This will lead eventually to large numbers of electric cars or gas-electric, hydrogen-electric hybrids.

Legislation may help to control emissions from the burning of fossil fuels, which raise the levels of carbon dioxide in the atmosphere, increasing global warming, and in the ocean, diminishing the ability of corals and plankton to create calcareous skeletons (e.g., Gattuso et al. 1998).

In the past few decades, tourism has become an important growth area, especially in coastal areas, for beach holidays. This has put pressure on the coastal zone, notably through the construction of marinas for pleasure craft, hotel and camping complexes, and related services. These leisure uses are increasingly likely to conflict with industrial use, especially in densely populated regions. The need for comprehensive coastal zone management programs is therefore going to increase.

Population growth and wealth both produce more waste. The present goal is to minimize pollution of the marine environment, for example, through international action to reduce the input of pollutants in runoff from land, and to phase out the dumping of waste at sea. But the availability of land for waste disposal is necessarily limited, so disposal at sea may again become acceptable (Angel 1996; Angel and Rice 1996). There is even talk of using the deep ocean to dispose of some of the excess greenhouse gas, carbon dioxide (CO_2), arising from human activities. CO_2 has become steadily more abundant over the past century, and international protocols are seeking to stabilize and reduce emission levels.

Exploitation of exclusive economic zones (EEZs) became more feasible when the United Nations Convention on the Law of the Sea (UNCLOS) came into force in November 1994. The nearshore mining of sand and gravel for construction is growing to meet the demands of expanding coastal populations (Cook 1996). Increasing demand for phosphate as fertilizer may cause some offshore phosphate deposits to become economic (Cook 1996). Deep ocean metal mining was once thought feasible in the short term. It is likely to be postponed to the long term because it is still not competitive with land mining.

Public concern about the environment is having an increasing influence on political decisions regarding the use of marine resources and the siting of industrial operations. When new technologies are to be implemented in the offshore industry, it will be increasingly necessary to carry out an environmental impact study first and build monitoring strategies into the final development. This will aim to satisfy legal requirements while increasing public understanding. Environmental protection measures, established on the basis of research, have to address a number of considerations:

- Potential environmental impacts and requirements for monitoring of effects (e.g., through appropriate preoperation impact studies of adequate scope and the development of better monitoring and observation technologies).
- The prevention of accidents and damage (e.g., through selection of safe sites, safe equipment, prediction of possible dangers, and improved emergency measures).
- Containment or reduction of chronic disturbances (e.g., through technical improvements to minimize physical disturbances and leakage of pollutants).

These changes are going to require technological research and development in the various industrial sectors as well as in the supporting marine sciences. Meanwhile, any expansion in offshore industry and new technologies will call for a parallel development of adequate environmental observation techniques and strategies. The need to keep an increasingly critical public informed about potential environmental effects is also likely to drive research to some extent. So, partnerships between industry, government, and academia will need to be reinforced.

THE ENERGY INDUSTRY AND ITS MARINE RESEARCH NEEDS

Offshore Oil and Gas

There are perhaps two "last frontiers" in the search for giant oil fields: the Arctic and the continental slope at depths of 1000 to 3000 m. In 1999, the deepest production well was located in 1853 m of water in the Roncador Field in the Campos Basin off Brazil (Anon. 1999a).

But to find and extract hydrocarbons in these hostile environments (figure 6.1) requires extensive developments in both the supporting science and the operational technology, since many of the operations increasingly take place on the seabed. Improved and more environmentally friendly exploration and drilling techniques are required, as are methods for decommissioning and disposing of old offshore platforms and facilities. Larger diameter and longer pipelines are needed, along with deep-water riser systems to circulate drilling mud from the drill platform to the borehole. The riser systems keep the borehole clean and stable and carry chips of drilled rock (cuttings) back to the drill platform so that progress can be evaluated (e.g., McKenzie 1997; DTI 1997, 1999b).

Figure 6.1. Claymore oil rig in bad weather. Courtesy of Talisman Energy.

Geological/Geophysical Exploration, Site Survey, and Maintenance

As oil and gas reservoirs are explored at greater burial depths, there is going to be a need to clarify images of the subsurface geology. This will call for continued improvements in seismic reflection and refraction techniques, combined with advanced geophysical data processing and supported by down-hole sampling and advanced geophysical logging. Virtual reality visualization technology is being used more and more to help interpret these complex data. These advances will in turn influence fundamental studies of the geological development of passive and active continental margins. But there may be a downside to some of these developments. For example, there is already concern about the impact that the use of increasingly high-energy sound for seismic exploration is having on marine life, particularly whales and dolphins.

Operations in deep water may be affected not only by ocean currents and weather but also by the instability of continental margins. Slumps and slides can affect large areas of the deep sea floor through turbidity currents (Masson et al. 1996) (figure 6.2). It will therefore continue to be important to map the seabed and its shallow subsurface characteristics both for geological evaluation as well as platform and pipeline siting. These data will be of particular value as a basis for understanding and forecasting slope stability (e.g., to assess the risk of tsunamis being generated by slope failures) as well as for detecting shallow gas hazards. Improved models of slope stability will need increasingly accurate topographic studies by swath-bathymetry and side-scan sonar systems carried on surface and deep-towed vehicles.

Robotics will play a much greater role in work in very deep water. Autonomous underwater vehicles (AUVs) are needed for deep-water site surveys for platforms and pipelines to cut costs and guarantee safe drilling (Anon. 1999b). Remotely operated vehicles (ROVs) controlled via cables are similarly required to carry out complex tasks, such as rig testing in deep water. As these technologies develop, costs will fall, making them more accessible to the scientific community for scientific experimentation and survey operations in deep water.

The Ocean Drilling Program (ODP) has considerably advanced knowledge of the fundamental geology of continental margins, thereby improving our understanding of the habitat of petroleum (ODP 1996). The ODP offers the opportunity to expand these fundamental studies by probing the deep structure of continental margins and ocean crust. However, to drill 3 to 4 km deep sections safely in deep water the ODP needs a riser system (see

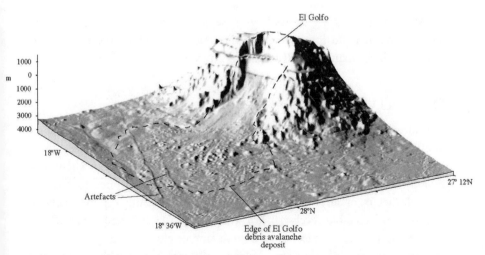

Figure 6.2. Three-dimensional image of the El Golfo landslide scar and deposit on the northwest flank of El Hierro, Canary Islands, viewed from the west. From Masson et al. 2002, with permission from Elsevier Science.

earlier discussion) along with the ability to seal the borehole with a blowout preventer to protect against hydrocarbon leakage. The passive continental margins formed where the Earth's tectonic plates move apart can throw light on the initial processes of stretching and breaking apart the continental crust, as well as on the history of crustal uplift and subsidence. At active margins, one plate is being subducted beneath another as the two converge, scraping sediments off the subducting one and forming an accretionary wedge. Models of how fluids and chemicals move through the wedge can be tested by drilling through the wedge, into the underlying subducting plate, using the boreholes as natural laboratories. This is already done where the wedge is thin, but riser drilling is needed to find out more about how the deformation system works at greater depths.

Physical Information for Offshore Operations and Design Criteria

Good information about currents, winds, waves, and sea-ice conditions is essential if offshore operations are to be cost-effective, safe, and environmentally acceptable. This need has become even more critical with the move to exploration in ever-greater water depths. For example, information on the vertical profile of currents is needed in deep water to predict shear effects on drill strings. For design purposes, the industry also needs to take into account seasonal and interannual variations, including the effects of global climate change. Box 6.1 shows how these various requirements continue to increase the demand for operational oceanographic information, using the Brazilian oil company PETROBRAS as an example.

OIL POLLUTION

Offshore oil and gas rigs are a source of chronic, low-level pollution, commonly caused by the disposal of drilling mud and drill cuttings, which smother the environment and may be contaminated with oil or chemicals. Biological evidence of significant disturbance is largely confined to within a few kilometers of any one rig, with damage decreasing exponentially to almost nondetectable levels 10 km away (e.g., Olsgard and Gray 1995). Changes in practice, for example, by reducing or eliminating the level of oil or chemicals in drill cuttings, help to alleviate this problem. Box 6.2 describes the effects on the marine environment of oil pollution from drilling and production. Research will continue to evaluate the extent of pollution and its effects over time.

Box 6.1. Contribution of Oceanography to Offshore Oil and Gas Activities: A Case Study from Brazil

Leonardo F. Souza, Jose Antonio M. Lima, M. Eulalia R. Carneiro, and Waldemar T. Junior

PETROBRAS has been producing petroleum from the Campos Basin on the continental margin west of Rio de Janeiro since the late 1970s. Initially, fields were developed on the continental shelf at depths of 90 to 170 m around 80 to 100 km from the coast. Oil production moved east to water depths around 500 m in the 1980s and to depths greater than 1000 m in the 1990s. Before 2020, production is likely from water depths greater than 2000 m, about 150 km offshore. Each progressive advance offshore has presented new challenges for oceanography.

During the 1970s, most of the meteorological and oceanographic (met-ocean) data needed to design safe offshore oil production platforms came from international databases and from ships' logbooks and naval databases, because local data on waves and winds were sparse.

During the 1980s, PETROBRAS worked with universities to improve knowledge of the marine environment and used simple numerical models to estimate tidal and wind-driven currents and extreme wind and wave conditions.

In the mid-1980s, with the discovery of the giant Marlim field (400 to 1000 m deep), it was noted that on-site measurements of current profiles at drilling platforms differed from those used in the design of platforms for operation on the continental shelf. So, new data were collected to provide greater reliability in the design of floating production systems for operation on the upper continental slope. To develop technology for deep-water production systems operating at 1000 m water depth, PETROBRAS created a technological development program that included the collection and analysis of met-ocean data in deep water. A buoy fitted with current meters throughout the 1250 m water column was installed over the Marlim field to collect data on surface waves, winds, and currents.

With the move toward production in water depths of 2000 m, a new technological program was created to develop technology for operation at those depths. One of the projects involved the use of moorings and hydrographic profiling with advanced equipment to characterize the subsurface, plus remote sensing and satellite imagery to map the surface features of the Brazil Current. Data were transmitted ashore to feed numerical models to make forecasts.

New production platforms are equipped with met-ocean instruments that collect data on waves, winds, and currents and transmit the data ashore for processing. Radar is used to measure wave parameters. Remote sensing is used to identify oceanographic features (eddies, internal waves, upwelling). The instrumented platforms form a data collection network that improves reliability. The enhanced data flow will make platforms safer and operations more cost-effective.

(continues)

> **Box 6.1. Continued**
>
> For the future, the growing complexity of offshore operations in deep water demands the growing use of advanced numerical models to forecast and simulate waves, winds, currents, and extreme events. Technological improvements are needed to reduce the costs of collecting met-ocean data. Several limitations must be overcome. First, field and remotely sensed data must be supplied in real-time to modeling centers to facilitate accurate mapping of present conditions (now-casting). Second, the predictability and reliability of meteorological models must be improved beyond the present 60 percent reliability for a five-day weather forecast and 80 percent for a two- to three-day forecast. Decision makers need more accurate information to enable appropriate operational responses to changing met-ocean conditions. Higher accuracy may come from improvements in the assimilation of real-time data by models, which is a global issue. In addition, some thought needs to be given to the possible effects of climate change on oceanographic variables and hence on platform design and offshore operations.
>
> Since 1989 PETROBRAS has used biological oceanography to assess impacts on the offshore environment. New developments are needed in studies of the bioaccumulation of pollutants in living organisms and in forecasting the state of biological communities.

To put these effects in perspective, it should be noted that drill cuttings affect around 1 percent of the total area of the North Sea, whereas, according to the OSPAR Commission's draft Quality Status Report for the Greater North Sea, trawling is seriously impacting 530,000 km^2 (i.e., about 70 percent) of the total area. At this time there seems to be no legislation underway to combat the effects of trawling. All efforts are focusing on oil.

We do not yet have an answer to the question as to whether it is environmentally better to clean up the cuttings piles as rigs are decommissioned or to leave them for natural processes such as biodegradation to clean up.

While pollution from drilling muds and drill cuttings is significant in the North Sea and other oil production areas, it is not the largest source of oil pollution in the ocean. The International Maritime Organisation estimates that nonpoint source discharge from land is now the main source of oil pollution. This pollution commonly passes unnoticed by the public, whereas oil spills from accidents, like that of the Exxon *Valdez,* that cause massive local damage attract media attention and divert attention from a very real problem. The other major sources of oil input to the marine environment include

Box 6.2. Effects of Oil Pollution from Drilling in the North Sea
Karl-Heinz Van Bernem

Drill cuttings are a large source of oil pollution and contamination in the North Sea. They are the end product of a drilling process, which starts with drilling muds being pumped down the well to lubricate the bit when a well is being drilled. The mud, which is under pressure so as to prevent a blowout when oil is struck, circulates back to the surface, carrying the cuttings chipped from underwater formations by the drill bit. The composition of the drilling mud depends on the nature of the strata being drilled. For easier drilling, the muds contain water. Harder drilling requires mixing the mud with up to 70 to 80 percent oil. Before 1985, diesel oil was primarily used, but now low-toxicity oil is usually used. Though there is some attempt to separate the drill cuttings from the oil-based muds before they are dumped on the seabed beneath the platform, they are inevitably still heavily contaminated. For some years, there have been attempts to ensure that oil-contaminated cuttings are brought ashore for disposal. There is no significant trend in the total amount of oil from these sources from 1984 to 1993, just fluctuations in response to economic factors.

The oil in cuttings and associated drilling muds may be accompanied by substances such as barite with variable amounts of toxic heavy metals, bentonite, inorganic salts, surfactants, organic polymers, detergents, corrosion inhibitors, biocides, and lubricants in the form of oil-in-water (oil-based mud) or water emulsions (water-based mud). The chemical composition and amount of contaminants depend on factors such as the type of rock being drilled, the depth of the well, and so on.

Factors such as the amount and type of cutting discharged, topography, and current regime dictate the size of the contaminated area around an oil platform. The total area of seabed contaminated with oil in the North Sea is about 8000 km^2. Typical concentrations of oil in sediments in remote (background) areas are normally in the range of 0.2 to 5 mg THC per kilogram of dry sediment (THC, total hydrocarbons). In contaminated areas of the United Kingdom sector, values of around 15 mg THC/kg are found. Within, say, 50 m of platforms, oil concentrations as high as 10 to 100 g THC/kg have been detected.

Drill cuttings dumped on the seabed have a profound impact on the benthic fauna. Close to source they blanket the seabed, create anoxic conditions that lead to the production of toxic sulfides in the bottom sediment, and so more or less eliminate the benthic fauna. Outside this area is a recovery zone, showing a succession of

- opportunistic species such as *Capitella capitata* (polychaete worm),
- species able to tolerate stressful conditions, and
- the less tolerant species, which gradually reappear farther from the center of disturbance.

(continues)

> **Box 6.2.** Continued
>
> This disruption extends up to 5 km, more in places, around North Sea drilling platforms. Where bottom currents are strong, as in the southern North Sea, discarded drill cuttings are more widely dispersed and do not blanket the seabed.
>
> Once drilling has ceased and oil production starts, the areas over which there are detectable biological effects or oil contamination decrease. It usually takes the macrobenthos two to three years to recover in the moderately affected zones.

- accidental or illegal discharges from ships at sea or in harbor, and
- deposition from the atmosphere as a result of the burning of fossil fuels (this input is the most difficult to estimate).

DEEP WATER CORALS

Despite exposure to agreed standards of operational discharges of oily water, drilling muds, and chemicals, apparently healthy colonies of the coral *Lophelia pertusa* have been found growing on oil platforms in the North Sea (Bell and Smith 1999). Such colonies could spread with time in the North Sea if the "footings" of large platforms were left in place at the time of decommissioning (Bell and Smith 1999).

DISPOSAL OF OFFSHORE OIL AND GAS FACILITIES

At the time of writing, the U.K. sector of the North Sea contained 219 platforms and 9870 km of pipeline. Once these and other similar facilities have come to the end of their useful life, they must be disposed of. This issue attracted considerable attention when the planned disposal of the Brent Spar in the deep Northeast Atlantic in 1995 was eventually abandoned after public pressure. At present the industry favors land disposal of steel platforms. There will have to be a full appraisal of the problems of platform disposal on land or at sea, because an increasing number of platforms will be decommissioned in the next few years.

Separate financial and strategic considerations made by individual operators for each installation will determine the timing of decommissioning and abandonment operations. As far as can be predicted at present, the bulk of decommissioning activity in the U.K. North Sea will fall in the period 1995 to 2025 (see the AURIS Decommissioning Report at www.ukooa.co.uk/

issues). Shepherd et al. (1996) noted that some of the 3500 platforms in the Gulf of Mexico are disposed of by being converted to reefs, which are reported to attract fish. But the authors accepted that there may well be reservations about translating a "rigs to reefs" policy to the North Sea. Research is needed to provide the basis for informed decision making.

When studying the use of the deep ocean as a place in which to dispose of used oil platforms, Shepherd et al. noted that the deep ocean environment is cold and dark, typified by abyssal waters at depths of around 4000 m. Biological productivity and standing stock are low, though biodiversity (in terms of species richness) can be quite high. Physical and biological transport processes from the seabed to shallow waters are weak, and there is little or no fishing. The environment is marked by processes that are generally slow compared with those on the continental shelf, apart from the occasional benthic storm caused by eddies reaching to the bottom, when currents may reach more than 25 cm/sec, and (rare) turbidity currents.

Shepherd et al. suggest that much may be learned about the possible effects of disposing of large structures in such environments by studying analogues such as shipwrecks. The deep sea is already disturbed by large-scale events, such as deep-sea hydrothermal vents (introducing metals), cold seeps (introducing hydrocarbons), and the arrival and decomposition of large dead animals (whales). The authors point out that the Brent Spar was equivalent to an oil tanker of 30,000 to 50,000 tons, with its tanks pumped out. So far, no effects have been detected on the deep ocean from the million tons of wrecked shipping scattered on the bottom. But these wrecks must create local disturbances, for instance, by smothering organisms. Shepherd et al. estimate that it would take two to ten years for the local environment to recover from disposal of a rig, provided that all toxic materials were first removed. The authors do not promote the deep-sea disposal of decommissioned offshore structures or of any other wastes. But they conclude that the environmental impacts of deep sea disposal of structures such as the Brent Spar are probably too small to be a crucial factor in the selection of the best disposal options or to exclude this option from consideration.

GAS HYDRATES: AN ALTERNATIVE ENERGY SOURCE?

Seabed gas hydrates may be a potential alternative energy source to fossil fuels (e.g., DTI 1997; McKenzie 1997). Gas hydrates (or clathrates) are solid, ice-like substances consisting of water and natural gas (methane), with the gas molecules trapped in cages of ice. They occur wherever temperature and pressure conditions are appropriate, for example, in permafrost in polar regions and beneath the seabed on continental margins in water deeper than about 450 m. They are the energy source for specialized bacteria that provide food for larger animals. When gas hydrates are located close to the seabed, abundant communities of mussels, worms, crabs, and fish develop, all indirectly sustained by methane released from the hydrates. These environments are special, productive habitats in the otherwise rather barren deep sea.

With increasing depth below the seabed, gas hydrates eventually become unstable under the influence of the Earth's internal heat. The base of the hydrate layer is a phase boundary between frozen gas and free gas. As the existence of clathrates speeds up the velocity of sound passing through seabed sediments, the change in velocity can be used to detect the hydrate layer. Below that layer, where conditions are appropriate, free gas may be trapped in potentially economic amounts.

Scientists from the U.S. Geological Survey have been mapping the thickness of the widespread marine gas hydrates using geophysical signals. They estimate that worldwide reserves represent an untapped source of hydrocarbon energy twice as great as that found in all known fossil fuel deposits on Earth (Cruickshank and Masutani 1999). Offshore gas hydrate deposits have been drilled by the ODP, which found them to be rather discontinuous. This could mean that such deposits are low grade. They may also be rather difficult to mine, and the gas (160 cubic meters of methane gas per cubic meter of hydrate) may prove difficult to extract. Modes of recovery are likely to be unconventional.

Nevertheless, some governments, notably those of Japan and India, still see continued research into their potential as a priority, with the first commercial production from hydrate deposits off Japan planned for 2010 (Cruickshank and Masutani 1999). Several other nations have also recently stimulated research into gas hydrates, notably Canada, the European Union, and the United States. Much research is needed to refine geological models of hydrate formation. This will help us to identify the geophysical signals correlated with hydrate concentration and mode of formation, and thus find the deposits with the greatest commercial promise.

Submarine slides are widespread on continental margins, but we do not fully understand what is happening. Improved understanding of the nature of gas hydrates should provide new insights and improve models of slope stability. We should also learn more about the possible role of gas hydrate stability under changing climatic conditions. Release of the greenhouse gas methane from decomposing marine gas hydrates is likely to have, or may in the past have had, a considerable influence on climate.

RENEWABLE MARINE ENERGY SOURCES

Greenhouse gas emission targets set at the Kyoto Meeting of the Conference of the Parties to the U.N. Framework Convention on Climate Change in December 1997 (the Kyoto Protocol) helped to accelerate the development of renewable energy sources as alternatives to the use of fossil fuels. Marine sources under consideration include offshore wind power and the extraction of energy from waves, tides, currents, and large ocean temperature differences (DTI 1997, 1999b; McKenzie 1997). Offshore wind farms already exist (e.g., at sites around Europe), while research and development in wave power are underway. The economic potential and environmental impacts of alternative offshore energy sources still have to be assessed.

Ocean thermal energy conversion (OTEC) uses the difference in temperature between warm surface water and cold deep water to power a turbine and generate electricity. The potential for applying this technology is greatest in the tropics between 10°N and 10°S, where warm surface waters of 25 to 30°C overlie cold water of 4 to 7°C at depths of 500 to 1000 m. The National Energy Laboratory of Hawaii has been operating OTEC plants (Matsuda et al. 1999). A near-term application may be land-based plants of 1 to 10 MW capacity designed for small island states.

OTEC may provide multiple byproducts as well as energy (Takahashi 1992; Nakashima 1995; Brown 1994; Matsuda et al. 1999). The cold water can be used for air conditioning, to encourage plants to grow that would not otherwise grow in the tropics, and to generate freshwater through desalination. Excess power could be used to generate hydrogen fuel. Matsuda et al. propose that OTEC plants on offshore floating structures could be used to generate artificial upwelling to raise nutrient-rich subsurface water to fertilize the upper ocean in what they call a Blue Revolution. The nutrients would stimulate the growth of plankton that would feed fish in a kind of ocean ranching operation. OTEC plants in Japan and in Hawaii have already used conventional pumps to supply ponds with nutrients to feed algae and fish, making the proposers confident that this process can be replicated offshore.

Some Possible Negative Effects

Little is known about the effects on the environment of large-scale extraction of energy by wave machines, offshore wind farms, or OTEC plants. The large scale of the plants required to produce alternative energy economically may have significant environmental impact. This is readily apparent in the extraction of tidal energy from the ocean, which usually requires the damming of rivers or estuaries and can lead to substantial environmental changes. But these changes may not all be for the worst. Mud flats exposed at low water may be less disturbed by currents and thus more productive. Calmer waters may provide excellent facilities for recreation.

Regarding artificial upwelling created by OTEC, any increase in production could be offset by increased oxygen demand in the bottom waters. Offshore wind farms are usually located in shallow waters with rich benthic life and can be valuable feeding grounds for fish and birds. The infrastructure for offshore energy plants includes the cables taking power from the site to the shore. The effect of large numbers of power cables on marine life is not known. In each case, research and development is needed, along with full environmental impact assessments.

OTHER MARINE INDUSTRY SECTORS

Submarine Cables

Advances in telecommunication services have stepped up the laying of marine fiberoptic cables. There are plans to construct a worldwide terrestrial and undersea fiberoptic cable network in the next few years to give twenty-five times present capacity. A Pan Pacific fiberoptic cable from Japan to the United States, plus a Trans-Atlantic cable and a Pan-European crossing, will link major business centers throughout Europe, Asia, and America.

Other telecommunication systems are in the design and construction phase, for example, to link offshore petroleum production platforms by submarine fiberoptic cables connected to land lines, to enhance oil platform automation and robotic capabilities, and to add emergency medical capabilities to remote sites.

There is a growing demand for submarine power cables to permit export and import of electrical power between continents and offshore islands (e.g., between mainland Europe and the United Kingdom) and to supply power to offshore installations, or to bring power ashore from offshore power sources, like wind farms.

High-resolution bathymetric maps are needed to help select the best route and design for these cables. There is also a need for detailed information about sediment characteristics and geological structure in the top few meters of the seabed, along with information about the strength of bottom currents, to see whether it is feasible to bury the cables. It will be important to evaluate slope stability and seismic risk, because slumps, slides, or turbidity currents can break or damage cables.

But before seabed surveys can be improved, some other developments will be needed:

- Side-scan sonar and other mapping systems need to be an order of magnitude more productive and capable of providing very high resolution at high speeds.
- Accurate environmental information must be extracted about the character of the seabed and subsurface from sonar signals.
- Data collection must be made a semiautomated activity integrated with digital navigation, charting, and geographic information systems (GIS) (Summerhayes et al. 1997).

Maps and sediment data from cable surveys may be useful for geological research on sedimentary processes. Cables offer the potential for monitoring seismic activity or observing benthic communities. This approach to data collection from the deep sea floor is very promising.

Seabed Mining

SAND AND GRAVEL

In many areas offshore sand and gravel (aggregate) extraction is an essential source of materials for the construction industry on land. This activity will spread, along with a growing demand for the use of aggregates for "soft" engineering solutions for coastal protection. Research is needed to find appropriate materials as well as to look at the cumulative and regional effects of aggregate extraction on ecosystems and sediment dynamics (Summerhayes et al. 1997). Since the exploited deposits are mainly in shallow coastal areas, there are large environmental impacts on bottom living fauna and on the fish and birds that feed on them.

SULFIDE

The ODP has already given a boost to our knowledge of the sulfide deposits on mid-ocean ridges, helping us to understand how sulfide ore bodies are formed (ODP 1996). ODP borehole observatories are collecting time-series measurements, studying circulating liquids, and monitoring active processes over the long term.

METAL DEPOSITS

Several offshore metal deposits have long-term potential:

- Manganese nodules and encrustations rich in copper, nickel, zinc, or cobalt
- Red Sea brines rich in metals
- Metal sulfide deposits on mid-ocean ridges

The depressed state of the metals market means they will not be exploited for some time to come. However, some nations retain an interest in research into deep-sea metal mining (notably Japan, Korea, China, Russia, France, and the United States) (e.g., see Cruickshank 1999). Tailings from manganese nodule mining have potential economic uses in building materials, drilling muds, ceramics, and coatings that could justify their exploitation (Cruickshank 1996). There is ample scope for research in resource evaluation.

Mining of metal-rich materials, such as Red Sea brines, manganese nodules, and crusts (figure 6.3), and sulfide deposits could have major environmental impacts, however. Mining manganese nodules or crusts will involve the transport or relocation of large amounts of material, and it will inevitably cause large disturbances in an otherwise quiescent habitat (box 6.3).

Phosphorites

Most phosphate fertilizer comes from onshore sedimentary phosphate rock deposits (phosphorites). Over the past 40 years phosphate production from these sources has grown by 500 percent. With growing population, these tra-

Figure 6.3. Sea floor of the deep Peru Basin (4150 m) covered by manganese nodules that are inspected by a rat-tail fish (*Coryphaenoides* sp.). Photograph courtesy of H. Thiel, DISCOL Project.

ditional sources may no longer satisfy the demand, leading to exploitation of offshore deposits, such as those on the Chatham Rise off New Zealand. Some developing countries may not be able to afford to continue importing if prices rise and may turn to offshore deposits (Cook 1996). Offshore

---

Box 6.3. Effects of Manganese Nodule Mining on the Deep-Sea Ecosystem
Karin Lochte

DISCOL, the disturbance and recolonization experiment in a manganese nodule area of the deep South Pacific, and ATESEPP, a study of the effects of technical impacts on the deep sea ecosystem of the southeast Pacific off Peru, have given us a great deal of information to guide the development of protective measures before mining is started. This new knowledge from small-scale studies will be put to the test when large-scale deep-sea mining operations start.

Deep-sea manganese nodules are likely to be mined as follows. At the seabed, a caterpillar-tractor moves forward carrying a system that flushes away the surface sediments with a water jet and collects the nodules. Most of the sediment and the smaller animals that are picked up by the collector are released quickly close to the seabed. Nodules and attached animals together with larger animals and some residual sediment are transported up to the mining ship through a suction pipe. Tailings consisting of residual sediment and nodule shavings are returned to the seabed through a waste pipe. Cargo vessels service the mining ship and carry the nodules to land for further processing.

The potential disturbances at the sea floor are obvious. The mining process is rather like strip-mining, with inevitable massive local disturbance to the deep-sea ecosystem. It would represent the biggest direct human impact yet seen in the deep sea.

By removing the surface sediment layer and manganese nodules, the substrate on which many organisms live will disappear, and the chemical environment will change. The mining process exposes deeper sediment layers, with low oxygen content, while releasing heavy metals from the sediments. This could harm the animals that remain. However, there is evidence that organisms living in polymetallic nodule fields are not very sensitive to higher metal concentrations. It appears that only the initial burst of dissolved metals may be difficult to tolerate. But this aspect has not yet been fully investigated and is not yet understood. Models simulating reestablishment of the original chemical gradients at the sediment surface have found it will take decades to centuries. However, settlement of the released fine material may speed up this process.

Large particles sink fast and are deposited close to the collector tracks or below

(continues)

Box 6.3. Continued

the waste pipes. This material will only affect the immediate surroundings of the impacted area. But fine particles, with their high content of organic matter, sink very slowly. How long they remain in the water column depends on the depths at which they are released, while their distribution is determined by the prevailing currents. Much of the sediment resuspended by mining will contain organic material whose decomposition is likely to increase the consumption of oxygen in bottom water over the impacted area.

Studies with pilot collector systems showed that some of the small organisms that lived off organic remains in the sediment survived the collection process and settled in the disturbed area. The mining process generally destroyed larger organisms that could not escape. Representatives of these species can resettle the disturbed area by migrating into it from undisturbed areas and by larval transport, provided there is enough food. Studies carried out seven years after a mechanical disturbance demonstrated that the fauna that had returned were more irregularly distributed, had different diversity patterns, and were less abundant than before. The greatest impact was on the fauna that previously settled on the nodules, but which had been removed with the substrate to which they had been attached. In this case, resettlement was impossible, because no hard substrate remained to settle on. Depending on the size of the operation, this could lead to species extinction. In order to preserve this special community, some areas of manganese nodules should be protected.

Studies suggest that the food web structure in the disturbed area is likely to be altered. Estimates based on the long-term resettling experiment (DISCOL) indicate that it will take decades before a balanced community is reestablished. The new community will most certainly be different from the original one, since the nodules, which structured the original environment, will have gone. It will probably resemble communities in the surrounding nodule-free soft sediment areas.

---

mining of phosphorite would probably release some phosphate into surface waters, adding to the runoff of nutrients from land, stimulating the formation of phytoplankton blooms, and thereby depleting oxygen levels. Phosphate inputs in excess of nitrate may furthermore encourage the mass growth of potentially toxic cyanobacteria (Graneli et al. 1999).

Biotechnology and Medicine

During the next twenty years we can expect to see spectacular growth in biotechnology. Automated DNA sequencing will unravel the DNA code of thousands of organisms, with profound implications for biological sciences and medicine (Kaku 1998). Genome sequencing and DNA chips, advanced computational analysis, and new biosensors will help take this field forward (Colwell 1999).

Marine organisms often possess unique characteristics to cope with their environment. The marine environment presents a vast range of conditions not found on land, such as deep hydrothermal vents, where life survives at over 100°C, high salt concentrations, and extremely high pressures. Sessile organisms have evolved mechanisms for firm attachment to substrates. Other physically vulnerable animals have developed highly potent toxins to deter predators, capture prey, and prevent overgrowths. Many of these adaptive characteristics are of pharmacological interest.

Several medicines are already produced from marine plants and animals. The great diversity of marine species makes it likely that many more have medicinal properties. Pharmaceutical companies are stepping up their studies of these organisms, but the rate of discovery of beneficial new compounds is still low, not least because of the relative inaccessibility of the marine environment.

Marine organisms are also the basis for many other kinds of products. For example, kelp have long been used as a source of fertilizer and food for livestock. Algin from kelp controls the properties of mixtures containing water, and it is used in many industrial applications in the paper, pharmaceutical, cosmetic, and food industries. There is considerable potential for using natural products from marine organisms instead of synthetic chemicals to prevent fouling. Antifoulants and bioadhesives are already being exploited. Other marine-based products include aquaculture pharmaceuticals and diagnostic systems; waste-stream degraders; oil and chemical spill treatments; enzymes and polymers; cosmetics; reagents for clinical diagnosis; fuel production from biomass; and pesticides.

There is considerable commercial interest in industrial uses for the hyperthermophilic bacteria of hot vents on mid-ocean ridges. Organisms surviving at the high metal concentration of hydrothermal vents may also present new solutions to heavy metal resistance and detoxification. These microbes may be abundant within the seafloor below hydrothermal vents and are likely to be the focus of flourishing research in the decades ahead.

There is also commercial interest in the psychrophilic bacteria that thrive in the extremely cold Arctic waters. Their biological products may be useful for defrosting cars or preventing frostbite (Colwell 1999). Bacteria useful for industrial processes under high pressure and low temperatures may be found in cold, deep-sea water.

Deep ocean drilling by the ODP has discovered live bacteria hundreds of meters beneath the seabed, suggesting that there is a deep and previously unsuspected biosphere that may be as large as the one above the seafloor (Parkes et al. 1994). Here surprising new discoveries are likely as the organisms of the deep biosphere must have been separated from the populations at the sediment surface for very long periods of time and may have evolved new biochemical properties. The challenges of extracting, cultivating, and handling the organisms from this extreme environment will require the development of new technologies.

Marine organisms are of interest to humans not only because of their exploitable properties, but also because they may cause disease. Colwell (1996) has drawn attention to the possible role of phytoplankton blooms off India, Bangladesh, and Peru in spreading cholera to humans and the apparent association of such events with ocean warming. However, the evidence of a link between algal blooms and cholera has been questioned (Gray et al. 1996).

WASTE DISPOSAL

Disposal of Solid and Liquid Waste

The space in the ocean is already being used as a resource for waste disposal. But contemporary legislation is tending to oppose this, reflecting the view that the sea is not a dustbin. Efforts are increasing to minimize or eliminate waste production at source or to reclaim, reuse, or recycle it (Hawken 1993). Nevertheless, as pressure grows on land space, waste disposal in certain parts of the deep ocean may emerge as a viable alternative if experiments can demonstrate that it is safe and containable (Angel and Rice 1996; Shepherd et al. 1996). Studies of the feasibility of disposing of organic-rich materials like sewage sludge, fly ash from municipal incinerators, and dredged material suggest that different disposal strategies are needed. Dredged material containing contaminants may be best suited to abyssal isolation, whereas sewage sludge may be better disposed of by dispersion. Meanwhile, fly ash, containing high levels of heavy metals, is a poor candidate for ocean disposal (Young and Valent 1997).

Three broad kinds of research can be envisaged:

- Analogy from previous unintentional and uncontrolled experiments
- Controlled experiments
- Modeling

Studies of past military and civil waste disposal in deep water as well as past shipwrecks in deep water could improve our understanding of the effects of the disposal of waste or structures in the ocean.

Large-scale controlled experiments provide another way of assessing the potential impact of industrial scale activities on the ocean, as in the projects described in box 6.3 on manganese nodule mining. Although these experiments themselves may substantially disturb the environment, the European Community's RISKER project (Environmental Risks from Large-Scale Ecological Research in the Deep Sea: A Desk Study) (Thiel at al. 1997) is backing them. RISKER argues that such experiments

- are essential before commercial scale deep-sea mining or deep-sea disposal (of sewage sludge, dredge spoil, or carbon dioxide) can be contemplated; and
- must be conducted early enough to ensure that adequate data are available to guide, and possibly restrain, a potentially rapid proliferation of commercial development in the deep ocean. Adequate automated in situ observation techniques need to be developed to improve the monitoring of impacted sites.

Finally, modeling is an essential tool in combination with observations to assess the probable fate of contaminants once deposited.

Nonendogenous Species

In many cases the effects of the introduction of alien substances on deep-sea life are not yet known. Low numbers of organisms and low biological activity in the deep sea does not implicitly mean that dumping of wastes has little impact. Due to the slow pace of deep-sea life, recovery from disturbances will take much longer than in shallow waters. Since we have little information about the geographical distribution of most animals in the deep sea, it is very difficult to assess which regions of the deep ocean are vulnerable and should be protected and which ones are perhaps better suited as potential dumping sites. Therefore, research is needed to investigate direct effects on deep-sea organisms as well as to map the biological landscape of the deep sea. Furthermore, the ways that different types of waste are deposited also determine the severity of their impact.

Deep Sea Disposal of Carbon Dioxide

In the 1970s, Italian engineer E. Marchetti (1977) had the idea of separating carbon dioxide from the other emissions from smokestacks and pumping it into the ocean off Gibraltar. Here, the reasoning went, salty Mediterranean water would sweep it out into the deep North Atlantic and thus keep it out of the atmosphere (Monastersky 1999). At the time, the idea was not followed up because it was considered to be too expensive. However, in recent years the notion of carbon sequestration, as it is called, has found more favor, not least because of political demands for a reduction of carbon dioxide emissions. Some disposal is already taking place in old petroleum reservoirs below the sea off Norway (Monastersky 1999). Current scientific progress on carbon sequestration in the ocean is described in box 6.4 and summarized in several recent papers (Brewer et al. 1999; Tamburri et al. 2000).

Box 6.4. Ocean CO_2 Sequestration

Peter Brewer

With the adoption of the Kyoto Protocol in 1997 the industrial nations committed themselves to numerical targets for reduced emissions of greenhouse gases by 2010. Direct reduction of emissions is possible through increases in efficiency or alternative energy sources. But progress may also be made by enhancing carbon sinks through changing land use and by taking advantage of the enormous chemical capacity of the ocean.

The surface ocean today takes up about 2 billion metric tons of carbon per year by gas exchange across the sea surface. This is driven by the winds forcing the physical turnover of the surface layers and by the chemical driving force provided by the CO_2 levels in the atmosphere. The ocean is a vast, slightly alkaline (pH ≈ 8.1 to 7.5) solution, with a buffer capacity so large that eventually some 90 percent of all the fossil fuel CO_2 emitted to the atmosphere will be taken up by the ocean with only small changes in its pH. This buffer capacity ensures that once it is in the ocean, little will return within the next few centuries.

This exchange, however, is so slow that CO_2 can build up in the atmosphere, with the effects on climate we are seeing today. So, is it possible to speed up the process? We cannot change the rate of ocean mixing or affect the global gas exchange rate, but the idea of capturing industrial CO_2 from power plants and disposing of it directly in the deep ocean has been considered. The ocean deep waters have a circulation age of about 500 years, so the CO_2 would be isolated for many centuries.

CO_2 has special properties that make this process even more feasible. Liquid CO_2 is highly compressible, so that at great depths (about 3000 m) it becomes denser than seawater and will sink to the ocean floor. It also reacts with seawater at low temperature and high pressure to form a solid hydrate, with six water molecules wrapping a cage around each CO_2 molecule to form a clathrate, a strange solid, very similar to ice. This solid "ice" dissolves only very slowly. Recently, small-scale field experiments have been carried out to examine what happens when CO_2 is put into the deep ocean (Brewer et al. 1999). The rate of release of CO_2 from the "ice" is so slow that marine animals even very close by apparently show no negative response (figure BX6.1).

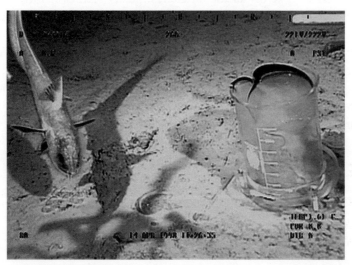

Figure BX6.1. Large Pacific grenadier fish (*Coryphaenoides acrolepsis*) inspects a blob of liquid CO_2 on the ocean floor at a depth of 3627 m. The liquid has been expelled from a nearby 4-liter beaker by the rapid reaction of CO_2 with seawater to form a dense mass of hydrate that has sunk to the bottom of the beaker, increasing the volume and forcing the remaining liquid to overflow. The experiment was carried out to test theories and models of ocean sequestration of fossil fuel CO_2 as a hydrate at great depth. Copyright 2001, Monterey Bay Aquarium Research Institute (MBARI).

This subject is ripe with exciting research prospects, such as finding out what exactly happens when you put carbon dioxide in the ocean and assessing impacts on the chemical environment and marine animals (e.g., Tamburri et al. 2000). At present, our understanding of the physicochemical behavior and the biological and geochemical effects of CO_2 disposed of in the deep sea is very limited, based on just a few small-scale experiments. A major challenge will be to demonstrate how much CO_2 could be disposed of in this way. This will then make it possible to evaluate the effects of disposal on that scale on atmospheric concentrations of the gas.

CONCLUSIONS AND RECOMMENDATIONS

When we were considering the probable developments of different industry sectors and their possible needs for research and development, we found it was also important to identify the most important scientific or technological advances needed in the industries addressed in this chapter. But two requirements cut across all the different industry sectors, namely, adequate environmental impact assessments and improved monitoring techniques, in particular automated systems. Both are demanded by legislation and to meet public concerns before industrial development can be accepted. Another cross-cutting demand is for advanced techniques for data information and management. These aspects are not specifically dealt with here, since they hold for future marine science in general, but they need to be taken into consideration in research related to offshore industries.

RESEARCH PRIORITIES

The following list of scientific or technological advances needed in the different industry sectors is not in any order of priority.

Energy

- Riser-supported ODP drilling of active and passive continental margins, supported by advanced multichannel seismic reflection profiling and seismic refraction, to improve understanding of the development of these systems, thereby aiding development of models for deep-water oil and gas exploration.
- Use of ODP holes as natural laboratories in which in situ measurements, logging, and down-hole geophysical experiments, including measurements of fluid flow and chemical fluxes, can be used to characterize the properties of sedimentary rocks and relate them to their deformational history, thereby aiding models of deep-water oil migration.
- Advanced study of the characteristics and causes of continental slope instability as the basis for improved prediction of slope failure.
- Advanced operational oceanographic information and models as the basis for improved forecasting of conditions affecting offshore oil and gas operations.
- Advanced study of seabed gas hydrates as a potential alternative energy source and a specialized habitat.
- Continued research on and development of renewable marine energy sources (wind, waves, tidal, and marine currents, and OTEC).
- Continued research on potential spin-off technologies associated with OTEC (e.g., algal and fish production, hydrogen production, etc.).

Cables

- Improved high resolution bathymetric survey capabilities for deep-water fiberoptic cable deployment.

- Studies of slope stability (see energy, above).
- Development of techniques to use undersea cables for environmental monitoring.

Minerals

- Expansion of fundamental studies of sulfide ore formation through use of ODP-based borehole observatories to collect time-series measurements of parameters, to study circulating liquids, and to monitor active processes over the long term.
- Improved understanding, characterization, and models of hot vent mineral deposition and biology and of the influence of vents on ocean chemistry.
- Research to find appropriate materials for construction and on the cumulative and regional effects of sand and gravel extraction on ecosystems and sediment dynamics.
- Better understanding of the impacts of extensive release of phosphate during subsea phosphorite mining.
- Improved understanding and modeling of sediment dispersal in relation to physical and biological processes in coastal seas and on continental margins to improve understanding and predictability of sediment entrainment and movement and its environmental impact.
- Better understanding of the resilience of marine habitats to human-made disturbances in order to protect biodiversity while allowing the (sustainable) exploitation of marine resources.
- Develop and establish automated in situ observation techniques to monitor impacted sites.

Biotechnology

- Improved molecular biology tools for characterizing pollution signatures, identification of animal populations, and screening for molecular products.
- Basic and applied studies of the potential of marine bacteria and higher organisms (including those from extreme habitats like hydrothermal vents, the deep biosphere, and gas hydrates) for the generation of beneficial new compounds for medicine, combating fouling, and other biotechnological applications.
- Biotechnological advances to identify new products and processes.
- Improved cultivation techniques and taxonomy to aid in identifying and analyzing marine organisms and their possible products.
- Better sampling and extraction facilities for gas hydrate communities and deep biosphere organisms.
- Improvement of techniques to access and cultivate organisms from extreme habitats (hydrothermal vents, deep biosphere).

Waste

- Fundamental research and modeling of the disposal of wastes, in particular CO_2, and its environmental impact.
- Better understanding of the potential mechanisms, effects, and impacts of the disposal of waste or structures in the deep ocean, including carbon dioxide.
- Better understanding of the biological and biogeochemical properties of different parts of the deep ocean to identify particularly vulnerable areas in need of protection.
- Better understanding of the risks associated with hormone disrupters and persistent organic pollutants.
- Improvement of environmental impact assessment methodologies.

ACKNOWLEDGMENTS

We gratefully acknowledge the help in particular of the participants in the Potsdam Working Group on Offshore Exploitation (non-Fishing), including David Rogers (Chair), Mary Altalo (Rapporteur), Rick Spinrad, Reg Beach, and Leonardo De Souza, Geoff Holland, and Ralph Rayner. In addition we gratefully acknowledge the writers of texts for boxes (Leonardo De Souza and colleagues from PETROBRAS, Ralph Rayner, Peter Brewer, and Carlo Van Bernem), and the reviewers of different drafts of the manuscript, notably Martin Angel, Julie Hall, Eduardo Marone, Gotthilf Hempel, and Colin Grant. Jaque McGlade provided information on biotechnology.

ENDNOTES FOR FURTHER READING

Angel, M. V., and A. L. Rice. 1996. The ecology of the deep ocean and its relevance to global waste management. *Journal of Applied Ecology* 33:915–26.

Brewer, P. G., G. Friederich, E. T. Peltzer, and F. M. Orr. 1999. Direct experiments on the ocean disposal of fossil fuel CO_2. *Science* 284:943–45.

Colwell, R. R. 1999. Marine biotechnology. *Sea Technology* January, 50–51.

Olsgard, F., and Gray, J. S. 1995. A comprehensive analysis of the effects of offshore oil and gas exploration and production on the benthic communities of the Norwegian continental shelf. *Marine Ecology Progress Series* 122:277–306.

Westwood, J., B. Parsons, W. Rowley. 2001. Global ocean markets. *Oceanography* 14(3):83–91.

WWF/IUCN/WCPA. 2001. *The Status of Natural Resources on the High Seas*. Gland, Switzerland: WWF/IUCN.

Chapter 7

Marine Information for Shipping and Defense

David P. Rogers, Mary G. Altalo, and Richard W. Spinrad

Ninety percent of world trade is transported by sea. In 1998, the world merchant fleet reached a record size of 85,828 ships exceeding 531.9 million gross tons (Lloyds Register 1999a). The estimated value of seaborne trade and shipping in 1995 was $155 billion (UNCTAD 1997; McGinn 1999). In addition, worldwide naval operations cost approximately $242 billion (Weber and Gradwohl 1995). Together, these activities have similar safety and efficiency requirements, which advances in marine science can help fulfill.

Trade by sea is increasing, with a demand for faster and more efficient cargo transport (figure 7.1). In coastal zones, where populations are growing rapidly, there is an additional demand for more and faster short-sea transport for both vehicles and passengers. In Europe, studies suggest that trade in bulk goods will double over the next ten years. Given the pressure on road and rail systems, much of the increase may go by sea. The increasing use of the ocean for transport, the development of larger and faster ships, greater competitiveness, and less tolerance for negative environmental impacts place enormous pressure on the marine transportation system to improve efficiency while maintaining or improving safety margins.

Transportation is not environmentally neutral. Maritime ports are located at the interface between the land and the sea and are in contact with important habitats, which are strategic components of the natural environment: the seabed, estuarine waters, rocky shores, coral reefs, mud flats, and wetlands. Adequate and continual assessment is essential to minimize the risk to the environment. The cost of environmental disasters is very high,

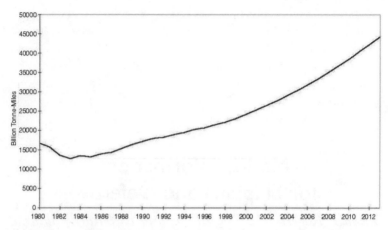

Figure 7.1. Projected growth in shipping (billion tons/miles) from 1980 through 2012. Reproduced with permission of the British Shipping Council.

both to society in general and to the human agents responsible; the Exxon *Valdez* disaster, for example, cost approximately $3 billion.

Despite stringent safety regulations, approximately 0.9 million tons of shipping and between 300 and 1000 lives are lost each year (Lloyds Register 1999b). About 3500 ships are involved in accidents in the United States alone, and about 6400 recreational boating accidents are reported each year. Human error is a significant factor in these losses, but they often occur in adverse environmental conditions. Using data from 1990 to 1996, the United States Department of Transportation estimates that human factors cause 75 percent of all commercial shipping accidents involving fatalities or significant financial cost, 16 percent are due to equipment failure, and 7 percent are due to weather (DOT 1998). Silting, shoaling, and debris are responsible for the remaining fraction of all accidents. The human factor includes operating in adverse environmental conditions without adequate situation monitoring. Adverse weather or seas were factors in at least 66 of the 175 ships lost in 1988 (Lloyds Register 1999b).

Improved ship design meets, in part, the demand for greater safety at sea and environmental security. Double-hulled vessels help minimize accidental oil spills, and better forensic methods developed from marine chemistry deter deliberate oil releases. International regulation helps the shipbuilding industry meet high engineering standards for ship design. But engineering solutions cannot solve all of the problems of safe navigation. There is also a need for up-to-date information about the operating environment in order to improve decision making and reduce risk. In particular, large vessels are vulnerable to rapid changes in water depth caused by wind fluctuations, poor predictions of worsening visibility increase the risk of collision, and heavy weather can cause delays and can damage cargo.

The technology exists, or will soon be available, to increase the quantity and quality of information obtained about the marine environment. There is an opportunity to provide benefits to a broad spectrum of users from marine operators concerned about optimizing vessel operations and routing to those involved in using ocean observations to improve climate forecasts or to sustain the health of oceans and coastal margins. This kind of service will need active partnering between the maritime industry, the military, international regulatory organizations, and research and development groups to develop a comprehensive observing and information system. The Global Ocean Observing System (GOOS) offers a strategy to accomplish this objective (IOC 1998a,b).

This chapter summarizes how environmental information has been used in the maritime industry and defense in the past, presents the current state-of-the-art technology, and suggests possible improvements that could be made over the next twenty years to improve decision making in commercial shipping, port operations, and defense.

PROMOTING SAFE NAVIGATION

Throughout the history of seafaring, weather and sea-state information have been critical to safe navigation. Severe storms, high waves, squalls, calms, sea ice, and icebergs have all taken their toll. Until the twentieth century, ocean-going shipping largely relied on climatological information compiled from past reports of weather conditions encountered by vessels in various parts of the world. In the nineteenth century, published atlases showed the recommended seasonal sailing routes around the globe and the effect on navigation of various wind regimes along the selected tracks (e.g., Maury 1860; Johnston 1856).

In 1853, Dr. Mathew Maury, then director of the U.S. Navy Hydrographic Office and later Superintendent of the U.S. Naval Observatory, introduced the concept of a uniform system for collecting meteorological and oceanographic data for the benefit of shipping. This was widely adopted and resulted in a standard form of the ship's log and a set of standard instructions for necessary observations. Maury combined 1,159,353 separate observations of the force and direction of the wind and 100,000 observations of surface pressure to establish the first global diagram of the winds (see plate 1 in Maury 1860). This was of great benefit both in practical navigation and for research into the physical processes that controlled global wind patterns.

The savings to commerce from knowledge of the climatalogical mean wind patterns was substantial. *Hunt's Merchant Magazine*, in May 1854, credited Maury with saving $2.25 million per annum by reducing the transit time of vessels outbound from the United States alone. The establishment of universal observations for meteorological application also owes its origins to the network of marine observations, which were first expanded to include land measurements by Vice-admiral Fitzroy, head of the Meteorological Department of the British Board of Trade in September 1860.

The practice of marine observations continues today as part of the World Meteorological Organization (WMO) Volunteer Observing Ship (VOS) program, with forty-nine participating countries and seven thousand vessels reporting data (Taylor et al. 2001). These reports include vessel position, surface pressure, wind speed and direction, air temperature, sea surface temperature, humidity, surface waves, and cloud amounts. In some regions, they are a major source of routine in situ marine meteorological and oceanographic data for forecasters and weather prediction modeling centers.

The VOS program is an effective system of cooperation on a nonprofit basis, devoted to the safety of life at sea. The cost to the weather services is merely the cost of the instruments and the transmission of coded weather messages. But there are disadvantages: the observations are recorded by crew members whose duties associated with the safe operation of vessels must often take precedence. But planned improvements in automation of shipboard observations will offset this problem.

Reliable forecasts rely on adequate initialization of numerical weather prediction models. The quality of the forecast product is directly related to the quality of the data entered into the model. In many instances, the uncertain quality of the VOS data means they are not used, while some modeling centers prefer to optimize model initialization to data from above the 500 mb pressure height. This in turn restricts the value of the forecast product for marine applications. There is an increasing reliance on satellite data for sea-surface temperature fields, surface wind velocities, and waves for model inputs. VOS data plays an important role in detecting biases in these remotely sensed data (Taylor and Kent 1999). These in situ data are also used increasingly for climate analysis and forecasting, loss adjustment in the insurance industry, and search and rescue. But careful correction of the data is essential for application in weather and climate studies (Jones et al. 1988).

Dissemination of forecast products to mariners has developed over the past decade or so, moving from the facsimile transmission of Difax weather charts generated by meteorological offices to full-fledged weather routing services. These are the contemporary equivalent of the nineteenth century navigational atlases aiding mariners in the selection of routes according to weather, sea ice extent, currents, and so forth. Companies such as Weather News Incorporated, which operates Oceanroutes and Weatherwatch, and national agencies, such as the Marine Weather Group of the U.K. Meteorological Office that operates MetRoute, provide these value-added services, tailoring products provided by various agencies to the specific and often unique needs of their customers. These agencies include the U.S. National Oceanic and Atmospheric Administration's (NOAA) National Weather Service, the U.S. Navy's Fleet Numerical Meteorology and Oceanography Center (FNMOC), European Centre for Medium Range Weather Forecasting, the U.K. Meteorological Office, and the Japanese Meteorological Agency. The aim is to help users make well-informed operational decisions that save time and money, while increasing the safe operation of vessels.

The development of these services has evolved in parallel with the growth in satellite communications. It is now possible to provide products through the Internet, exchanging information via personal-computer–based systems. Ships participating in these services are provided weather, optimum routing, and fuel consumption information. In the case of ocean routing, the ship's master will supply information about the vessel's performance, engine capacity, speed, fuel consumption, and handling characteristics along with port of origin, destination, and waypoints. The routing service will then supply environmental information and guidance, which can be used to select a route to minimize the effects of adverse weather and optimize fuel consumption. This is becoming increasingly important in the context of intermodal transport, where rail, road, and sea transport systems are integrated. Delays in one sector directly affect another; these must be factored into the optimization of each component.

COMMUNICATION

A limiting factor in the use of these types of services remains the cost of communication. Greater bandwidth satellite communication is desirable, but the cost of airtime is still prohibitively expensive. Advances expected in data compression for transmission will partially solve the problem. Even so, communication between the open ocean and the shore is an expensive undertaking.

The advent of wireless communication enabled ships to receive routine reports from other vessels and marine safety organizations and aided navigation through direction finding in coastal waters. But dependence on high-frequency (HF) radio for long distance communication is cumbersome, limiting ship-to-shore information exchange. This began to change in 1982 with the introduction of services provided by the International Marine Satellite Organization (Inmarsat), which was formed through intergovernmental agreement to provide communications for commercial, distress, and safety applications for ships at sea.

Inmarsat grew out of an initiative of the then International Maritime Consultative Organization, now the International Maritime Organization (IMO) under the auspices of the United Nations. At the time, mobile satellite communication was an unexplored technology and the industry still embryonic and untested. A joint cooperative venture of governments was required with contributing capital and expertise provided by national post and telecommunication providers, who bore the high risk of the venture. In April 1999, Inmarsat became the first international treaty organization to become a commercial company.

Today there are several competing service providers. Orbcomm provides a messaging service, similar to e-mail, but limited to less than 2000 characters. Global Star provides a full mobile telephone service, which will also be able to carry Internet services. As more communication providers enter this market, the cost of transferring data between ship and shore will come down.

COASTAL INFORMATION SYSTEMS

Coastal waters are dangerous areas for shipping—from the breakwater to the open ocean. On the one hand, water depth varies enormously, and traffic is at its most dense. Indeed, the most frequent causes of environmental disasters are groundings and collisions in coastal waters. Meanwhile, the revenue generated for both port operators and shipping companies is directly related to the maximum draught available in their respective port or waterway. An increase in draft for large containerships or bulk carriers can be worth as much as $12,000 in revenue per centimeter, per voyage. But deeper drafts and larger ships also increase the risk of grounding. United States national economic models indicate that if port shoaling or uncertainty about water levels only added 1 percent to the cost of crude petroleum imports, the United States would lose $3.1 billion from its gross domestic product, along with 61,000 jobs. It is therefore essential to be able to determine real-time changes in water depth that affect vessel safety in restricted waterways.

In many countries, coastal meteorological and oceanographic data are routinely available from buoys and other fixed platforms. National meteorological agencies provide forecasts tailored to shipping in coastal waters. But these forecasts need to be focused increasingly on regional scales and often do not contain sufficient information to provide much more than general synoptic conditions. Similar problems confront navies engaged in littoral operations.

In response to the need for high-resolution regional weather forecasts, the U.S. Navy developed the Coupled Ocean-Atmosphere Mesoscale Prediction System, operated by FNMOC. They recognize, however, that model predictions are only as accurate as the data used to initialize the models. There is a need to be able to define the operating environment as completely and immediately as possible (box 7.1). Rapid assessment of the environment will require a more comprehensive observing system than is presently available to either military or civil sectors. There is a need for very high temporal and spatial resolution observations of weather, bathymetry, and ocean conditions. These will depend on the development of new automated sensors and observing systems (see, e.g., IOC, 1998).

In the United States, the NOAA has developed a system to provide real-time information near selected ports, which helps to overcome this problem. The purpose of the NOAA Physical Oceanographic Real-Time System (PORTS) is to support safe and cost-efficient navigation by providing ships' masters and pilots with accurate real-time environmental information to avoid groundings and collisions. This type of system has the potential to save the maritime insurance industry from multi-million dollar claims resulting from shipping accidents.

PORTS includes centralized data acquisition and dissemination systems that provide real-time data on water levels, currents, and other oceanographic and meteorological variables from bays and harbors to the maritime user community. The system uses a variety of formats, including telephone voice response and the Internet. PORTS also provides "nowcasts" and predictions of these parameters with the use of numerical circulation models (figure 7.2). Telephone voice access to accurate real-time water level information allows United States port authorities and maritime shippers to make sound decisions regarding loading of tonnage (based on available bottom clearance), maximizing loads, and limiting passage times, without compromising safety.

Box 7.1. The Role of Marine Science: A U.S. Navy Perspective
Richard W. Spinrad

U.S. naval oceanography has evolved to meet the changing nature of the navy mission. Between 1950 and 1990, the major emphasis was on the open ocean, resulting in naval oceanographic research focusing on basin-scale and mesoscale oceanography, in conjunction with studies of underwater acoustics. Products of this research include the Modular Oceanographic Data Assimilation System (MODAS), developed for the U.S. Navy. MODAS was designed to provide the fleet with improved acoustic prediction. In effect, without MODAS, the fleet only has a climatology of ocean conditions (temperature and salinity fields) for their operating area. Such a climatology (if it even exists, given the paucity of historic oceanographic data) offers, at best, a very coarse definition of existing conditions in one's operating area. Given, for example, the highly variable nature of Gulf Stream meanders and warm core rings, a climatology is virtually useless at some times and locations within this part of the North Atlantic. MODAS, however, starts with existing climatologies, assimilates data from expendable bathythermographs, satellite derived sea-surface heights and temperatures, and produces a three-dimensional, detailed depiction of the local thermohaline structure.

In the past decade, the emphasis of the U.S. Navy has shifted from the open ocean to the littoral zone. Now naval oceanography emphasizes more meteorology, small-scale physical oceanography, nonacoustic technologies (e.g., optics, magnetics, and radar), and high frequency acoustics. One particularly powerful product is the Coupled Ocean-Atmosphere Mesoscale Prediction System, which has been developed to provide accurate forecasts of environmental conditions over relatively long periods (many days to weeks) and for all regions throughout the world with a spatial resolution of less than 10 km. As with all numerical models, the forecasts are only as accurate as the data used to initialize them. Thus, there is a need to define the environment within an operational area using observations, which depend on the application of next generation sensors and observing systems to obtain timely weather, bathymetry, and ocean conditions on a fine scale.

These advanced observations will be obtained in the context of a clearly defined strategy for research and development (available at www.oceanographer.navy.mil), battlespace meteorology and oceanography data acquisition, assimilation, and application. Acquisition means both sensing of data as well as "finding" (i.e., accessing) them through optimized database management. Assimilation includes the ingest and upgrade of data into models and simulations as well as the representation of the output through advanced visualization techniques. Application of the data will be for two fundamental functions: (1) prediction of system performance because of environmental variability and (2) decision-support regarding tactics that are affected by the environment.

> Because of the expected changes in naval operations and technical capabilities, there are specific challenges for the oceanographic community:
>
> - First, perhaps most fundamentally, is the ability to develop fully integrated, seamless representations of the environment from below the seabed to beyond the upper atmosphere.
> - Second is the challenge to build systems that are capable of fusing information and sampling in a sophisticated adaptive fashion.
> - Third is the need to deliver environmental knowledge, not just information, anywhere, and at anytime, regardless of bandwidth, in support of operations.
> - Fourth is the challenge to design architectures, which allow straightforward "plug and play" capabilities for inserting new environmental models and simulations.
>
> Research and development in a broad range of areas will address these challenges. Sensor concepts (materials and methods) and techniques for data assimilation will be important. Research in areas that are not specific to the environment, such as decision support and communications will also serve our oceanographic challenges well.
>
> In summary, the trends of oceanographic research as applied to naval operations have shown us that the efficiency of the transition of science to applications depends as much on the operational "pull" as it does on the technological "push." The exercise of speculating about trends in naval missions, when coupled with thinking about science and technology, suggests gaps in our future capabilities. Within the oceanographic community, these gaps will be filled, at least for the U.S. Navy, through a combination of research efforts dedicated to environmental understanding and research efforts common to many other technical arenas.

This type of system is critical to environmental protection, since marine accidents can lead to hazardous material spills that can destroy coastal ecosystems and the tourism, fishing, and other industries that depend on it (see, e.g., Petrae 1995). The human, environmental, and economic consequences of marine accidents can be staggering. Many of those losses are avoidable with better environmental information aids to navigation such as those provided by systems like PORTS. The original prototype for this system was installed in Tampa Bay, Florida (figure 7.3), in response to the tragic loss of thirty-five people who were killed in the May 1980 ramming of the Sunshine Skyway Bridge. This system is now operational in five high-traffic U.S. port areas.

Other prototypical operational systems include Coast Watch (e.g., Johannessen et al. 1997a), which has been tested along the coast of Norway. This

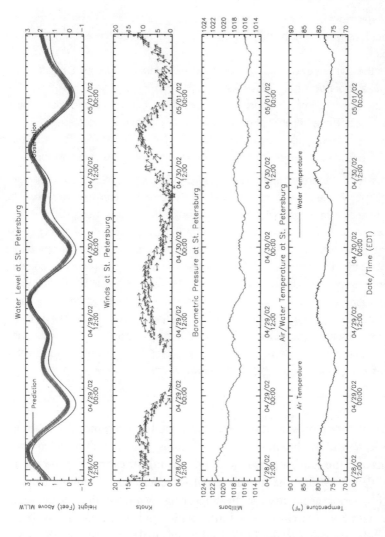

Figure 7.2. Observations and model predictions provided by NOAA National Ocean Service PORTS, for one specific location: Old Port Tampa, Florida. Courtesy of NOAA. MLLW = mean lower low water. (http://www.co-ops.nos.noaa.gov/d_ports.html)

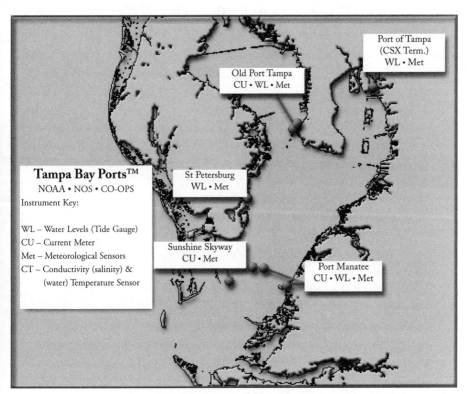

Figure 7.3. Regional Tampa Bay PORTS observation network of the NOAA National Ocean Service. Courtesy of NOAA. (http://www.co-ops.nos.noaa.gov/d_ports.html)

system combines satellite synthetic aperture radar (SAR) with in situ measurements to monitor coastal currents, winds, surfactants, and oil spills. These data can be input to data assimilation and numerical prediction models to provide forecast products for ship traffic control centers. European Space Agency SAR data are also used in a project called Ice Watch to monitor sea ice operationally along the northern sea route between the Bering Strait and Barents Sea (Johannessen et al. 1997b). These data have been combined with Russian Okean satellite side-looking radar, aircraft, coast stations, ice breakers, and drifting ice stations to improve the forecast system for ice navigation conditions. SAR can detect ice edges, boundaries between different ice types, and identify floe types for classification. Repeat SAR passes can detect the motion of ice features. Systems such as Ice Watch will become increasingly important as faster polar routes from Japan to Europe are developed.

ELECTRONIC CHARTS AND POSITIONING SYSTEMS

The single most important development in navigation in the modern age has been the Global Positioning System (GPS). This allows us to know exactly where we are on land, at sea, and in the air in all weather, anyplace in the world. Combined with an electronic chart display and information system (ECDIS) that meets the IMO Safety of Life at Sea (SOLAS) Convention requirements, it is possible to know the exact position of a vessel relative to known hazards to navigation. This makes it possible to issue automated warnings to bridge officers and pilots (DTI 1997; Andreassen 1999). The electronic nautical chart (ENC) provides the reference data for the ECDIS, written in an exchange standard of the International Hydrographic Organisation. The IHO is responsible for coordinating the activities of national hydrographic offices, the uniformity of nautical charts, the adoption of reliable and efficient methods for hydrographic surveys, the development of sciences in the field of hydrographic charting, and ocean mapping. GPS also makes the task of automating environmental observations from ships much more reliable.

New rapid survey technologies and methods have enabled better detection of underwater wrecks and other hazards to navigation. Advances in side-scan sonar and swath multibeam technology, using 450 kH sonar, have improved the accuracy and speed of bathymetric surveying for charting activities. With this technology, a square meter of the bottom is measured twice, with no need for interpolation between points. This provides unsurpassed accuracy for seafloor mapping. With the advent of this technology, the IHO is setting new international standards. The U.S. Army Corps of Engineers and Coast Guard routinely use these instruments in their waterways maintenance. Improved bathymetric maps will facilitate optimal utilization of ports, improve modeling of tides on continental shelves, and improve modeling of ocean circulation. This in turn will improve ocean-atmosphere models for weather and climate forecasting.

However, while the technology exists, its application has been very limited. At the current rate of funding, it will take an estimated twenty years at least to resurvey major U.S. waterways, for example, to create sufficiently accurate electronic nautical charts.

MARITIME PORTS, INTERMODAL TRANSPORT, AND THE COASTAL ENVIRONMENT

The demand for more efficient cargo transportation and tracking systems results in routing strategies in which freight shipping is optimized for decreased costs and decreased time to destination. Integrated intermodal

Box 7.2. Ship Routing and Real-Time Weather Forecasting
David P. Rogers and Mary G. Altalo

Time is money, and the efficient and safe delivery of cargo translates into dollars. Port selection and carrier selection are determined on the basis of time to marketplace and proximity to or ease of transport from the production area. In coastal waterways and in the littoral zone, local weather events can drastically alter the transit time of a vessel as well as threaten the safety of the cargo. Information from tide tables, local bathymetric charts, and current predictions is not sufficient to maximize the efficiency of transport. Short-term wind events may have a significant impact on the depth of water in coastal waterways. Coastal pilots have reported concern that sudden change in wind induces change in water depth, increasing the potential for grounding.

To combat the problem, an on-bridge display is needed combining real-time information on water depth in relation to vessel draft, with accurate short-term weather forecasts for local regions. Depending on the area, high-quality coastal navigation also requires forecasts of such phenomena as storm surges and their interaction with tides, unusually high and steep waves, icing due to local winds, and sandstorms that may reduce visibility.

transportation is becoming the preferred way to do business, using a combination of carriers, including ship, rail, trucking, and airplane (box 7.2). The infrastructure to accommodate such elaborate transportation schemes must be solidly in place to maintain efficiency. There must also be competent and cost-effective "gateways" between the different modes for the system to work well.

Advances in information technology have enabled the smooth transition and precise tracking of vessels and cargo throughout the handling chain. Well-developed and maintained transit corridors of roads, railways, and waterways make up the backbone of the intermodal system. Terminals and ports are the critical links at the land/water transportation interface. As demand grows, there will be a need to examine the capacity of these entities to handle the increasing traffic.

There is constant pressure on ports to augment the landside access infrastructure in order to improve productivity, throughput capacity, mobility, and accessibility. The increasing growth of container ship capacity and size will add to rail and highway capacity problems. As more governments and shippers endorse this intermodal method of cargo transit, significant public and private capital investment is needed to meet the emerging demands. Inefficiencies at any point in the intermodal pathway can cause bottlenecks in the system, leading to transit slowdowns and increased costs.

Increasing recreational and commercial activity is also putting a strain on the capacity of waterways leading to the ports. In addition, the use of deeper and larger tankers and container ships has led to the need for channels over 20 m deep. And this requires dredging and widening of the channels and port facilities. In the United States, approximately 400 million cubic meters of material are dredged from harbors to deepen federal access channels, to provide turning basins for ships, and to maintain deep water along waterside facilities. Advances in channel survey technology in the past decade have greatly aided the effectiveness of the dredging operations. All new surveys are using swath multibeam technology, which is capable of surveying the entire width of the channel at seven times the water depth (see collection of articles in JGR 1986). In addition, new vessel monitoring technologies allow for better tracking of dredge disposal operations for compliance verification.

All aspects of the shipping operation place some stress upon the environment. With the projected increase of activities in the next decade, the potential for negative environmental impacts will also grow. A better understanding of the impacts of dredging and disposal activities on the water column and benthic communities has led to more stringent requirements. Dredging causes resuspension of sediments in harbors, releasing buried toxic materials and degrading water clarity and destroying bottom communities. But improved physical circulation and chemical dispersion models for bays and harbors have made it possible to predict the effects of dredging, helping to reduce the environmental impacts.

Recent studies on toxic algal blooms and on harmful invaders transported from elsewhere have prompted a new look at the way that ships handle their ballast water. The accidental introduction into port and coastal areas of nonindigenous species, which are contained in the ballast water of ships and living on vessel hulls, may impact the local economy and health of the human population. Moreover, these species compete with the resident populations of the area and often have serious impacts on existing populations of commercially and environmentally important communities. And by introducing toxic red tide organisms into the food chain, they can cause significant seafood safety concerns.

There is a need to quantify the introduction of and damage attributed to nonindigenous or exotic species into foreign waters (box 7.3). The U.S.

Box 7.3. Ballast Water
Julie Hall

The unintentional introduction of nonindigenous marine organisms has resulted in the establishment of many species outside their native range. It is impossible to predict the effects that such introductions will cause to local ecology through competition with and replacement of native species, and to local economies through harmful organisms threatening aquaculture sites, damaging port installations, causing diseases, or reducing aquaculture production. Among the main vectors of organism transfer in the marine environment is transmission via ships. Since the development of steel-hulled vessels in the late nineteenth century, ballast water discharges have increased considerably throughout the world, and the probability of successful establishment of self-sustaining populations of nonindigenous species has increased with greater volumes of ballast water as well as with reduced ship travel times. It has been estimated that the major cargo vessels of the world, in particular bulk carriers and containerships (total number, 70,000 [Stewart 1991]), are transferring 10 billion tons of ballast water globally per year, indicating the significant size of this problem.

The impact of a significant number of nonindigenous species introduced to nonnative ecosystems throughout the world has resulted in a call for treatment of ballast water in order to minimize introductions. Desk studies revealed 53 nonindigenous species of macrofauna and macroflora in British waters, 24 exotic organisms in Cork Harbour (Ireland), more than 100 species in German waters (North Sea and Baltic), and about 70 species along the Swedish coasts. More than 300 species are known to have been introduced and established in the Mediterranean Sea. Nearly half of the total numbers of these nonindigenous species are believed to have been introduced by shipping. Investigations by Hallegraeff and Bolch (1992) between 1989 and 1991 showed that viable toxic dinoflagellate cysts were found in up to 6 percent of the vessels entering Australian ports. The list of organisms reported to have survived ship voyages in ballast-water is being extended after each sampling program worldwide. The presence of human disease agents such as cholera bacteria in ballast also underlines the need for ballast-water treatment. A study of recent aquaculture development in the coastal zone showed a high risk of disease transfer from ballast water in cases where aquaculture facilities and fishing areas are located near shipping routes.

Eradication of introduced species once they have become established in a new marine environment is either very expensive or impossible. Therefore, efforts to prevent or minimize introductions should be given high priority. The uncontrolled discharge of untreated ballast water is a major international problem. It is up to governments, environmental agencies, and the shipping and fishing industries to make commitments with a view to identifying a solution to this complex problem. Ignoring the problems that may be caused by introduced species with ballast water could have a significant impact on aquaculture, other economically important activities, and the ecology of many coastlines.

Environmental Protection Agency (EPA) has recently published guidelines, setting limits for aquatic nuisance species. For example, the San Francisco Regional Water Quality Control Board declared the San Francisco Bay an impaired water body, interpreting exotic species as a pollutant under the Clean Water Act. In May 1999, the EPA approved the State of California list of impaired water bodies. The San Francisco Regional Water Quality Control Board is now required to develop a total maximum daily load (TMDL) for exotic species. By establishing this TMDL, the regulatory agency is setting compliance measures under which all ships must operate or be fined. This may be the first setting of such a limit for a living aquatic species.

Ballast-water research technologies need to be fully developed and legally binding. The IMO is formulating international regulations and standards in order to manage the issue. In the case of San Diego Bay, the state of California has begun exploring ballast-water treatment technologies that may be effective in eradicating ballast species, cost-effective to the port and shipping industries, and safe to the aquatic environment. On-board ballast technologies include the use of chlorine, high salinity treatments, heat treatments, biocides, and centrifuge.

IMPACT OF ADVANCES IN MARINE SCIENCE ON INTERNATIONAL REGULATIONS FOR THE SHIPPING INDUSTRY

Marine science data have underpinned the establishment of regulations and enforcement of policy since the beginning of the twentieth century. In addition, advances in marine science and technology have contributed to the shipping industry's capability to meet these enhanced standards. Traditionally, new information obtained from research on the oceans has been rapidly incorporated into the regulations for the safe operations of the shipping industry and for the safety of the marine environment. For example:

- The data on currents and winds in areas hazardous to shipping are continuously incorporated into warnings and charts.
- Advances in the chemical and biological characterization of the ocean have led to major international regulations and the guidelines and manuals of MARPOL (International Convention for the Prevention of Pollution from Ships, 1973, 1978) and OILPOL (International Convention for the Prevention of Pollution of the Sea by Oil, 1954), which are amended as new information is obtained.

- Regulations concerning pollution prevention and potential for spill damage have led to restrictive routing of oil tankers near reefs and sensitive areas, along with modified ship routings and operations.
- Chemical signature analysis is useful in determining the source of deliberate spills and in developing an appropriate hazardous material response to a spill (Petrae 1995).
- New regulations on vessel emissions result from research on greenhouse gases and damage to the environment.
- Fishing restrictions and regulations on bycatch reductions are evolving to rebuild overharvested fish stocks. The requirement for environmental impact statements or reporting regulations for operation in certain areas minimizes the impacts to endangered or sensitive species, such as the right whale and several species of marine turtle.

The links between advances in marine technology and policy setting for the environment are clear. Advances in ocean engineering and instrumentation, ocean modeling and forecasting, ocean animation and environmental simulations, and physical and biological or chemical coupled models are critical to the "smart" management of the seas. Advances in remote sensing methods have led to better marine communications as well as vessel monitoring for compliance. Vessel monitoring systems (VMS) are readily available for installation on fishing and dredging vessels to track their movements precisely as well as record their trawling and hauling operations.

Better forecasting information from more comprehensive observations has improved emergency response measures, including Notice to Mariners and more frequent and precise storm tracking and warning communications.

CONCLUSIONS AND RECOMMENDATIONS

The enormous growth in marine commerce over the next twenty years will place increasing pressure on the marine environment. Ship safety must increase significantly to avoid environmental disasters. The safe operation of bigger and faster ships will demand greater knowledge of changing environmental conditions that affect vessel performance.

Despite recent advances, there is still a lack of reliable weather and seastate information offshore, resulting in poor coastal forecasts of limited use to coastal vessels, fishing boats, and recreational boaters. In the United States, one of eight recreational boating accidents are attributable to adverse weather (DOT 1998). Globally, one of three total vessel losses are related to adverse weather (Lloyds Register 1999b). Innovations such as NOAA

PORTS are cost limited to a relatively few observations in each port or waterway that the system serves. Relative to the value of commercial shipping, it is staggering how few resources are available to improve safe navigation and increase the efficiency of maritime commerce.

The difficulty of sustaining in situ observation systems and the small number of potential users of ocean data are often cited as reasons for the absence of a concerted effort to provide high-quality marine weather and ocean forecasts. However, when one is maneuvering a deep-draft vessel in coastal waters, operating an oil platform in deep water, conducting naval operations, or depending on the health of the ocean for mariculture or fisheries, the absence of adequate ocean observations can have dire consequences.

The role of the ocean in the global climate system; the migration of people to coastal communities; the increasing volume of commodities transported by sea; the exploitation of oil reserves in deeper water; the creation of offshore facilities, like fish farms and wind farms; national security; and the need to sustain and protect marine fisheries will force us to pay much more attention to the open ocean and coastal seas over the next twenty years. Minimizing loss of life and property and avoiding environmental disasters requires significant improvements in the information available to decision makers.

This need for comprehensive measurements in the marine environment led to the formation of GOOS, which is designed to produce economic and social benefits from ocean observations on a global scale at an acceptable cost and risk. The focus of GOOS is on climate, coastal seas, living marine resources, the health of the ocean (e.g., pollution), and marine meteorological and oceanographic services (IOC 1998a,b). Although initiated by national and international operational agencies, its success depends on the involvement of commercial companies, research organizations, national security, and aid agencies. It is impossible to imagine a single entity with sufficient resources capable of providing all of the necessary information to provide adequate forecasts of the marine environment.

The advent of new, cheaper observing systems, better global satellite communication, and the Internet, however, will improve the quality of VOS observations and lead to increased use of the data from the maritime industry. It is now possible to equip vessels with automated meteorological and oceanographic systems that can broadcast data almost continuously from ship to shore via any appropriate communication method. In turn, the data transmitted can be combined with similar information from other platforms and then rebroadcast to provide vessel operators with current spatial information on sea state and weather within their operating area. For example, wind can force rapid changes in water depth and currents, which can severely restrict the operation of deep-draft vessels, causing delays and increasing the risk of grounding.

Data broadcast from other vessels, combined with some fixed monitoring stations, could provide up-to-date information on the sea conditions and alert the vessel operator to any rapid changes that may influence vessel operations. Data from these regional vessel data networks (RVDNs) would also be used by national agencies with operational forecast responsibilities for marine and coastal regions. If the quality of shipboard data can be assured, improved regional maritime and coastal forecasts are certain. These model fields and data also provide the necessary boundary conditions for the specialized limited-area forecast models, which might be run locally by a value-added service provider to give additional guidance to the vessel operator and to company or port operatives scheduling vessel discharge and loading. Delays to port operations could be anticipated and avoided.

The success of RVDN initiatives requires a high level of cooperation among the end users, who are providing a large fraction of the data; the value-added service sector, who are using these data to provide information tailored to specific user needs; and the national agencies entering these data into quality-assured databases for model initialization. All must work together to make maritime services work effectively. The provision of real-time or nowcast data on an operationally useful spatial scale requires the free provision of data from end-user platforms. These data must be quality assured to be useful in forecast models, and the national agency forecast products must provide reliable boundary conditions for limited-area models developed for specific commercial applications. National governments can provide the data assembly and forecast infrastructure, and the growing value-added service sector can extend this information to develop economically beneficial products for end users, while helping to improve maritime safety.

International commitments to maritime services are needed as new initiatives for marine forecasting are developed. There is an opportunity for governments and the private value-added service sector to provide direct economic benefit to shipping and a higher level of marine safety, while also providing the information necessary to support global climate services, which will benefit the world community. Taylor and Kent (1999) have recommended ways to improve the accuracy of basic meteorological and oceanographic observations aboard VOS. In particular, there is a need for more complete information about the vessel to enable corrections to be made to the observations for height above sea level, the overall shape of the ship, and the position of the instruments. Errors in the ship's weather report need to be decreased by calculating true wind accurately, correcting pressure for changes in the draft of the vessel, minimizing position errors, and ensuring proper exposure of temperature and humidity sensors. Automatic sensing and coding of the reports will avoid most of these problems (Taylor et al. 2001).

Persuading the commercial sector to cooperate in the provision of data understandably depends on demonstrating that there is indeed an economic and safety benefit. End users must be willing to supply the necessary data. If the benefits are sufficient, it is likely that automatic weather and oceanographic systems could be installed on many more vessels than currently contribute to the WMO VOS program, although costs will have to be reduced well below those currently proposed (Taylor et al. 2001).

The concept of rapid environmental assessment for strategic decision making is a relatively untested tool in commercial shipping, and much can be gained from the approach that is being developed for naval applications. In naval parlance, the operating environment is the battlespace, in which meteorological and oceanographic conditions must be characterized as completely as possible using data acquisition, assimilation, and application techniques (see box 7.1). Similar commercial techniques, which provide information on the state of the marine environment, should feature in any integrated transportation system that connects road, rail, and sea.

The task is daunting, but the technology is at hand, and the potential economic and environmental benefits are large. What must be undertaken first is a series of pilot studies, in cooperation with end users, to determine the scope and feasibility of an information system dedicated to marine shipping needs.

RESEARCH PRIORITIES

Key Scientific and Technological Challenges

- Seek implementation of new VOS technology to increase the value of the data collected from this source.
- Expand VOS program by demonstrating the value of the data collected to ship operators.
- Develop a coastal equivalent of VOS: a regional vessel data network.
- Expand programs such as NOAA PORTS to a global network.
- Improve sensors for undersea communications.
- Develop new installation techniques for deep-sea networks.

- Enhance the reliability and redundancy of surface communications.
- Improve the speed and bandwidth of global telecommunication system (GTS) or replace with an Internet-based system.
- Extend the geographic coverage and bandwidth of communications.
- Develop affordable precision clocks through acoustic chip technology for precise navigation.
- Optimize environmental information from platforms through new data assimilation techniques.

Key Socioeconomic Challenges

- Seek greater cooperation between the IOC, IMO, ship registry organizations, pilots, shipbuilders, ship owners, and harbor authorities to test new information systems.
- Establish the utility of enhanced environmental information to the commercial sector through socioeconomic models.
- Improve the rapid transition of technology between sectors through appropriate partnering mechanisms, including strategic alliances and joint ventures.
- Define the relative roles of legislation and voluntary standards in driving the utilization of marine products.
- Develop appropriate mechanisms for the exchange of data.

ENDNOTES FOR FURTHER READING

Baker, D., and P. Graykowski. 1998. Oceans and commerce. In *Oceans of Commerce, Oceans of Life*. Proceedings of the National Ocean Conference, 11–12 June 1998, Monterey, California. Silver Springs, Md.: U.S. Department of Commerce, Office of Public and Constituent Affairs.

Dalton, J. H., and M. Kimble. 1998. Oceans and global security. In *Oceans of Commerce, Oceans of Life*. Proceedings of the National Ocean Conference, 11–12 June 1998, Monterey, California. Silver Springs, Md.: U.S. Department of Commerce, Office of Public and Constituent Affairs.

Pink, A. (ed.). 2000. *Integrated Coastal Zone Management: The Review of Strategies and Technologies for the Management of EEZs and Coastal Zones Worldwide*. London: IGC Publishing Ltd.

Stopford, M. 1997. *Maritime Economics*. London: Routledge.

Thornes, J. E. 1997. Transport systems. In R. D. Thompson and A. Perry (eds.), *Applied Climatology: Principles and Practice*, 198–214. London: Routledge.

PART II
TOOLS AND APPROACHES

Chapter 8

Operational Oceanography

Colin P. Summerhayes and Ralph Rayner

Over the next twenty years, the need to understand and forecast the oceans and their resources is going to increase significantly—and on time scales that permit relevant and effective management decision making. Safe and sustainable navigation, the exploitation of marine resources, and the safeguarding of both local and global marine environments will all depend on this enhanced capacity (OECD 1994; OOSDP 1995; Stel et al. 1997; DTI 1999a,b). We can expect the new field of operational oceanography (box 8.1) to grow to meet this need, providing operationally useful information for a wide range of users and customers about the present state of the sea, and about its future states for as far ahead as possible.

END USES AND USERS

Interest in operational oceanography starts with private and public concerns about the oceans and related systems—and ways to address these issues (box 8.2). Many of these concerns are the responsibility of government agencies whose job is to meet statutory national and international obligations. These "public good" issues are the major motors for the development of operational oceanography.

There is a need for long-term ocean observations to address these issues (Nowlin 2001). The data will be used for numerical weather prediction and to improve the safety and efficiency of marine operations. They will help to monitor and predict climatic variability, preserve and restore healthy marine ecosystems, and manage living marine resources for sustainable use. The data will also help to mitigate the effects of natural coastal hazards, while being

Box 8.1. A Definition of Operational Oceanography (based on Woods et al. 1996)

Colin P. Summerhayes and Ralph Rayner

Operational oceanography is the activity of routinely making, disseminating, and interpreting measurements of the seas and oceans and atmosphere, so as to

- provide continuous forecasts of the future condition of the sea as far ahead as possible;
- provide a description of the present state of the sea, including living resources, with optimal accuracy; and
- assemble long-term climatic data sets to describe past states and time-series showing trends and changes.

Operational oceanography relies on measurements that are systematic, routine, cost-effective, high quality, sustained for the long term, available in a timely manner, and relevant to users' needs. Several examples of operational oceanography are given by Stel et al. (1997).

Operational oceanography proceeds usually, but not always, by the rapid transmission of observational data to computerized data assembly centers, where the data are processed through numerical forecasting models (Woods et al. 1996). Outputs from the models are used to generate secondary data products that have special applications, often at the local or regional level. Final data products and forecasts must be distributed rapidly to users in industry and commerce, government agencies, and regulatory authorities.

Operational oceanography already exists at local levels and for a limited number of factors (Woods et al. 1996). Forecasts regularly provided at present include wind velocity and direction over the sea; wave height, direction, and spectrum; and surface currents, tides, storm surges, floating sea ice, and sea surface temperature. There are great advantages in making operational oceanography global so that all parts of the system can be analyzed and forecast simultaneously with greater accuracy and further into the future. There are many products of value to industry and government agencies that can be made available soon or for which forecast periods and accuracy can be increased (Woods et al. 1996). These include indicators of marine pollution and contamination, movement of oil slicks, prediction of water quality, concentrations of nutrients, primary productivity, subsurface currents, temperature and salinity profiles, sediment transport, and erosion.

Observing systems are maturing. Rates of capture of data from both satellites and in-the-water instruments are rapidly increasing. Value is being added to these data by storing, retrieving, managing, and manipulating them digitally to derive products tailored to the needs of different customers, including policymakers.

Much operational oceanographic information meets national strategic and tactical defense needs as well as civilian requirements. This synergy offers the prospect of coordinated civil and military research projects and priorities.

Box 8.2. Matters of Public Concern about the Oceans
Ralph Rayner and David Szabo

- Climate variability including El Niño and other oscillations
- Global warming, sea level rise, and small island states
- Frequency and intensity of hurricanes and storm surges
- Potential melting of Antarctic ice sheets
- Pollution (e.g., oil and other industrial chemicals from land affecting water quality and public health)
- Oil spills and other marine accidents
- Dumping and waste disposal, including radioactive waste
- Loss of amenities due to coastal development and urbanization
- Coastal erosion
- Loss of coastal ecosystems and fragile habitats
- Preservation of coral reefs and mangrove forests
- Nutrient runoff leading to eutrophication
- Toxic algal blooms
- Invasive species
- Exhaustion of fish stocks
- Biodiversity in the ocean
- Wildlife conservation
- Mass mortalities of marine mammals, including damage to whales and dolphins
- Artificial fertilization of the ocean
- Safety of passenger and cargo ships, ferries, and offshore operators

used to monitor the supply and effect of pollutants and their effects on water quality. National surveys carried out across Europe by EuroGOOS have amply confirmed the growing demand for this kind of a broad range of services (Fischer and Flemming 1999).

The user community for operational oceanographic data is rather broad. The data are essential not only for offshore and coastal activities, but also to underpin weather and climate forecasts used to plan supplies of water, food, and energy (box 8.3).

An example of operational oceanographic applications in support of industry is given by the North Brazil Current Rings Experiment (box 8.4 and figure 8.1 in the color section), which shows how different observing practices may be used to track rings as they move. In this case, operational results were obtained by exploiting ongoing research activities.

Box 8.3. Broad Classes of Users and Customers
Ralph Rayner

- Government agencies, regulators, public health, certification agencies
- Environmental management, wildlife protection, amenities, marine parks
- Operating agencies, services, safety, navigation, ports, pilotage, search, rescue
- Small companies, fish farming, trawler skippers, hotel owners, recreation managers;
- Large companies, offshore oil and gas, survey companies, shipping lines, fisheries, dredging, construction
- The single user, tourist, yachtsman, surfer, fisherman, scuba diver
- Scientific researchers in public and private institutions

Box 8.4. Case Study: North Brazil Current Rings Experiment
Ralph Rayner and David Szabo

For part of the year, the North Brazil Current normally flows northwest along the northeast coast of South America. During the second half of the year it turns back to the east, in what is called the North Brazil Current (NBC) retroflection (see figure 8.1). When the current bends back on itself in this way, at about 50°W, masses of warm water separate off, forming clockwise rotating eddies or rings. The rings drift slowly northwest along the continental slope for several months before reaching Trinidad, where the continental slope then guides them to the north. Rings may form as early as August and as late as April. They can be tracked from their well-defined velocity structure, their sea surface elevation signature, and the differences in water temperature between the core and the exterior.

Information about these rings and their associated currents is potentially useful to companies exploring for oil on the continental slope off the east coast of Trinidad, where the challenge for the industry is to determine operational and extreme conditions for currents. Because the rings migrate into the area, it is not sufficient to collect just local measurements of current velocity, although such measurements are important. What is required as well is the ability to forecast the arrival of rings in the drilling area. The NBC Rings Experiment (funded by the National Science Foundation and the National Oceanic and Atmospheric Administration) is providing the regional picture that is needed for forecasting. Details can be found at www.aoml.noaa.gov/phod/nbc.

This experiment maps NBC rings by using a combination of inverted echo sounder moorings, current meter moorings, altimeter data, near-synoptic hydrographic surveys of the velocity and water mass structure, surface drifters, and subsurface floats. These data complement those collected by the industry in the drilling area, confirming the role of rings in strong currents measured off Trinidad and demonstrating the value of regional observations in understanding local processes. Academic

> research on the NBC rings has provided the scientific basis for a commercial operational forecasting service based on satellite altimetry and drifters. This commercial service now provides offshore operators with advance warning of high current events associated with NBC rings, allowing them to take operational precautions well in advance.

EMERGING TRENDS

In its early years, operational oceanography—and the marine information market that it serves—relied primarily on observations of physical parameters such as waves, tides, currents, temperature, and salinity. Three trends are now becoming evident:

- An oceanographic and marine meteorological services sector is growing to meet the needs of the offshore and coastal industries and other customers for information to improve the effectiveness and safety of operating conditions and to contribute to sustainable development of the environment.
- The further development of operational ocean forecasting services will depend on improvements in several areas: (1) numerical modeling and simulation of oceans and shelf seas; (2) increases in data acquisition, management, and assimilation into models; and (3) improved understanding of the ocean's role in climate variability.
- The information market is expected to change from one dominated by physical measurement, with the oil and gas and shipping sectors as major customers, to one with more environmental measurements and forecasts for people managing coastal zones and monitoring pollution. This third area calls for progress in biological and chemical oceanography, including:

    - monitoring of pollution by assessing its biological effects on specific organisms (customers include local authorities, tourism, fisheries, aquaculture);
    - monitoring eutrophication by in situ chemical analysis or by changes in phytoplankton mass or species (customers: local authorities);
    - monitoring changes in coastal communities by observing benthic communities by optical methods (video etc.) (customers: local authorities, tourism, fisheries);
    - monitoring fish stocks through improved acoustical methods (customers: fisheries agencies and fishing community); and
    - monitoring development of harmful algal blooms through refined observation of harmful algae (optical-fluorometric identification) and

forecasting of their occurrence (customers: aquaculture, fisheries agencies, tourism, health authorities).

Some of the improvements required in data collection will come from the further instrumentation of ships. In the recent past, voluntary observing ships have collected marine meteorological data. Similarly, the ships of opportunity have used disposable bathythermographs or disposable conductivity/temperature/depth devices to collect subsurface marine data. The Intergovernmental Oceanographic Commission (IOC) and the World Meteorological Organization (WMO) have now merged these activities under a new Joint Technical Commission for Oceanography and Marine Meteorology (JCOMM). This will facilitate the collection of both marine meteorological and subsurface data through increasingly automatic systems from the same ships. The improved data-gathering network and data flow will lead to more efficient services.

Improved bathymetric maps are also needed, especially on continental shelves and along coastlines, because the modeling of circulation, waves, tides, and storm surges is critically dependent on the shape of the bottom topography used in numerical models.

OPERATIONAL NUMERICAL MODELS

Advances in operational oceanography in recent years are largely a reflection of the substantial increases in computer power. Numerical ocean models can now even be run on desktop personal computers to simulate the way the ocean works and to forecast how it may change in response to external forcing. The application of numerical models is now an essential part of the toolkit of the manager of ocean or coastal zone operations (Flather 2000). Switching to model-based generation of products would also greatly improve the performance of local operational oceanographic services (Woods 2000). This normally involves redesigning the sampling strategy for observations and may turn out to cut costs by requiring fewer observations than before.

Numerical models can add enormous amounts of information to data (see figure 8.2 in the color section). These models can be initialized using climatological data based on observations collected in a nonsynoptic manner over a period of years and forced by meteorological data. The Fine Resolution Antarctic Model (FRAM) of the Southern Ocean, for instance, was constrained with 100 years worth of ocean data (FRAM Group 1991). The hand- or machine contouring of such data can provide broad fields for different parameters, such as sea surface temperature. But they cannot produce synoptic interpretations other than in a seasonal sense, nor can they provide much detail of ocean structure. In complete contrast, numerical models can convert such original nonsynoptic data into richly textured fields of synoptic information, including the probable locations of fronts, the simulation of eddy fields, and the likely locations of currents and other important structural features of the water column.

Even where the present observing network is sparse, the combination in a numerical model of (1) historical in situ ocean data sets with (2) available remotely sensed data from satellites and (3) available meteorological data can produce quite a realistic synoptic representation of ocean circulation on regional or local scales. Local ocean data, meteorological data, and satellite data can be assimilated into such models to provide continual updates and projections. Bearing in mind the caveats mentioned below, this approach can help developing countries to manage their coastal zones in a sustainable way.

An ideal observational network for accurate operational forecasting will comprise a whole array of observations from different sources:

- Remote sensing from satellites
- Meteorological stations on land
- A mixture of fixed and floating data-gathering arrays fitted with (relatively) cheap and disposable instruments
- Observations from ships
- Careful collection at selected sites of long time-series records demonstrating the scales of natural variability

The observing array should be designed to meet the demands of the model in order to provide output that is useful to the customer. Simulation experiments or pilot observing system projects can be used to discover the minimum set of data that is required, which variables to measure, and which regional forcing factors to consider to improve regional or local forecasts. This will guide the development of an observational network capable of providing new in situ data to feed to models and so improve their accuracy.

Coastal zone managers are now becoming more aware that open ocean information, in the form of observational data and model output, is essential to provide them with the boundary conditions they need for the operation of local- and regional-scale numerical models of shelf seas. It is no longer sufficient to collect information just from the local (usually coastal) area of interest, if accurate forecasts of waves, tides, currents, and other parameters are required. Instead, local-scale models with high resolution output, on say a 0.5 km grid, must be nested within regional models, operating on a somewhat coarser grid. These, in turn are nested in basin scale or global models with a much coarser grid, each model providing boundary conditions for the next level down.

In box 8.5 we give an example of the use of operational oceanographic

---

Box 8.5. The Atlantic Margin Metocean Project
Ralph Rayner

Oil is now being produced from fields along the northeast Atlantic margin (off the coast of Scotland) where environmental conditions are severe, complex, and difficult to predict. Providing reliable meteorological and oceanographic (met-ocean) environmental information to back up hydrocarbon exploration and production is a lengthy business and needs to be planned and implemented well before drilling starts.

The Atlantic Margin Metocean Project brings together research institutions and commercial oceanographic service providers to produce this information. The project team includes experts in oceanographic measurement, numerical modeling, data interpretation, and criteria development to meet practical industry requirements and carry out important new research. The project comprises a year-long measurement campaign in parallel with high-resolution numerical modeling. This regional appraisal of oceanographic processes creates a unique data set from which site-specific met-ocean criteria can be derived to meet industry needs.

Eight comprehensive oceanographic moorings have been deployed west of Ireland, and three in Faeroese waters, to gather data for a synoptic period of twelve months. Upward-looking Doppler current profilers and conventional current meters and temperature sensors are spaced through the water column to determine water column structure.

More than 2000 research and commercial archives of in situ measurements of oceanographic parameters exist for the European continental slope and shelf seas in the project area (20°W to 13°E, 42°N to 65°N). These, plus remotely sensed data including sea surface temperature from advanced very high resolution radiometry, ocean color from SeaWiFS, and ocean height from altimetry, are held in various data banks to which the project has access. The project will develop a Web-based gateway to all met-ocean data for the region, building on existing archives, developments, and ongoing measurements. The gateway will combine distributed Web and database technology. Relevant data will be reanalyzed to assess the decadal and interannual variability of slope processes and ocean-shelf exchanges.

Local circulation in the region is determined by remote processes, such as variability in the transport in the Gulf Stream and the North Atlantic Current, as well as local topography. A large area model is therefore required to give realistic circulation in the study region. Previous attempts to model the observed current structure have involved relatively simple 3-D representations. The project will extend this work using two independent, complementary, high-resolution, eddy-resolving models to establish water-column current and density structure over the project area for the year-long duration of the measurement campaign.

Remotely sensed data will be used to provide a powerful synoptic visualization of surface processes and a detailed time-series picture of the movement of dynamic features such as fronts or eddies even in cloudy regions. The synoptic time-series observations of current velocity and density structure will be used to describe ocean behavior, to provide a reference data set against which to validate numerical model performance, and to help to place existing data sets in a regional and temporal context.

A major component of the study will be the verification and calibration of the models, based on comparisons of model output with remotely sensed data and in-water measurements. The validation will focus on the capability of representing water masses and transport. Given the higher resolution of the model outputs compared to previous studies, the study will place emphasis on achieving an accurate representation of the slope current and the variability in the surface circulation, including mesoscale eddies, fronts, and currents.

The project will deliver a set of criteria tables and figures representative of the current, wind, wave, water level, sea temperature, and salinity characteristics for each specific exploration location.

Increased skill in medium-range forecasts and the development of ensemble-averaged forecasts (based on the combination of outputs from a number of forecast runs) will help improve weather-routing services for ships. Better evaluation of possible errors in wave and wind forecasting can be used to provide routes optimized in relation to wind and wave statistics. This may be useful not only for reducing or planning journey time, but also for reducing hull fatigue in very large tankers. Forecasts should include extreme values, for example, warnings of gales, high swells, storm surges, and floods. Nowadays there is a requirement to include wave forecasts in weather bulletins for mariners. In the future, forecasts may also be required to include surface currents. Recognizing that the most dangerous waves for shipping are generated through interaction with surface currents, wave-current interaction is increasingly taken into account in sea-state forecasting, requiring the application of sophisticated numerical techniques.

No model is perfect. It is wise to run different models to ensure that predicted features are not artefacts. The forecasting skill of any given model will vary depending on data input and needs constant testing against reality. At present, we cannot easily provide numerically testable estimates of confidence on model output, for which precise

(continues)

Box 8.5. Continued

determinations of error level are required. The accuracy of forecasts from numerical models depends not just on the model formulation. It also depends on having access to a continuous stream of real-time data that can be assimilated in the model.

Depending on requirements, many models, especially for coastal seas, can be run with relatively modest (desktop) computing resources. But, for high resolution at the basin-wide or global scale, it essential to have access to the most powerful supercomputers in order to run advanced coupled ocean-atmosphere models. Very high-performance computers will be required for models of water quality with equations for physics, chemistry, biology, and sedimentology (Woods 2000).

---

data and models to produce accurate regional meteorological and oceanographic information to support offshore oil and gas exploration. This particular case study brought together academic institutions and commercial oceanographic service providers.

ACCESS TO DATA

Widespread access to publicly acquired data will be needed for maximum benefit. In any regional sea area bordered by two or more countries, neighbors have more to gain by sharing their data than they do by keeping it to themselves, since the waters washing one coast today will wash another coast tomorrow. Without complete knowledge of the system, any one country's forecasts of its behavior will remain incomplete and less than ideal. Free exchange of such data is to be encouraged.

Although access to real-time data is increasingly important for operational purposes, access to historical data is essential for understanding regional patterns and changes with time. Often these historical data are to be found in obscure files, in analog form, and not in any one place. But efforts to retrieve them are highly advisable, as recommended through the IOC's Global Oceanographic Data Archaeology and Rescue Project.

REMOTE SENSING DATA

Remote sensing technologies can supply more data at faster rates and will be increasingly in demand for marine forecasting services (see figure 8.3 in the color section). Satellite-borne sensors are needed that can measure salinity, surface air temperature, all-weather sea surface temperature, precipitation, and organic and other pollutants (see, e.g., OOSDP 1995; DTI 1997; Summerhayes et al. 1997). Data gathering by satellite needs to be developed further, especially in:

- gravity missions for better altimetric data;
- algorithms for the use of ocean color data; and
- salinity sensors to characterize ocean surface properties fully.

Remote sensing by aircraft can provide essential increases in resolution in critical areas. Land-based high-frequency radar is becoming an indispensable tool for mapping wave and current fields out to considerable distances in coastal seas, and it may be supplemented by sophisticated systems on ships and offshore platforms. Remote sensing within the water mass, using acoustic tomography, is also becoming more frequent, especially for detecting significant changes in bulk water mass properties such as temperature.

There is a growing market for "value-added" products derived from remotely sensed images from space. Trained professionals provide the interpretation that adds the value. In many developing countries, there is a need for professionals trained to interpret remotely sensed data from space for end users.

THE GLOBAL OCEAN OBSERVING SYSTEM

Needs, Benefits, and Costs

Most operational oceanography is carried out locally to solve local problems, for instance, to provide information for oil platform operators in a specific area or to monitor and model water levels in a particular port and its approaches. However, local conditions are always subject to regional controls, set in a global ocean-atmosphere-ice system with teleconnections between far-flung areas (e.g., Woods 2000). For example, physical processes far out in the Atlantic and in the Gulf Stream, driven by the North Atlantic Oscillation, affect the oceanography, water quality, and biology of the northeast Atlantic and the adjacent North Sea (figure 8.4). Similarly, events in the eastern Indian Ocean and western Pacific influence the depth of the thermocline off Peru, while weather in the Southern Ocean partly determines swell heights off South Africa.

There is a need to improve the accuracy of local forecasts and other local services as well as climate prediction. A global view of ocean behavior is

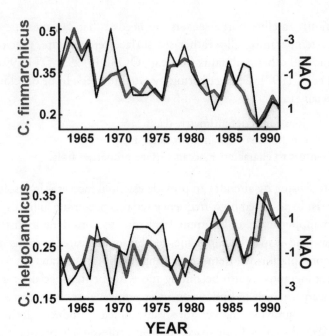

Figure 8.4. Continuous plankton recorder data from the North Atlantic show clear correlations between the populations of different species of the plankton *Calanus* (log abundance) and the North Atlantic Oscillation (NAO) index measured as the difference in normalized sea level pressure between Akureyri, Iceland (the Iceland "low"), and Ponta Delgada (the Azores "high"), which demonstrate the control of ocean biology by climate and the need to continue observations over the long term in order to develop a predictive capability. Figure based on Fromentin and Planque 1996. Courtesy of SAHFOS and Inter-Research.

needed to provide the input conditions for this; hence, the Global Ocean Observing System (GOOS). By providing the local user with an accurate regional framework, GOOS will help to improve predictability. It will complement, but not replace, the collection of data and application of models for specific local applications.

Improving predictability demands routine, systematic, long-term measurements of relevant ocean properties on a global basis or, at the very least, on a basin-wide scale. The El Niño forecasting system is a good example of GOOS operating at the basin-scale to forecast climate change (figure 8.5). It offers roughly nine-month warnings of impending change, helping countries to plan ahead to mitigate impacts (e.g., Weiher 1999).

The world community called for GOOS at the United Nations Confer-

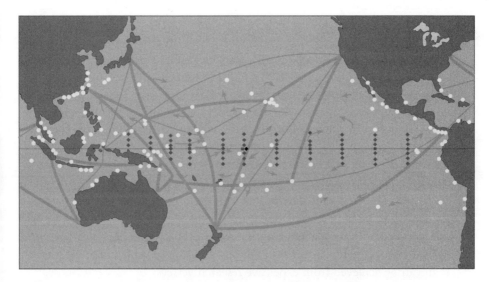

Figure 8.5. The El Niño Southern Oscillation (ENSO) observing system in the tropical Pacific, which uses moored buoys (stacked diamonds), drifting floats (arrows), tide gauges (white circles), observations along ship's tracks (lines), and satellite data (not shown) to monitor atmospheric and ocean conditions to help forecast climatic changes associated with El Niño and La Niña. Courtesy of NOAA.

ence on Environment and Development. It is being developed by nations working together through its main sponsors, the IOC, WMO, UNEP (United Nations Environment Programme), and ICSU (International Council for Science). It will meet the demands of operational oceanography (see box 8.1), but it is not solely operational. It includes sustained observations needed for research (e.g., on climate variability), and it stimulates research needed to improve observations and forecasts.

GOOS is like a spatially distributed large facility, and it needs the same kind of international management as the European Organisation for Nuclear Research (CERN) or the Ocean Drilling Program (ODP). If it is designed and managed well, it should enable nations to convert research results into useful products to meet societal needs.

Several studies have been published on the costs and benefits of making global ocean observations (e.g., Flemming 1994, 2001; Weiher 1999). Where calculations have been made, a return on investment of significantly greater than 10 percent seems commonplace. In fact, the benefit from forecasts is often felt not so much in the marine sector as on land. The forecasts can be used to improve prediction of the water cycle and precipitation on adjacent landmasses. This information can help manage supplies of water and food. The forecasts also help to anticipate and mitigate floods and droughts or to improve predictions of thermal conditions on land, thus contributing to the management of energy supplies.

The Conference of the Parties to the UN Framework Convention on Climate Change now requires improved monitoring of the climate system and the ocean's role in it. At their meeting in Buenos Aires in November 1998, they noted a need to increase the number of ocean observations, particularly in remote locations.

While the benefits of developing GOOS are becoming more apparent, the costs of making observations are dropping. Satellites are becoming cheaper, and measurements of upper ocean properties are beginning to be made by profiling floats and autonomous underwater vehicles more cheaply than can be achieved from expensive research vessels.

The GOOS Design

The GOOS design has now emerged in broad terms for application by individual countries in the form of the GOOS Strategic Plan and Principles (IOC 1998a) and the GOOS 1998 Prospectus (IOC 1998b) (see also Nowlin 2001; Nowlin et al. 2001; Woods 1997, 2000). Detailed designs have been published for GOOS in relation to

- coastal seas (IOC 2000a),
- pollution and the health of the ocean (IOC 1996), living marine resources (IOC 2000b), and
- climate (OOSDP 1995; IOC 1999; Needler et al. 1999).

A key feature of the design is that certain core measurements will meet the needs of a wide range of end users. Design information is obtainable through the GOOS Web site (www.ioc.unesco.org/goos). GOOS should be fully implemented in the period 2010–2020, once the system has been tested via pilot projects.

A new design of an ocean observing system for climate emerged from OceanObs99, the First International Conference for the Ocean Observing Systems for Climate (held in St. Raphael, France, 18–22 October 1999). This meeting sought consensus between the research and operational communities regarding the most appropriate blend of observations required to satisfy the collective needs of research and operational applications (box 8.6).

Box 8.6. New Design for Ocean Observing System for Climate, Endorsed by the OceanObs99 Conference

SPACE-BASED OBSERVATIONS OF THE OCEAN

OceanObs99 gave high priority to:

- low resolution (100 km) satellite sampling for sea surface temperature;
- one precision altimeter and one high-resolution altimeter to measure sea surface height for ocean dynamics;
- two dual-swath scatterometers to observe surface wind vectors, daily;
- measuring ocean color by satellite as a proxy for ocean productivity;
- passive microwave satellite systems for observing sea ice;
- synthetic aperture radar (SAR) for surface wave and ice data;
- a satellite gravity mission to improve estimates of the geoid and, hence, the accuracy of altimetric measurements of sea surface height; and
- development of global remote sensing of salinity.

IN SITU OBSERVATIONS FROM THE OCEAN

OceanObs99 gave high priority to:

- maintaining the El Niño–Southern Oscillation (ENSO) observing network;
- improving coverage of sea surface temperature by ships of opportunity and drifting buoys;
- the Argo Project for measuring global temperature and salinity in the upper ocean;
- repeat sampling along selected hydrographic lines;
- time-series measurements at a selected number of stations to resolve complex interactions; and
- collection of surface wind data by dedicated surface moorings.

In remote sensing, the key message for the space agencies is the need for continuity of certain kinds of observations from one mission to the next, as well as a sufficient number of similar sensors being flown at the same time (see box 8.6).

GOOS: Implementation

Implementation will be incremental, building on a GOOS Initial Observing System (GOOS-IOS) that unites the main global observing subsystems supported by the IOC, WMO, and (in the case of coral reefs) the IUCN (listed in box 8.7). It will include measurements from ships, buoys, coastal stations, and satellites. Other operations will be added in due course. In addition to these international elements, many nations are now contributing substantial parts of their national observing systems to GOOS. Implementation will be aided by the formation of the new JCOMM, which integrates the marine meteorological and oceanographic data-gathering communities.

The Tropical Atmosphere Ocean (TAO) array of buoys in the tropical Pacific (see figure 8.4) proved its worth during the 1997–98 El Niño by

---

Box 8.7. Components of the GOOS Initial Observing System

- Upper ocean measurements made by the Ship of Opportunity Programme
- Marine meteorological measurements made by the Voluntary Observing Ship program of the WMO
- Observations of sea level made by the tide gauges and pressure gauges of the Global Sea Level Observing System
- Observations from fixed and drifting buoys coordinated by the Data Buoy Cooperation Panel
- Meteorological and oceanographic observations from the operational El Niño observing system in the tropical Pacific, including the Tropical Atmosphere-Ocean and TRITON arrays of buoys
- Ocean surface and marine meteorological measurements made by NOAA's operational satellites
- Plankton data from the Continuous Plankton Recorder Survey, managed by the Sir Alister Hardy Foundation for Ocean Science
- Time-series data from time series stations "S" (Bermuda) and Bravo (Labrador Sea)
- Physical, chemical, and biological data from the International Bottom Trawl Survey of the North Sea, managed by the International Council for the Exploration of the Sea
- The monitoring activities of the Global Coral Reef Monitoring Network;
- High-quality data from the upper ocean provided by the Global Temperature and Salinity Profile Programme
- Ocean data collected by the Global Data Centre managed by NOAA
- Transmission of information through the Global Telecommunications System of the WMO
- GOOS products available on the electronic Products Bulletin at http://iri.ldeo.columbia.edu/climate/monitoring/ipb/.

providing even more accurate forecasts than before (Leetma et al. 2001; McPhaden et al. 2001). Retrospective analyses show that the first indications of the event appeared in subsurface data from the buoys. Continued investment in Pacific observing systems is needed to improve El Niño forecasts.

GOOS is also being implemented through pilot projects that test aspects of the GOOS design. The main GOOS pilot project is GODAE, the Global Ocean Data Assimilation Experiment, designed to demonstrate by 2005 the power of integrating satellite and in situ data, the power of assimilating the integrated data into numerical models, and the value of a global system capable of working in real time. This experiment calls for a global network of surface and upper-ocean temperature and salinity data that can be integrated with the surface ocean data provided by remote sensing from satellites. To provide the experiment with the first-ever global coverage of upper ocean temperature and salinity, the Argo Pilot Project is underway to deploy 3000 profiling floats. These will cycle through the upper 2000 m every fourteen days (figure 8.5), returning 75,000 to 100,000 temperature and salinity profiles per year during their anticipated four-year lifetime (Roemmich et al. 2001). For details on GODAE and Argo, see www.BoM.gov.au/bmrc/mrlr/nrs/oopc/godae/homepage.html.

Another major GOOS pilot project is PIRATA (Pilot Research Moored Array in the Tropical Atlantic). This extends the TAO array to monitor ocean and atmospheric variables and upper ocean thermal structure at key locations in the tropical Atlantic region, and it will improve climate prediction in West Africa and South America. For details on PIRATA see www.ifremer.fr/orstom/pirata/pirataus.html.

GOOS is also being planned and implemented at the regional level. The two main regional programs of GOOS are EuroGOOS in Europe and NEAR-GOOS in the Northeast Asian region. EuroGOOS (see Woods 1997; Flemming 2001) is attracting resources from the European Commission for preoperational research projects to develop the skills and capabilities to implement GOOS. Significant operational developments are expected in the five EuroGOOS subregions: the Mediterranean, the Arctic, the Baltic, the northwest shelf, and the wider Atlantic. NEAR-GOOS is increasing its constituents and contributors and will move toward ocean forecasting (IOC 1998c).

Several newly created GOOS regional bodies are expected to increase their capabilities and implement operational activities during the next five to ten years, thus helping the further development of GOOS. They include

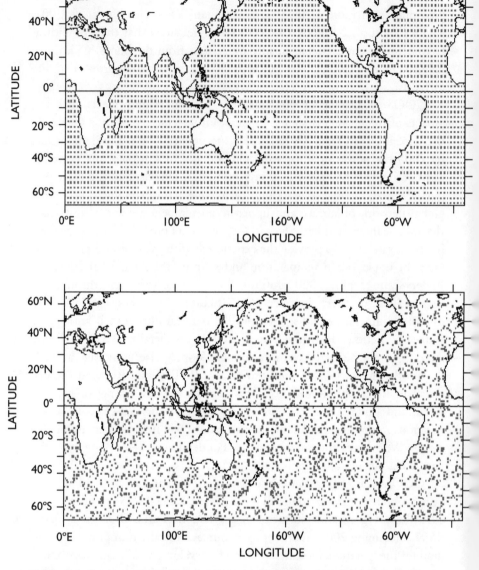

Figure 8.6. Computer-projected distribution of 3000 Argo profiling floats at 2000 m depth, after three years, from initial sites evenly spaced 300 kilometers apart. Courtesy of NOAA.

PacificGOOS, covering southwest Pacific island states; MedGOOS for Mediterranean countries; IOCARIBE-GOOS for Caribbean states; an Indian Ocean program; Black Sea GOOS; and a North Atlantic program.

The range of observations needed to understand and monitor Earth system processes and to monitor and assess human and societal impacts cannot be satisfied by any single program, agency, or country. Effective monitoring on the global scale demands international cooperation. Recognizing this requirement, GOOS in 1998 became part of the Partnership for an Integrated Global Observing Strategy (IGOS), which brings together the main organizations involved in global observations of the Earth. The IGOS provides a framework for integration, enabling better observations to be derived in a more cost-effective and more timely fashion by building on the strategies of existing international global observing programs.

CONCLUSIONS AND RECOMMENDATIONS

Looking ahead to the next twenty years, there will be some tough scientific and technical challenges, especially concerning remote sensing of the ocean from space, and research for and the development of operational oceanographic information services. But the ratio of benefit to costs is potentially high.

More than just advances in science and technology are needed to advance operational oceanography in the cause of sustainable development. In this context, key socioeconomic challenges are seen as:

- Achieving widespread and timely dissemination and exchange of ocean data
- Developing the value-added service sector through synergy between the customer and the basic sciences
- Providing incentives for participation by industry in sharing information
- Persuading navies to declassify data and to provide data in real time
- Developing a framework for the transition of research-based observing networks to operations
- Reducing vandalism of buoys by fishermen
- Obtaining commitment to continuity of key satellite missions
- Increasing national participation in GOOS and expansion of the observing network
- Building the capacity of developing countries to contribute to and benefit from operational oceanography and GOOS, for example, through education and training
- Integrating traditional ocean and meteorological data-gathering systems
- Expanding operational oceanography (including GOOS) as the basis for improved operational services.

RESEARCH PRIORITIES

Key challenges

- Improving forecasting and information services for safe and efficient marine operations
- Improving practical climate predictions and ocean state estimates and data assimilation (e.g., through GODAE)
- Integrating interdisciplinary analysis and modeling for improved environmental understanding and forecasting
- Increasing the provision and use of high-quality data in real-time
- Filling data gaps, for example, in the South Pacific, Indian, and Southern Oceans
- Obtaining global coverage of the upper ocean through profiling floats (e.g., Argo Project) and by other means (e.g., using autonomous underwater vehicles or gliders for full ocean depth hydrographic sections)
- Developing new remote sensors for salinity, precipitation, surface currents, ice thickness, and biology
- Developing new in situ acoustical, optical, chemical, and biological sensors
- Improving technologies for data management
- Increasing the bandwith of communication systems for transmitting and disseminating data
- Improving maps of the seabed to improve the constraints on ocean circulation and tidal models

ACKNOWLEDGMENTS

We gratefully acknowledge the help in particular of the participants in the Potsdam Working Group on Offshore Exploitation, including David Rogers (Chair), Mary Altalo (Rapporteur), Rick Spinrad, Reg Beach, and Leonardo De Souza, and Geoff Holland. In addition we gratefully acknowledge the writers of texts for boxes and the reviewers of different drafts of the manuscript, notably Neville Smith, Worth Nowlin, Geoff Brundrit, Vladimir Ryabinin, Geoff Holland, Gotthilf Hempel, Julie Hall, Nic Flemming, Reg Beach, and Eduardo Marone.

ENDNOTES FOR FURTHER READING

Flemming, N. C., S. Vallerga, N. Pinardi, H. W. A. Behrens, G. Manzella, D. Prandle, J. H. Stel. 2002. *Operational Oceanography: Implementation at the*

European and Regional Scales. Second International EuroGOOS Conference. Amsterdam: Elsevier.

Rayner, R. 2002. *Operational Oceanography: A Perspective from the Private Sector.* Anton Bruun Memorial Lecture, 2001. IOC technical series 58. Paris: UNESCO.

Stel, J. H., H. W. A. Behrens, J. C. Borst, L. J. Droppert, and J. v.d. Meulen (eds.). 1997. *Operational Oceanography: The Challenge for European Co-operation.* Proceedings of the First International Conference on EuroGOOS. *Elsevier Oceanography Series 62.* Amsterdam: Elsevier.

Woods, J. D. 2000. *Ocean Predictability.* Bruun Memorial Lecture, IOC 20th Assembly. IOC technical series 55. Paris: UNESCO.

Chapter 9

A Vision of Oceanographic Instrumentation and Technologies in the Early Twenty-first Century

Tommy D. Dickey

Observations based on developments in instrumentation and technologies have led to most of the major advances in ocean sciences. Modeling has also benefited from new technologies and is playing an increasingly important role (e.g., Robinson and Dickey 1997; Le Traon et al. 2001; Stammer et al. 2001). Undersampling is the main limitation on our understanding and modeling of problems such as global climate change as affected by and affecting the oceans, variability in biomass and fish abundance and regime shifts, and reduction of ocean forecasting error (described in other chapters). A major challenge is therefore to massively increase the variety and quantity of ocean measurements. These measurements are expensive, but vital for effective stewardship, preservation, and utilization of the oceans and atmosphere. Fortunately, many innovative technologies involving computing, robotics, communications, space exploration, and physical, chemical, biomolecular, and biomedical research are being developed at unprecedented rates for a great many applications (Kaku 1998). Many of these will be very beneficial for oceanography in the early twenty-first century.

The general aims of this chapter are:

- to present a brief summary of the challenges of observing the ocean environment;
- to describe a variety of observing platforms;

- to introduce emerging interdisciplinary sensors and systems as well as data telemetry methods; and
- to discuss data dissemination and utilization issues.

Following the introductory section, each subtopic is subdivided into two components: present and near-future capabilities and a vision toward capabilities within the next two decades.

A brief summary of challenges for developing and utilizing new technologies concludes the chapter.

Because of the broad scope of the chapter, it is not possible to develop the ideas and concepts in detail, so several important observational tools can only be mentioned in passing, while others must be omitted. Further, the treatment is not exhaustive and the focus is deliberately on oceanographic instruments. Marine geology and geophysics are touched upon briefly, but not in depth; however, box 9.2 concerns seafloor observatories. Several recent review papers and reports are cited within the chapter, providing starting points for particular interest areas. A few are included in the Endnotes for Further Reading. Finally, an overarching goal of this chapter is to stimulate new and creative ideas concerning ocean technologies and their applications to societal problems.

OBSERVATIONAL CHALLENGES

Factors such as rapid population growth, expanding use and abuse of the oceans, and increasing awareness of environmental change have produced a sense of urgency to understand and minimize human impact on the oceans and atmosphere. In particular, it is vital to improve measurements of critical variables if we are to be able to distinguish natural from anthropogenic changes. It should be emphasized that virtually all important environmental problems require interdisciplinary approaches and necessarily atmospheric, physical, chemical, biological, optical, acoustical, and geological data sets. Ideally, these data should be collected simultaneously (concept of synopticity) and span broad time and space scales to observe the processes of interest (figure 9.1).

Detection limits, precision, and accuracy of ocean measurements are important. However, the oceans are naturally dynamic, with large-amplitude periodic and episodic variability, which is especially confounding for quantifying long-term trends and changes. Oceanography is confronted with challenges beyond those faced by laboratory scientists, in that it is important

9. Oceanographic Instrumentation and Technologies in the 21st Century | 211

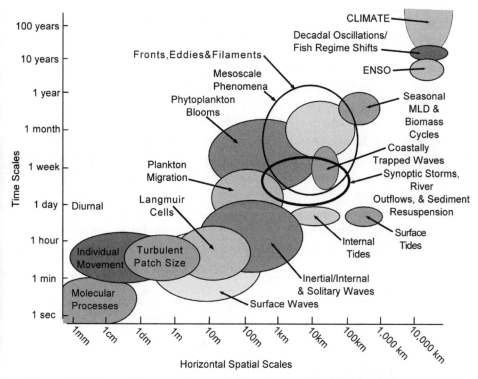

Figure 9.1. Time and horizontal space scales of some of the important oceanic processes. MLD, mixed-layer depth. ENSO, El Niño–Southern Oscillation. After Dickey 1991.

to collect large volumes of data from an uncontrolled and typically harsh environment. And data volume is often more important than precision and accuracy. Presently, limited numbers of variables can be directly measured in the marine environment, so it is important to carefully select those that are most critical. Because of the paucity of data, there will be a need for interdisciplinary numerical models capable of synthesizing observations and predicting variability over broad time and space scales (e.g., Robinson and Dickey 1997).

Many of the necessary observations and instrumentation systems are common to the topics described in the previous chapters. For convenience, the following sections provide information according to platforms, sensors, and systems (and data telemetry) as well as data dissemination and utilization.

PLATFORMS

Multiplatform Approach

Several major interdisciplinary oceanographic programs have adopted multiplatform approaches as conceptualized in figure 9.2. These include the World Ocean Circulation Experiment (WOCE), the Tropical Ocean Global Atmosphere (TOGA) program, the Joint Global Ocean Flux Study (JGOFS), the Global Ocean Ecosystems Dynamics (GLOBEC) program, and the Climate Variability and Prediction (CLIVAR) Study. These programs have utilized mooring arrays, drifters, voluntary observing ships (VOS), and satellite data. The Global Ocean Observing System will also follow this approach, allowing studies of El Niño–Southern Oscillation (ENSO) and interdecadal phenomena such as the North Atlantic Oscillation (NAO), the Pacific Decadal Oscillation (PDO), and the Arctic Oscillation (AO). Further, numerical modeling

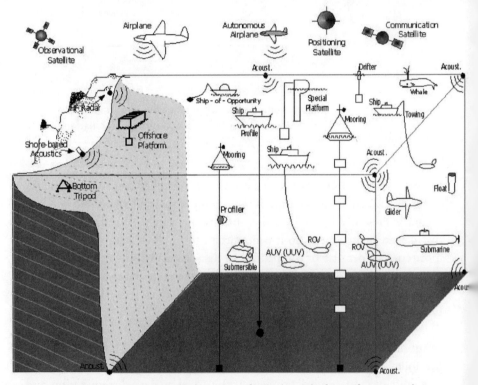

Figure 9.2. Schematic illustrating a variety of sampling platforms for ocean observations. AUV, autonomous underwater vehicle; ROV, remotely operated vehicle; UUV, unmanned underwater vehicle.

is central to these collective programs. Many of the societally important oceanographic problems, like their atmospheric counterparts, require forecasting and rapid information dissemination to decision makers and the public. Thus, two important aspects are near real-time data telemetry and data assimilation modeling. It is evident that oceanographic technologies face challenges not only in the natural scales of environmental variability, but also in particular applications that require rapid response as well as climatic scale problems.

The schematic shown in figure 9.2 illustrates a variety of platforms, several of which can utilize physical, chemical, bio-optical, acoustical, and geophysical sensors or systems. The time-space diagram shown in figure 9.3 provides a rough means of estimating the utility of different platforms in space (horizontal aspect depicted in the figure) and time. It also reemphasizes the need for deploying sensors from both in situ and remote platforms. Nesting of platforms can optimize the use of these observational assets.

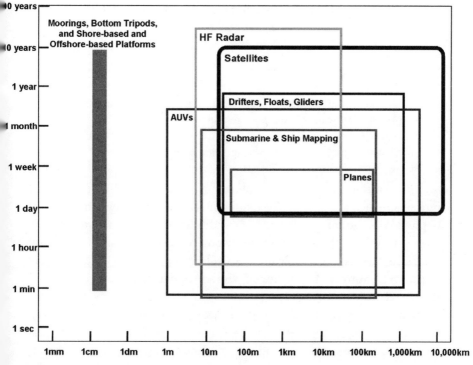

Figure 9.3. Time and horizontal space domains accessible with different observing platforms. After Dickey 1991.

Below, brief summaries of capabilities and future directions using different platforms are presented.

Ships and Submarines

PRESENT AND NEAR-FUTURE CAPABILITIES

Ships have played important roles in providing access to ocean observations since early expeditionary voyages beginning several centuries ago. Today they are important for:

- direct observations and data collection and
- deployment of other sampling platforms, such as moorings, drifters, and others described below (e.g., Knox and Wallace 1999).

One of the advantages of ships is that advanced analytical instrumentation, which cannot presently be routinely deployed in situ from other platforms, can be utilized, often with real-time data analysis. Examples of such instrumentation include flow cytometers, mass spectrometers, radioactivity measurement systems, and turbulence sampling systems. In addition, large volumes of water and net towing are still required for some applications.

Four useful modes of ship sampling include:

- on-station profiling of instruments;
- underway sampling of surface waters using flow-through systems;
- underway sampling using towed undulating or fixed depth bodies or chains, which act as platforms for sensor suites; and
- underway acoustical measurements (e.g., using acoustic Doppler current profilers [ADCPs], hydrophone arrays, and side-scan sonar).

These various modes of sampling are especially useful for regional process-oriented studies and for long transect sampling programs designed to provide important spatial maps. It should also be noted that commercially operated VOSs or ships-of-opportunity observational programs (e.g., Smith et al. 2001) are especially valuable, particularly for obtaining data in remote oceanic regions where few dedicated research sampling programs can be routinely executed.

Some interesting examples of data collection from submarines have been reported. In particular, they have been used for making various physical measurements (sometimes including turbulence probes and biological samplers as well) and bubble measurements (using acoustics). Nuclear submarines have been used by civilian scientists under Arctic ice since 1993.

9. Oceanographic Instrumentation and Technologies in the 21st Century | 215

TOWARD THE FUTURE

Future ships, submarines, submersibles, and other instrument-carrying platforms will benefit from rapid advances in composite materials (e.g., high-performance fibers) whose properties are superior to those presently used (see box 9.1). Ships will likely continue to serve as essential oceanographic platforms. Their utilization for process-oriented and transect sampling will continue, but probably with less weighting because of increasing autonomous sampling platform deployment and recovery operations (e.g., Knox and Wallace 1999). Ships used for direct sampling are expected to use more advanced instrumentation, often using fiberoptics for greater data bandwidth. Newly designed ships are likely to be faster and have improved sea-keeping characteristics in order to make them more cost-effective. Lightships and specialized manned platforms with unique sampling capabilities (e.g., such as Research/Platform Floating Instrument Platform [R/P FLIP]) can play important roles as well.

Ships have served the oceanographic community well. But their limitations in terms of cost, availability, limited synoptic sampling, sample degradation, and contamination, etc., have stimulated development of other complementary platforms as described below. In addition, it is likely that international cooperation through sharing of ships and other seagoing assets will be accelerated in order to make the best and most economical use of ships. Most submarines are used for military purposes, but more may become available to the research community in the future.

Moorings, Bottom Tripods, Offshore, and Shore-Based Platforms

PRESENT AND NEAR-FUTURE CAPABILITIES

Interdisciplinary moored and bottom tripod measurement systems and sensors are being used primarily by the research community to study environmental changes in the ocean on time scales from minutes to years. An increasing number of bio-optical, chemical, geological, and acoustical parameters are being measured from moorings (e.g., Dickey 1991, 2001; Tokar and Dickey 2000; Varney 2000). This work has led to discoveries of new processes, such as:

- primary production variability associated with ENSO and equatorial long waves (e.g., Foley et al. 1997; Chavez et al. 1999);
- sediment resuspension through internal solitary waves (e.g., Bogucki et al. 1997);

- cloud-induced and diel fluctuations in phytoplankton biomass (e.g., Stramska and Dickey 1992);
- phytoplankton blooms associated with incipient seasonal stratification (e.g., Dickey et al. 1994);
- blooms related to frontal- and eddy-trapped inertial waves (e.g., Granata et al. 1995).

Measurements of nitrate, partial pressure of carbon dioxide (pCO_2), dissolved oxygen (DO), and primary production rates (see Taylor and Doherty 1990; Jannasch et al. 1994; Friederich et al. 1995; Merlivat and Brault 1995; DeGrandpré et al. 1997, 2000; Tokar and Dickey 2000; Varney 2000) have enabled new insights into primary and new production and gas exchange across the air-sea interface.

Interestingly, several diverse and often adverse oceanic regions have been studied using interdisciplinary moored systems (e.g., Dickey and Falkowski 2002). These range from the equatorial Pacific (e.g., Foley et al. 1997; Chavez et al. 1999) to the Arabian Sea (e.g., Dickey et al. 1998b) to high latitude areas south of Iceland (e.g., Dickey et al. 1994) and in the Southern Ocean (Abbott et al. 2000). Interdisciplinary moorings and bottom tripods have been used in both open ocean and coastal settings (e.g., Dickey et al. 1993, 1998d; Bogucki et al. 1997; Chavez et al. 1997; Dickey and Williams 2001; Chang and Dickey 2001; Chang et al. 2001).

So-called biofouling degrades sensors, so useful data from moorings have often been limited to a few months in the open ocean and less in coastal waters. But work is underway to mitigate this problem (e.g., Chavez et al. 2000). Moored systems have proven their value in the research realm and need to be deployed in critical regions for studies of seasonal through decadal variability and longer-term monitoring purposes (e.g., Send et al. 2001; Weller et al. 1999; Glenn et al. 2000). High temporal resolution data will continue to be needed to minimize sampling induced uncertainties (via undersampling and aliasing).

Instrumentation deployed on bottom tripods can be used to study and monitor benthic processes (e.g., Chang et al. 2001). Bottom tripods and their instrumentation may be placed in virtually the same environments as moorings. In fact, essentially the same suite of sensors and samplers that are deployable from moorings can also be used on bottom tripods. The chemical species and geological parameters of interest will vary depending on the type of environment (e.g., harbor, coastal, or open ocean).

Offshore platforms, including dedicated and oil production platforms, provide unique opportunities for conducting oceanic and meteorological research and monitoring (e.g., Dickey 1997b). These often large and very stable platforms typically have space and facilities for manned research laboratories and are equipped with adequate power and other services that are needed, making them ideal for oceanographic observations. They offer several advantages over shipboard platforms, including absolute stability in high sea states, suitability for time-series measurements, and capability for housing personnel. It should be possible to launch autonomous underwater vehicles (AUVs) and other mobile sampling devices from these platforms for spatial sampling as well. Active platforms would not be preferable for all types of measurements because of possible chemical contamination and nonrepresentative biology that may result from drilling operations.

TOWARD THE FUTURE

Time-series observations in the coastal ocean and at selected sites of expected high environmental consequence (e.g., equatorial Pacific, high latitude sites of deep-water formation, and/or CO_2 uptake or release) or special long-term monitoring value (e.g., oligotrophic areas of the gyres of the Pacific and Atlantic Oceans, the Arctic region, the Southern Ocean, and near Antarctica) will require the platforms mentioned above (e.g., Dickey and Falkowski 2002). The need to consider costs and return of investment will make it essential to optimally select locations. Increased multiuse of platforms for interdisciplinary sensors and systems is imperative. For example, a mooring designed for a tsunami warning system has been used in the Pacific for measuring upper ocean parameters relevant to global climate change (e.g., Dickey et al. 2001b). There is also a need for essentially expendable moorings, which can be deployed in remote areas and would not require excessive ship time for recovery, and for special observational programs (e.g., in paths and wakes of hurricanes [e.g., Dickey et al. 1998d,1998e] and typhoons and in harmful algal blooms). Only a few studies have utilized mooring arrays (e.g., McPhaden 1995; Dickey et al. 1998b; Abbott et al. 2000); however, this approach has been effective for studying evolving spatial features and long waves.

Novel uses of the platforms will evolve. For example, moored profilers (see box 9.1) have been and can be used to good advantage for situations requiring high vertical as well as temporal resolution data (e.g., including temperature, salinity, current, and bio-optical variables [Marra et al. 1990; Provost et al. 1999]). For the open ocean these may be buoyancy or mechanically driven (e.g., by traction drive) and wave driven in the coastal zone. A variant of moored profilers is the "pop-up" system, which could be deployed as an expendable system with a telemetry module (Dan Frye, personal communication). The package would be dropped from a ship or airplane and rest on the ocean bottom until it rose to the surface either on command or at a prespecified time. If several of these systems were co-located, a time-series record for the site could be obtained.

Increased bandwidth for telemetry of data (using communication satellites and fiberoptic cable) should enable transmission of multifrequency acoustical and multiwavelength optical as well as video data (e.g., Dickey et al. 1993, 1998a; Detrick et al. 1999; Frye et al. 2000). Utilization of the data sets will also be enhanced through synthesis with data collected from other in situ and remote sensing data sets. Use of offshore platforms will require formation of partnerships between government agencies, private industry, and academia. Shore-based instrumentation is being used to obtain surface current and wave data in some coastal regions (e.g., Glenn et al. 2000). Coastal sites and piers may be used for other ocean observations (e.g., sea level, acoustics, horizontally oriented ADCPs, etc.). New acoustic instruments are also becoming available for measurements of wave directional spectra, bottom pressure, and currents in the surf zone.

Drifters and Floats

PRESENT AND NEAR-FUTURE CAPABILITIES

Whereas the platforms described in the previous section can be used to provide high temporal resolution, long-term measurements at fixed locations, drifters, and floats may be used to provide spatial data by effectively following water parcels (e.g., Griffiths et al. 2001). Physical oceanographers have used these methodologies for several decades, most recently to great advantage for WOCE studies. Drifters and floats can collect data in regions of the world oceans, which are rarely visited by oceanographic vessels. These collective devices now use the Global Positioning System (GPS), which gives very accurate position data (down to a few meters). Surface drifters (e.g., Swenson 1999) provide important upper ocean data (e.g., temperature, salinity, and meteorological variables). Early floats typically moved at predetermined depths. More recently, profiling floats (using buoyancy changes to move vertically) have been used to provide data (e.g., temperature, salinity, and reference velocity) during rise and descent through the water column as part of their function to telemeter data back to the scientific community (Roemmich and Owens 2000; Argo Science Team 2001). Present projections suggest that roughly 1000 surface drifters and 3000 profiling floats (near surface to 2000 m) will be in operation annually within the next few years.

TOWARD THE FUTURE

The costs of drifters and floats are expected to decrease as more are produced and used. Emerging systems will be simplified and made more rugged to enable easy deployment from ships-of-opportunity and aircraft. Alternative deployment modes include preprogrammed and on-command releases of floats. New GPS and telemetry (including two-way duplex information exchange) capabilities will continue to improve position accuracy and increase the daily number of reported positions, resulting in greater computed current resolution and accuracy along with much greater volumes of data throughput. Within the past decade, a few oceanographers have begun to deploy optical and/or chemical sensors from drifters and floats (Abbott et al. 1990; Chavez et al. 1997; James Bishop, personal communication; Greg Mitchell, personal communication). An increasing number of interdisciplinary variables are expected to be sampled from these platforms in the future as sensor or system size, weight, power, etc., become less limiting. Biofouling of conductivity as well as chemical and bio-optical sensors will require special measures, as mentioned earlier.

Manned Submersibles

PRESENT AND NEAR-FUTURE CAPABILITIES

Manned submersibles have been in regular use for ocean exploration since the 1960s. Human observations have been important for scientific and industrial activities as well as for engaging the public in ocean exploration (see Hawkes 1997). These craft have been equipped with a variety of in situ sensors and allow on-site decision making by scientists and operators. Investigations of hydrothermal vents, mid-ocean ridges, deep-sea biology and bioluminescence, exploration of archaeological sites and shipwrecks, and pipeline inspection are only a few of the many diverse examples of the usefulness of manned submersibles. Manned submersibles have also been used recently for exploring and collecting baseline data in national undersea marine sanctuaries and parks in the United States (e.g., Earle and Henry 1999).

TOWARD THE FUTURE

New designs of manned submersibles will probably enable deeper dives. Some efforts are being directed toward lightweight "microsubs," that could be operated economically as independent vehicles from research and commercial vessels (Hawkes 1997). The use of microsensors and microprocessors and new hull materials facilitates these approaches. The use of real-time video and audio transmissions during scientific dives should be used to allow other scientists, students, and the public to share the experience of deep-sea exploration.

Remotely Operated Vehicles

PRESENT AND NEAR-FUTURE CAPABILITIES

The industrial ocean community has used remotely operated vehicles (ROVs) for a variety of applications, including underwater inspection and mechanical activities (e.g., Gilman 1999). They have also been used for exploration and science (e.g., Dawes et al. 1998), sometimes with connection to manned submersibles. In particular, they have played major roles in studying hydrothermal vents, observing animal life and behaviors, and for discovering shipwrecks, including the famous *Titanic*.

TOWARD THE FUTURE

ROVs will continue to be needed for many undersea activities and in some cases will supplant work that is now being carried out with manned submersibles. It will be possible to instrument ROVs with many of the small sensors described later, thus increasing their utility. The use of fiberoptics will make it easier to transmit video imagery.

Autonomous Underwater Vehicles

PRESENT AND NEAR-FUTURE CAPABILITIES

Major advances in the development of autonomous underwater vehicles (AUVs) have occurred during the past decade (e.g., Curtin et al. 1993; Griffiths et al. 2000, 2001). Box 9.1 provides a description of the history and present and future capabilities of AUVs. Several forms of AUVs are being developed. These include autonomous surface craft, moored profilers, gliders, and propelled AUVs. Most of the activities have been confined to engineering design and some practical field activities. However, these vehicles are beginning to be exploited for a variety of scientific studies, thanks to the development of new, relatively small sensors and systems that consume moderate power and that can be interfaced to the vehicles (e.g., Griffiths et al. 2001).

Some of the key advantages of the autonomous platforms include:

- cost per deployment;

Box 9.1. Autonomous Underwater Vehicles
Gwyn Griffiths

Autonomous underwater vehicles (AUVs) have a long heritage. Their ancestry can be traced back to the military torpedo and drift bottles of the nineteenth century. But, by harnessing the tremendous developments in miniature and low-power electronics, in reliable software and high-capacity data storage, and in data communications we can now produce a very capable family of AUVs for ocean science. Microelectronics, software, and communications are no longer the limiting factors in AUV design. Performance limits are now set by material properties such as weight to displacement ratio and by the limited amount of energy that can be economically and safely carried on board the vehicles. Radical improvements can be expected in these areas within the next decade. Exciting developments in fiber and ceramic composites promise a new generation of low-weight, high-strength materials for use in the deep ocean. New-generation batteries and particularly fuel cells under development for the automobile and portable computer markets will offer an order-of-magnitude increase in energy density at affordable cost.

These generic technologies form the building blocks for a variety of vehicle types, each with a set of advantages and disadvantages. There will be no single universal vehicle suitable for all purposes. The key vehicle types are discussed below.

AUTONOMOUS SURFACE CRAFT

Surprisingly, this class of vehicle has seen less research and development effort than the fully underwater vehicle. Surface vehicles have the great advantages of simple high-bandwidth communications using radio or satellite links and a far greater potential range using energy generation via combustion engines. In ship-like and semisubmersible formats, they form ideal platforms for autonomous near-surface mapping of physical and chemical parameters and as platforms for sonars imaging the seabed. By adding autonomous winches, their capability can be extended to profiling, similar to a research ship on station.

MOORED PROFILERS

Moored profilers are AUVs constrained to traveling along a mooring line. Traveling at vertical speeds of less than 0.5 ms^{-1} and carrying a micropower chemical detection system and current sensors, they are very energy efficient, with endurance in excess of 1 million meters. Onboard processors may enable their use as adaptive or conditional sampling tools. Through inductive, acoustic, and perhaps optical data telemetry, the profiler may communicate with a surface buoy to enable near real-time data transfer.

GLIDERS

Free drifting neutrally buoyant acoustically tracked floats have a heritage dating back fifty years. The glider concept takes the most recent version of the variable buoyancy

(continues)

Box 9.1. Continued

profiling drifting float, equips it with lift surfaces (wings), a hydrodynamic shape, and trajectory control (internal moving mass). These additions enable the vehicle to glide while diving and while surfacing along slopes as steep as 2:1 or as gentle as 1:5. With forward speeds of typically 0.25 ms^{-1}, gliders may be used as virtual moorings or on long purposeful transects. As with moored profilers, gliders will benefit from new-generation micropower sensors and from new energy sources. It is feasible that global initiatives such as the ARGO array of drifting floats may one day be replaced with an array of gliders operating as virtual moorings and on transects.

PROPELLED AUVS

Presently, propelled AUVs tend to be far larger than moored profilers, floats, or gliders (figure BX9.1). Consequently, they are more expensive to build and operate. However, because they can carry considerable energy (perhaps over 150 MJ) and have large payload spaces (up to 1000 L), they form valuable platforms for multidisciplinary process study experiments. In particular, they can carry large power-hungry sensors such as sonars and flow cytometers as well as new versions of traditional water samplers. With proven ranges in excess of 250 km, such vehicles have already made significant contributions to high-resolution marine geoscience, to fisheries research, to the measurement of oceanic turbulence and mixing, and to understanding small-scale coastal processes.

By deploying AUVs shuttling between docking stations included on deep ocean moorings, key sections may be monitored in the future without needing a ship. Propelled AUVs are very likely to be used for data gathering from otherwise impenetrable environments, for example, under Antarctic sea ice and under the floating ice shelves of Antarctica and Greenland.

Figure BX9.1. The Autosub autonomous underwater vehicle being recovered from a trial deployment. Courtesy of S. P. Hall, Southampton Oceanography Centre.

> Cost will remain a major issue. Whereas AUVs may provide an acceptable cost per profile over five years, the need for significant up-front capital investment will need to be justified through more rigorous cost-benefit analyses.
>
> The ocean-science community will undoubtedly develop and use autonomous vehicles of many different types over the next decade. Many will be able to share sub assemblies and design philosophies, and engineers will continue to seek cost-effective solutions by adapting modules and components imported from the consumer and industrial sectors. It will become even more essential for engineers and scientists to maintain dialogue on requirements and specifications. As the vehicles become more capable, the legal aspects of their operations will need careful thought in national and international forums. More information concerning AUVs may be found in Griffiths et al. (2001).

- capability to sample in environments generally inaccessible to ships (e.g., in hurricanes or typhoons and under ice) (see Bellingham et al. 2000);
- good spatial coverage and sampling over repeated sections;
- capability of feature-based or adaptive sampling; and
- potential deployment of several vehicles from moorings, mother ships, offshore platforms, and coastal stations.

An important factor for new platforms will be sampling flexibility. In particular, it should be possible to direct AUVs to critical sampling areas based on other remote and in situ data and model predictions. It is likely that AUVs will be increasingly called upon to replace some of the present ROV work efforts.

TOWARD THE FUTURE

At present, several specialized groups are developing and using autonomous vehicles (Griffiths et al. 2001). Numbers are expected to grow as mission length capabilities increase, costs decline, reliability improves, operation becomes more routine, and more sensors become available for various sampling needs. Creative uses of the vehicles will involve networking and information feedback loops to guide sampling programs (in some areas involving predictive models) and responses to extreme natural and anthropogenic driven events.

Remote Sensing from the Air and Space

PRESENT AND NEAR-FUTURE CAPABILITIES

Humankind's views, perceptions, and understanding of the Earth and its oceans have been dramatically affected first through images obtained from aircraft and more recently from space. The thinness of the atmosphere and the vastness of the oceans as well as the beauty and complexity of the atmosphere and oceans are only a few of the impressions provided in photographs by space explorers (e.g., Apt et al. 1996). While the visual impacts of space-based images are profound and in themselves informative, the next difficult step has been to extract quantitative information.

Satellite-based sensors are now capable of providing nearly global and, in some cases, "snapshot" or synoptic views—and, importantly, data—over much larger areas of the ocean surface (and ice) than possible from any other platforms (see figure 9.4 in the color section). The data are typically empirical inferences of surface signals (e.g., either passive or active electromagnetic radiation) and are often based on groundtruth data sets obtained from ocean-based platforms (e.g., Koblinsky and Smith 2001). Because electromagnetic radiation only penetrates to very shallow ocean depths (e.g., infrared to millimeters and visible to meters), satellite information must be complemented with in situ observations to characterize important subsurface ocean properties. Considerable research effort is being devoted to extracting subsurface data using remote sensing, in situ data sets and models.

Some of today's oceanographically important remotely sensed variables include solar radiation, wind stress and direction, rainfall, surface heat fluxes, sea surface temperature, ocean color (e.g., pigment concentrations) (IOCCG 1999), and sea surface height (several references in Koblinsky and Smith 2001).

A few of the interesting applications of remote sensing have included studies of mesoscale features, seasonal evolution of temperature and phytoplankton (via ocean color), El Niño and La Niña, equatorial waves, planetary scale waves (Kelvin and Rossby waves), wakes of ships, hurricanes and typhoons, coastal upwelling, storm runoff, surface and internal gravity waves, bottom topography, island wakes, and ice age, extent, thickness, and motion.

TOWARD THE FUTURE

Aircraft-based systems will be able to provide very high spatial resolution data for altimetry, color, temperature, salinity, and other variables. Aircraft-borne lidar has been used to detect mixed layer thickness as well as optical properties from backscatter profiles. Work is underway to implement similar methodology with satellite sensors. Autonomous aircraft technologies are progressing and operating ranges of 2500 km at high altitude with diverse sensor packages (e.g., for meteorology, sea surface temperature, hyperspectral color, altimetry for tides and currents, etc.) are projected for unmanned aerial vehicles (UAVs) (Paul Bissett, personal communication). UAVs could, in principle, fly over a site for a day or longer to enable collection of high spatial resolution time series.

The temporal and spatial resolutions of satellite remote sensing systems are also likely to improve for many parameters. Multisatellite missions are already collecting sea surface temperature data capable of resolving the diurnal cycle (sampling at hourly time scales) on cloud-free days. Clouds will continue to be an obstacle to temperature and color measurements. But, by using more satellites, the time gaps will be decreased and data sets more fully completed. Altimetry measurements are essentially unaffected by clouds, but present altimeters have good spatial resolution only in a narrow beam directly under the satellite's flight path. However, concurrent and coordinated sampling with multiple satellite missions (e.g., Raney and Porter 1999) and new wide swath systems (Pollard and Martin 1999) will make it possible to collect essentially two-dimensional altimetry data. Another important application of altimetry is to determine sea-level time series in conjunction with traditional tide-gauge-network data sets and benchmarked using GPS data (e.g., Nerem and Mitchum 2000). This methodology has led to observations suggesting that the globally averaged mean sea level increased by more than 10 cm during the 1997–1998 ENSO event. In addition, a new satellite mission is planned to improve the accuracy of the global marine geoid. This will allow considerable improvement in the altimetric calculation of ocean circulation. Planned missions for color (e.g., IOCCG 1999) and temperature can be expected to achieve spatial resolutions of tens of meters or less (probably over limited selected areas) as well as increased optical spectral resolution down to a few nanometers and less (Davis et al. 1998).

Work is progressing in remote sensing of other important oceanic variables. In particular, studies are underway to measure salinity from satellites (e.g., Njoku et al. 1999; Lagerloef and Delcroix 2001). Some promising results have already been obtained using aircraft-based sensors in areas with strong salinity gradients. Other variables are likely to include colored dissolved organic materials (e.g., Chang and Dickey 1999) and signatures of different phytoplankton groups, including those associated with red tides or harmful algal blooms (HABs) (e.g., Carder and Steward 1985).

Remote sensing of many important chemical and biological variables (e.g., zooplankton and fish) remains as a major research challenge. Regular observations of organisms from satellite platforms are not presently feasible. However, ocean color, temperature, and current data can be valuable for identifying features (e.g., fronts, eddies, upwelling areas, red tide blooms, etc.) where high biological activity may be located. Further, extremely high-resolution imagery may eventually be available for sensing surfacing mammals and large schools of fish. Also, studies of larger organisms, such as marine mammals, have used satellite radio tracking, but this approach requires initial tagging. In some cases, tagging instrument packages have included sensors for temperature and depth as well as positioning.

Possibilities for event-triggered sampling using sensors placed on special satellite platforms (e.g., steerable instruments in geostationary orbit) are being considered. This approach is most attractive for responses to disasters, directing field and other remote sensing observations to key locations and providing data, which would otherwise be unattainable. Advanced analytical and modeling activities will be required to optimize the use of present and future remote sensing data sets. Examples include removal of tides from altimeter data and incorporation of remote sensing and in situ data into models for three-dimensional spatial descriptions and predictions.

IN SITU SENSORS AND SYSTEMS

The emergence of more capable sensors and systems for oceanographic applications can be attributed to several factors. These include:

- technology transfer in the areas of measurement and analysis techniques originating in the medical, engineering, microelectronics, microprocessor, data communication, and global positioning research communities;
- support of projects devoted to development of both fundamental and societally relevant ocean technologies; and
- the formation of functional partnerships among academia, government laboratories, and private industry (e.g., Dickey et al. 2001b). Many of the requirements for deep-space measurement systems are similar in nature to those of oceanographic studies, so future synergistic partnerships between ocean and space technologists are attractive.

Considerations for future sensors and systems to be deployed from the platforms described above include response time, size, power requirements, durability, reliability, stability/drift, susceptibility to biofouling, data storage and telemetry, and cost. One of the problems facing sensor and system developers and users is the proper interpretation of the instruments' signals (e.g., Dickey 2001). The variability of parameters is usually well depicted, but it is often difficult to obtain and interpret absolute values. For this reason, testbed sampling programs that use multiple platforms for intercomparison and groundtruthing are critical elements for optimal use of ocean technologies (Dickey et al. 1998a, 2001a).

For convenience, in situ sensors and systems have been subdivided below in terms of their primary disciplinary use: physics, chemistry, optics and biooptics, biology and bioacoustics, and marine geology and geophysics. Interdisciplinary measurement suites are desirable because of the need for simultaneous, complementary observations and cost-effectiveness in shared platforms and telemetry systems. A final section briefly discusses this aspect.

Physics

PRESENT AND NEAR-FUTURE CAPABILITIES

Measurements of atmospheric variables above and at the sea surface, currents, and physical water properties are of great importance for many of the other subdisciplines of oceanography. This is partly because of the processes of air-sea gas exchange and other interactions as well as advection and mixing of chemical and biological species. New meteorological measurement systems developed for ships and buoys as part of WOCE (e.g., Weller et al. 1998; Taylor et al. 2001) now allow net surface heat fluxes to be made with an accuracy approaching 10 W/m^2.

Novel uses are being explored for underwater sound, using ADCPs and other instruments to estimate rainfall rate and wind speed. In addition, ocean temperature and salinity measurements have improved significantly. Temperature resolution and accuracy of 0.01°C is now possible with salinity drift of less than 0.015 to 0.055 psu over six-month observational periods from moorings. Current measurements have likewise advanced, both in terms of quality and quantity (Dickey et al. 1998c). Some current measurement devices (e.g., mechanical) must be deployed from moorings or fixed platforms. However, others such as ADCPs can be used for measurements from most platforms, including commercial ships, moorings, and AUVs.

Current and turbulence measurements from bottom tripods have improved greatly (Williams et al. 1987) and are now often made in conjunction with optical and acoustical measurements to study sediment resuspension and transport (e.g., Chang et al. 2001). Bottom pressure sensors are being effectively used for many applications (sea level, tides, waves, and currents). Expendable devices for measuring temperature, conductivity (salinity), currents, and turbulence have become available and can often be launched from ships and aircraft. In principle, optical, acoustical, chemical, and other probes can be added to expendable packages. Shore-based high-frequency radar systems are being used at some geographic sites to great advantage as two-dimensional surface currents in areas of hundreds of square kilometers can be sampled synoptically (e.g., Paduan and Rosenfeld 1996).

Drifters and floats have become more important research and monitoring tools for reasons described earlier (e.g., Griffiths et al. 2001). An important new method involves tracers that have been exploited for upper ocean, bottom boundary layer, and deep circulation studies (e.g., Fine et al. 2001; Watson and Ledwell 2000; Ledwell et al. 2000). A variety of tracers are being exploited. These have been introduced either through anthropogenic activity (radioactive tritium and carbon, chlorofluorocarbons, etc.) or by experimenters (e.g., fluorescein dye, sulfurhexafluoride, etc.). Different decay time scales of tracers are used to advantage.

TOWARD THE FUTURE

The use of remote sensing will need to be expanded to increase numbers of meteorological measurements over the world oceans. In situ measurements from buoys and drifters will continue to be needed to provide groundtruthing data and to provide local high-frequency time series data, which cannot be obtained from satellite sensors. For example, localized weather, storm, hurricane, and typhoon systems cannot be adequately sampled from satellites alone. Some in situ measurements (e.g., rainfall and relative humidity) remain difficult, and more research will be needed to improve these capabilities.

The cost of measurement systems is a primary concern because large numbers of measurements are needed. Multiple users and uses of data can reduce expenses. In particular, ADCPs can continuously sample vertical profiles of horizontal current and acoustic backscatter (related to zooplankton concentrations) (Smith et al. 1992; Roe et al. 1996). Shipboard profiling or tow-yo operations are underway, and many commercial ships now routinely measure currents using ADCPs.

One of the important physical variables for biological and chemical as well as physical oceanographers is vertical velocity. Very few examples of accurate measurements of vertical velocity have been reported (e.g., Dickey et al. 1998c). However, specially designed mechanical and electromagnetic instruments and acoustic systems using both the Doppler effect and backscattering have been used to investigate vertical motions (e.g., in strong upwelling and convective regimes and associated with Langmuir cells). Near surface-scanning Doppler sonar also has the potential to provide simultaneous measurements of wind direction, wave directional spectra, Langmuir circulation, and mean near-surface flow.

Another important parameter is turbulence. In the past, only a few specialists have been able to measure ocean turbulence successfully. More routine and cost effective turbulence measurements, which can be made from a host of platforms, are needed for many types of applications. It is expected that high-frequency radar will continue to be refined and exploited (e.g., Paduan and Rosenfeld 1996). Most systems are shore-based at present. However, offshore platforms, buoys, etc., will also be used to expand the spatial coverage (e.g., Glenn et al. 2000).

Acoustic tomography shows great promise for two important applications: for long-term ocean-basin-scale integrated temperature change (acoustic thermography) and for measuring ocean circulation patterns, especially at the mesoscale (e.g., Send et al. 1999b; Orcutt et al. 2000; Dushaw et al. 2001). This methodology is especially valuable for demanding areas such as boundary currents, straits and through-flows, under ice, and in regions of active convection and bottom water formation. Importantly, some existing hydrophone arrays may be used, and efforts are underway to create user-friendly software to make acoustic tomography more accessible and widely used.

There is a fundamental need to collect and transmit higher volumes of data (e.g., higher temporal resolution and longer duration). One of the important goals of physical oceanography is to optimally use remotely sensed data (e.g., altimetry, wind scatterometers, sea surface temperature, etc.) and in situ data (from all available in situ platforms, especially moorings with fixed depth and profiling instruments, AUVs, etc.). Sea-level measurement is one of several important examples.

Chemistry

PRESENT AND NEAR-FUTURE CAPABILITIES

Scientific areas of particular relevance to ocean chemistry include global warming due to the greenhouse effect, nutrients and their role in primary productivity, and the biological pump (for transporting carbon to the deep sea) as well as coastal eutrophication, ocean pollution, and hydrothermal vents. Natural and artificial chemical tracers are important tools for studying circulation, mixing, and dispersal, as mentioned earlier. Some of the important advances in chemical sensors and potential future applications are described below (also see reviews by Varney 2000; Tokar and Dickey 2000; Fine et al. 2001).

In the past, most ocean chemistry was done using water samples collected with bottles at sea. Some of the lab-based measurement methodologies presently used in chemical oceanography include colorimetric chemical analyses, gas chromatography, high-performance liquid chromatography, mass spectrometry, and infrared spectrometry. These instruments do not always respond favorably when installed aboard research vessels.

Methods using wet chemical techniques often require that samples be collected, stored, preserved, and transported from the field into the laboratory for chemical analyses. Solvent extractions and separations are usually required before actual chemical analyses can begin. This poses problems in that biological, chemical and photochemical, and physical processes can cause changes in the chemical properties of interest. The ability to identify and quantify trace metals, synthetic organics, and toxic substances in the field lags behind the ability to analyze them in the laboratory. The high costs associated with sophisticated laboratory analyses of environmental samples preclude their widespread use in many monitoring programs.

While analyses of ship-derived water bottle samples are still important, other methods are gaining increasing attention and are being put into use. A few examples follow (see also Varney 2000):

- Underway surface water sampling is carried out using shipboard water intake systems and automated laboratory chemical analyzers (e.g., for DO, pCO_2) and plant nutrients such as nitrate, phosphate, silicate, and ammonia.
- Pumping systems are used to bring subsurface samples onboard ships for similar analyses and water collection.
- Specially designed water samplers have also been used to sample key chemical components of venting hydrothermal fluids (i.e., DO, pH, Fe^{2+}, Mn^{2+}, and H_2S) from manned submersibles and ROVs.
- Moored serial water samplers are being developed and used to obtain discrete preserved samples periodically over the course of months to create chemical time-series (e.g., Wu and Boyle 1997; Dickey et al. 1998a). Similar systems have also been developed for drifters (e.g., Abbott et al. 1990). Trace metals, as well as macronutrients, can be analyzed using the water samples because of improved laboratory analytical systems.

As stated earlier, in situ measurements have major advantages in sampling for ocean chemistry because these samples do not suffer degradation and are representative of the local environmental conditions at depth. Several new sensors and systems are capable of making in situ time-series measurements with sampling intervals of a few minutes and durations of months. Measurements include nitrate and other nutrients, DO, pCO_2, pH, and alkalinity (e.g., Merlivat and Brault 1995; Friederich et al. 1995; DeGrandpré et al. 1997, 2000; Varney 2000; Byrne et al. 2000). The various measurements use a variety of methods, including polarographic electrodes, colorimetry (multiple reagents used for different analyses), and spectrophotometry. The latter method has taken advantage of long path-length absorbance spectrometry enabled by using fiberoptic liquid core waveguides (Byrne et al. 2000). It should be noted that some chemical sensors and analyzers have been successfully deployed from moorings (e.g., Dickey et al. 1998a, 2001b; McNeil et al. 1999), drifters, and AUVs. Real-time and near real-time telemetry of chemical variables has also been demonstrated, and there are a few examples of actually modifying sampling (gain changes, etc.) using two-way or duplex data telemetry systems.

In situ chemical detection systems have also been deployed using manned submersibles. These virtually eliminate the need to return to the surface to perform the analysis. One such system was designed for vent studies and can measure four chemical species, including Mn^{+2}, Fe^{+2}, Fe^{+3}, and H_2S (Coale et al. 1991). It is also equipped with a CTD and light-scattering sensors.

Finally, the importance of atmospheric input of dust and aerosols into the ocean has been recognized (e.g., role of iron fertilization). Samplers for dust, aerosols, and particular chemical species are being developed for deployment from surface buoys to avoid land contamination (Sholkovitz et al. 1998).

TOWARD THE FUTURE

Several other chemical methodologies can be used (the following examples were provided by Peter Brewer). For example, laser Raman scattering systems, which can be deployed from ROVs and manned submersibles, are planned for use to examine the time course of hydrate formation for deep sea CO_2 disposal (Brewer 2000). These systems can also be used for gas hydrate geochemistry, bacterial mats, particulate geochemistry, and sulfate profiling in sediments. Nuclear magnetic resonance (NMR) technology has been used by oil companies to detect the presence of water and to find the porosity of marine sediments and rocks in boreholes. This technique can provide a spectrum of pore sizes, but to date has not been available for academic science. Large and small borehole tools offer powerful opportunities to examine gas and hydrate sediment geochemistry experimentally under deep-ocean conditions using ROVs. The use of NMR to directly measure advection rates of fluids in sediments is another possibility.

Also, the radium isotope series is being shown to be useful for studying the important problem of saline groundwater intrusion into coastal sediments. Such methodologies can be powerful for quantifying nonpoint source invasion of pollutant species from terrestrial systems.

Chemical sensors and analyzers will probably become more readily available for users of the various platforms. Verification of data using discrete water samples will be necessary during the developmental phase. There is a need to increase capabilities for sampling broader suites of chemicals autonomously. Applications to problems of pollution, with diverse chemical species of concern (e.g., PCBs, DDT, toxic metals, etc.) will be quite demanding. It is expected that chemical sensors rather than analyzers will be preferable, if not required, for some platforms such as towed bodies, floats, drifters, and AUVs. It is also anticipated that artificial neural networks will be used in conjunction with some chemical and perhaps biological measurement systems.

A new technology mentioned earlier uses fiberoptic sensors for ocean applications (see Tokar and Dickey 2000). An optical sensor or modulation device is one element of an integrated system that also includes an excitation light source, optical fibers, a photodiode detector, and other associated components such as connectors, couplers, and signal processing/data logging equipment. Different light sources can be used. The modulated optical parameters of light may include amplitude, phase, color, state of polarization, or combinations of these. The fiberoptic chemical sensor is usually made up of analyte-specific sensing reagents immobilized on the side or located at the tip of an optical fiber. To construct an effective sensor, it is necessary to immobilize suitable and sufficient sensing reagents (i.e., fluorescent dyes) next to the sensing area of the fiber.

Fiberoptic sensors fall into two major categories: intrinsic and extrinsic. Intrinsic sensors use the optical fiber itself to sense the parameter being measured. This becomes possible when the fiber is altered by the physical or chemical external variable being sought. These external variations cause an alteration of the optical properties (total internal reflection) of the fiber, resulting in measurable light intensity, phase, or polarization changes.

Intrinsic refractive index sensors have been used to measure hydrocarbons in water. A number of extrinsic fiberoptic sensors have been designed and tested both in the laboratory and the field to measure selected analytes in water. Examples of successfully measured chemical species using fiberoptic sensors include ammonia, methane, and dissolved carbon dioxide. Several pCO_2 and pH systems have been developed recently (e.g., Friederich et al. 1995; DeGrandpré et al. 1997, 2000; Hopkins et al. 2000). One type of pCO_2 fiberoptic sensor for seawater is based upon fluorescence using a combination of dyes and is coupled to a commercial fiberoptic fluorometer. A renewable-reagent fiberoptic sensor has also been developed. Currently, there are a limited number of fiberoptic sensors available for use in seawater. Additional research and development is needed to improve response time, reproducibility, and long-term reliability.

Another promising type of sensor is the microelectromechanical system (MEMS) (see Tokar and Dickey 2000). MEMS is based on a relatively new technology, which is used for making and combining miniaturized mechanical and electronic components out of silicon wafers using micromachining. MEMS have shown encouraging results for sensing physical parameters, but work is needed to realize their full potential for chemical sensing. Most work with MEMS and other so-called nanotechnologies has been done in laboratories. But, transitioning to in situ applications is feasible. Potential advantages of MEMS include autocalibration, self-testing, digital compensation, small size, and economical production.

Optics and Bio-optics

PRESENT AND NEAR-FUTURE CAPABILITIES

The term ocean optics refers to studies of light and its propagation through the ocean medium. Bio-optics is a more recent, but very commonly used, term, which invokes the notion of biological effects on optical properties and light propagation and vice versa. These related disciplines have gained increasing attention in part because of new technologies and community realization of their central importance to several ocean problems. These involve biological-optical-physical interactions, such as ocean primary productivity, upper-ocean ecology, biogeochemical cycling and the biological pump, sediment resuspension and transport, ocean pollution, and bio-optically modulated variability in upper-ocean heating rates (e.g., Dickey and Falkowski 2002). Studies of phytoplankton now use optical measurements for size ranges depicted in figure 9.5. In the open ocean, optical properties are typically biologically modified, whereas in the coastal oceans terrigenous input and resuspended sediment also play major roles.

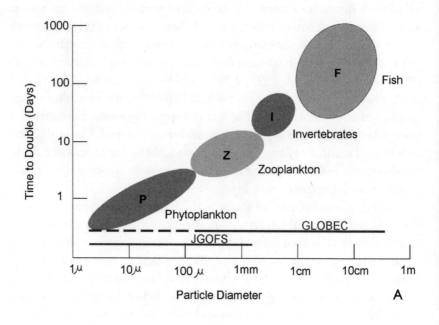

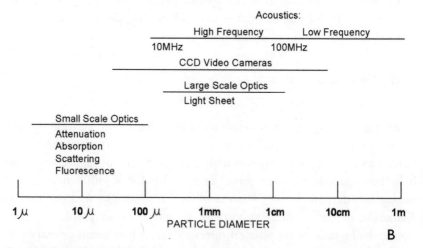

Figure 9.5. A: Rough scales of doubling of particular groups of organisms versus size. Based on Sheldon et al. 1972. B: Organismal sampling capabilities of various technologies. CCD, charged coupled device.

The relationships between physical and biological processes can often be seen by comparing satellite maps of sea surface temperature and ocean color (see figure 9.4b). Subsurface light fields are affected by both incident solar radiation and properties of ocean waters, which are highly variable in space and time because of a host of complex physical, chemical, and biological processes (e.g., Dickey 1991; Dickey and Falkowski 2002). Particulates, both organic and inorganic, and dissolved materials play key roles in the variability of ocean optical properties. The spectral quality of light is often fundamental to understanding processes.

Two operational classifications of bulk optical properties are useful for the following discussion: inherent optical properties (IOPs) and apparent optical properties (AOPs) (see review by Dickey 2001). IOPs depend only on the medium and are independent of the ambient light field. AOPs are those properties depending on both the IOPs and the geometric structure of the subsurface ambient light field. Instruments designed for underwater light observations are usually described as measuring either IOPs (e.g., spectral beam attenuation, absorption, and scattering coefficients) or AOPs (e.g., spectral diffuse light attenuation coefficients). Reviews of recently developed IOP and AOP measurement techniques are presented in Dickey (2001) and Dickey and Falkowski (2002).

Until quite recently, direct in situ measurements of IOPs were generally limited to single wavelength (usually 660 nm) beam attenuation. However, some instruments are now capable of measuring light absorption, scattering, and attenuation at multiple wavelengths (from 9 to about 100 different wavelengths) (e.g., Moore et al. 1992, 2001; Twardowski et al. 1999). Likewise, measurements of diffuse light attenuation have increased spectral resolution to a few nanometers in the visible. The power of these types of instruments lies in their ability to distinguish phytoplankton from detritus and colored dissolved materials (e.g., Chang and Dickey 1999), and potentially to identify phytoplankton (perhaps including harmful algae) at least by community groups.

An important goal is to estimate primary productivity; several different optical measurements have been used for these determinations (e.g., Dickey and Falkowski 2002). Examples include the use of chlorophyll fluorescence and photosynthetically available radiation (PAR) measurements with empirical models and more sophisticated measurements using "pump and probe" fluorometers (e.g., Kolber et al. 1998). The latter instruments have the important advantage of providing information about biophysical parameters related to photosynthesis (e.g., quantum yield as affected by nutrient and light conditions). Many of these optical instruments have been deployed from ships, moorings, profiling floats, and drifters. Thus, sampling with both high spatial and temporal resolution, comparable to those of physical parameters, is now possible.

One of the common uses of in situ optical measurements has been for groundtruthing and algorithm development of ocean color imagers (e.g., Coastal Zone Color Scanner, Sea-viewing Wide Field-of-View Sensor [Sea-WiFS], and Ocean Color and Temperature Scanner [OCTS]) (see IOCCG 1999). In addition, variability of phytoplankton biomass and primary productivity and upper ocean radiant heating rates (and penetrative component of solar radiation) have been estimated using both in situ and remotely sensed data (see Dickey and Falkowski 2002). Determination of bottom bathymetry using optics in coastal areas is another important applied research objective.

TOWARD THE FUTURE

A major thrust for in situ optical instrumentation is in the area of scattering of light (particularly, in terms of angular dependence (volume scattering function) and backscattering). These data types, along with absorption, are important for fully characterizing the underwater light field, which has great implications for underwater visibility and remote sensing as well as for fundamental optical radiative transfer problems. Concurrent measurements of key IOPs and AOPs are critical for developing inversion models so that IOPs can be determined from AOPs and vice versa. Another important research goal is to estimate the vertical structure of IOPs given remote sensing measurements of water-leaving radiance (Dickey 2001).

Spectral fluorescence is used in another new type of instrument. One such instrument, which uses six excitation wavelengths and sixteen emission wavelengths (e.g., Desiderio et al. 1997), has been used to measure dissolved organic materials as well as other dissolved substances associated with aging sewage discharge waters (Petrenko et al. 1997). Other optical instruments, using the Fraunhofer effect, have been used successfully to obtain particle size distributions, primarily in bottom boundary layers (Agrawal and Pottsmith 1994). A different and extremely powerful optical technology, flow cytometry, has been successfully used onshore and onboard ships for counting and distinguishing particles and phytoplankton as well as for characterizing their optical properties (e.g., DuRand and Olson 1996).

Work is progressing to miniaturize and ruggedize flow cytometers for deployment at-sea from buoys (e.g., Rob Olson, personal communication) and AUVs. The time series obtained from many of the aforementioned in situ optical systems show remarkable variability, associated with high frequency and episodic events as well as longer term processes (e.g., McGillicuddy et al. 1998; McNeil et al. 1999; Dickey et al. 2001a). The interpretation of these time series is often difficult because of the complex nature of the observed medium and organismal physiology. Thus, more effort in understanding these types of data will be essential. Nonetheless, some remarkable and somewhat unexpected correlations have been noted (e.g., very high correlations between beam attenuation (660 nm) and particulate organic carbon and optical backscatter and particulate organic carbon (see Bishop 1999).

Presently, most bio-optical sensors are deployed from ship-based profilers. With advances in microprocessor technologies, data processing and storage are not generally limiting. Thus, a growing number of observers are moving toward autonomous sampling from moorings, AUVs, drifters, and profiling floats. Capabilities for the telemetry of optical data are also increasing rapidly.

Biology and Bioacoustics

PRESENT AND NEAR-FUTURE CAPABILITIES

One important goal of current research is to understand and ultimately predict how populations of marine animal species (from zooplankton through fish) respond to natural and anthropogenic changes in global climate (e.g., the GLOBEC program) (Dickey 1993). This challenging research is driven in part by the waxing and too frequent waning of fisheries. Biological diversity is another topic of great interest. There remains no comprehensive census of marine organisms (although plans and efforts are underway for this activity), and some estimate that more than 5000 marine species remain to be identified. Massive volumes of data will be required along with interdisciplinary models capable of extrapolating and integrating these data sets (see *Oceanography*, vol. 12, no. 3, 1999). Because of the variety of marine life, key indicators and ecologically relevant variables must be carefully chosen. These will often be dictated by special regional aspects (e.g., target species, habitats ranging from coral reefs to the abyss). Fundamental physical (e.g., meteorology, currents, temperature, and salinity), chemical (e.g., plant nutrients and trace metals), and biological (e.g., phytoplankton and zooplankton biomass distributions) observations will form the core measurements and must be modeled with excellent fidelity.

Early studies of organisms, including phytoplankton, zooplankton, and higher trophic level organisms, typically relied upon net sampling. The advantage of this method is that an individual organism can be captured, numbered, analyzed, and studied. One of the interesting variants on this method involves the towing of continuous plankton recorders, which use progressively exposed portions of mesh netting, behind either research vessels or ships of opportunity. Long transects across the North Atlantic with these systems have provided valuable phytoplankton and zooplankton data for several decades (Continuous Plankton Recorder program, Reid et al. 1998). Modified versions of these systems are presently being developed to use new sensors. Large amounts of ship time and great numbers of personnel are typically required for net-based sampling, leading to high costs and thus limiting the geographic and temporal coverage. However, one novel approach is to tow nets from the ocean bottom to the surface. Net release is activated either on command or with a timer and the net is brought to the surface using a positive buoyancy element. Unfortunately, nets do not necessarily capture representative specimens (e.g., problems of net avoidance, bias toward capturing unhealthy organisms, and other sampling biases).

While studies of phytoplankton have benefited from advances in bio-optical instrumentation, studies of higher trophic level organisms (e.g., zooplankton and fish) are using emerging video and acoustical techniques (Jaffe 1999). The size ranges of these organisms and relevant sampling methods are illustrated in figure 9.5. One method of sampling zooplankton employs light sheets (optical plankton counters). These systems are usually profiled or towed behind ships and provide zooplankton biomass and size distributional data. Calibration and interpretation of the data relies upon occasional in situ net sampling. Video imaging systems (video plankton recorders) (Davis et al. 1996) provide more detailed organismal information. Image analysis is a demanding aspect of this approach (e.g., Tang et al. 1998); however, progress is being made and the informational value is high as statistical methods can be applied to examine important questions involving patchiness, co-location of species, and predator-prey interactions. Proto-type systems have been profiled and towed from ships. Work is underway to deploy them from AUVs.

The instrument packages described thus far typically include complementary sensors for measurements of temperature, salinity, pressure, chlorophyll fluorescence, and PAR. These measurements are most useful for determining the environmental conditions of the sampled volumes. Some platforms have included turbulence sensors in order to examine relationships between organismal patchiness scales and small-scale hydrodynamics. A key

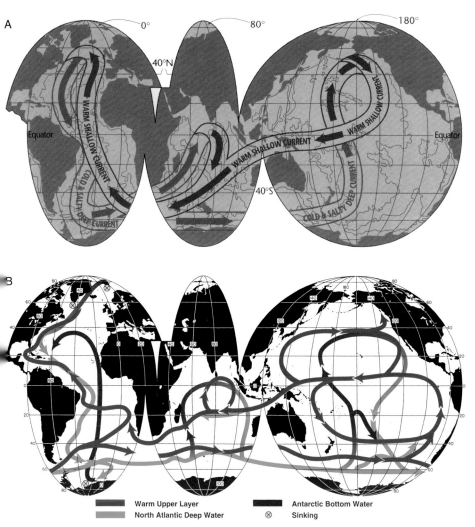

Figure 2.1. A: A simple schematic of the concept of the thermohaline circulation: the "conveyor belt." See text and chapter 4 for details. From Broecker 1991. B: A somewhat more detailed diagram of the thermohaline system showing the sinking zones in the North Atlantic and Antarctica. However, the thermohaline circulation is known to be variable and more complex than shown here. See text and chapter 4 for details. Adapted from Schmitz 1996.

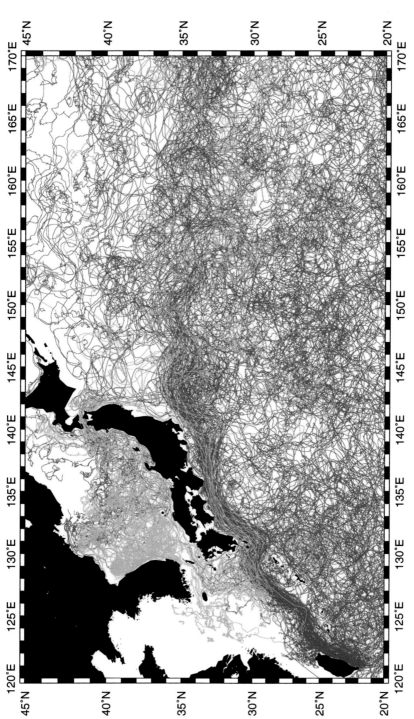

Figure 2.6. Tracks of individual drifters in the western North Pacific. Drifters were introduced in the Sea of Japan (green), the southern Kuroshio (red), and other parts of the Pacific Ocean (lavender). Jets, meanders, and eddies are clearly evident as is the Kuroshio extension and mixing of Sea of Japan waters into the Kuroshio and North Pacific. Courtesy of P. Niiler, personal communications.

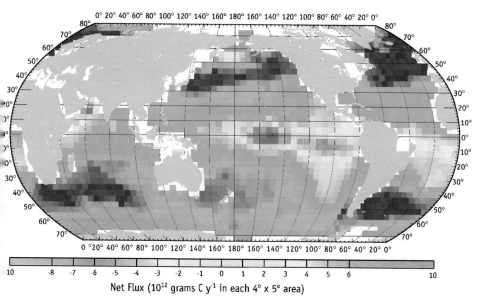

Figure BX2.2. Map showing the annual flux of CO_2 into (blue purple) and out of (yellow red) the ocean surface, corrected to 1995 atmospheric CO_2 levels. Note the strong ocean uptake (sink regions) in the high latitudes, and the strong source region in the equatorial Pacific. Takahashi et al. 1999.

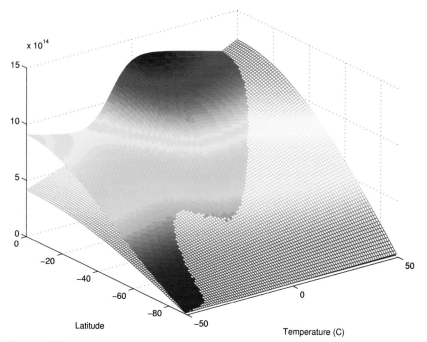

Figure BX2.3. Sketch of the climate system: the balance of incoming and outgoing heat fluxes.

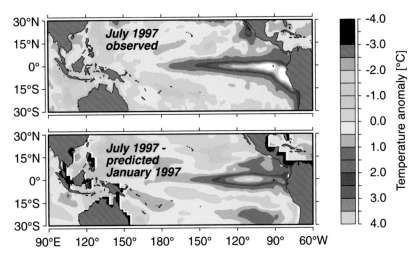

Figure 4.1. Observed (top) and predicted (bottom) Sea-surface temperature in the equatorial Pacific for the major El Niño event of 1997–1998. Courtesy of M. Latif, Max-Planck Institute for Meteorology, Hamburg.

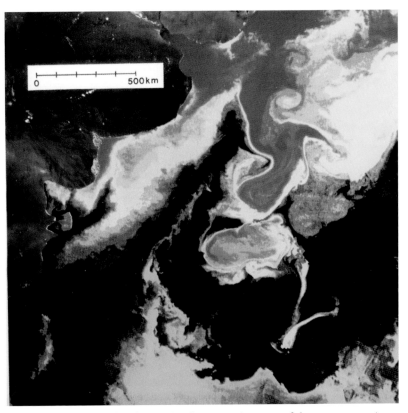

Figure BX4.2. The Brazil Malvinas Confluence region, one of the most energetic areas of the world ocean, is characterized by strong thermohaline gradients and intense mesoscale activity. In this satellite image of sea surface temperature, red through yellow colors represent relatively warmer temperatures, while greens through blues indicate colder waters. From Gordon (1988), with permission from Elsevier Science.

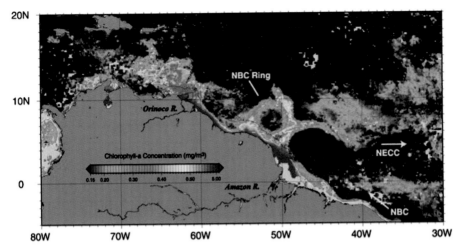

Figure 8.1. SeaWiFS satellite ocean color image converted to chlorophyll, showing a ring spinning off the North Brazil Current (NBC) at the point of its retroflexion to become the North Equatorial Counter Current (NECC). Image courtesy of David Fratantoni of Woods Hole Oceanographic Institution.

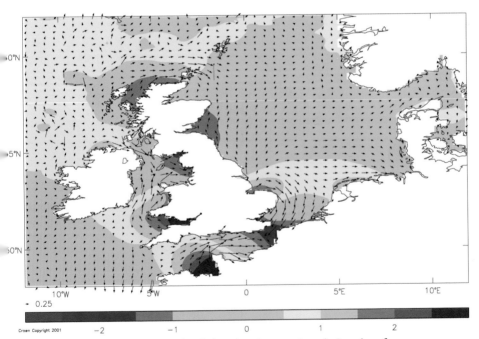

Figure 8.2. Instantaneous sea level elevation (meters, in color) and surface current (meters/second, as arrows), including both tidal and meteorological forcing, at 1800 hours local time on 29 October 2001. The example is taken from the POL-COMS circulation model of the NW European Shelf, run daily at the U.K. Met Office (as described by Holt 2002). Arrows are plotted every third gridpoint. Courtesy of Met Office, U.K.; copyright Her Majesty's Stationery Office, Norwich, England.

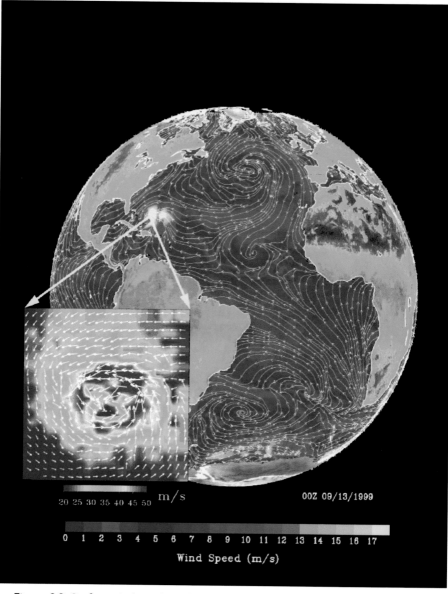

Figure 8.3. Surface winds in the Atlantic Ocean, as viewed by the QuickSCAT scatterometer, with the detailed structure of Hurricane Floyd (August 1999). Courtesy of NASA.

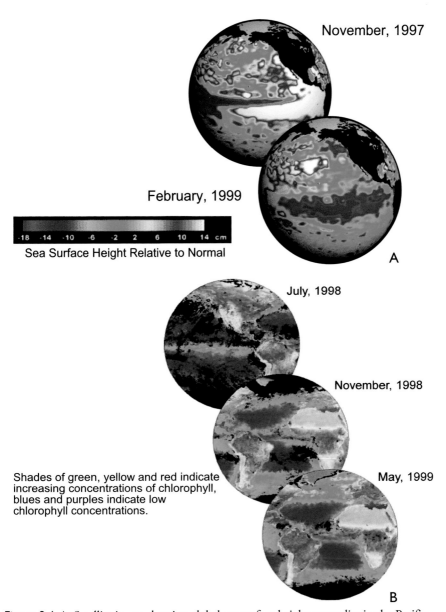

Figure 9.4. A: Satellite image showing global sea surface height anomalies in the Pacific Ocean for the months of November 1997 (El Niño; note elevations due to warm waters in equatorial Pacific) and February 1999 (La Niña; note depressions due to cool waters in equatorial Pacific). Images provided courtesy of the TOPEX/Poseidon and Jason-1 programs, the Jet Propulsion Laboratory, and NASA. B: Satellite ocean color images. Top image shows a ribbon of chlorophyll-rich waters in the equatorial Pacific in July 1998, a La Niña period with relatively great primary productivity. Center image shows the Atlantic Ocean in November when chlorophyll is relatively low in the North Atlantic and relatively high in the South Atlantic and Southern Ocean. Bottom image shows the Atlantic in May with high chlorophyll levels in the North Atlantic and relatively low values in the South Atlantic. Images provided courtesy of the SeaWiFS project, Goddard Space Flight Center, NASA, and ORBIMAGE.

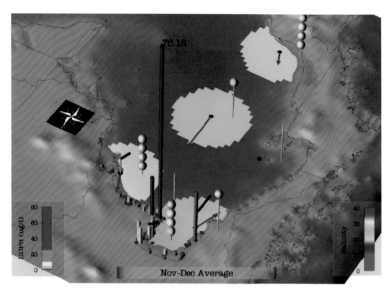

Figure BX9.4. A snapshot from a computer animation of the Gulf of Thailand showing observed net currents (blue arrows), wind (red arrows), the salinity at four depths from top to bottom (balls), the surface salinity (colored disks), and the dissolved and dispersed petroleum hydrocarbon concentration (DDPH, vertical bars). Adapted from Wattayakorn et al. 1998.

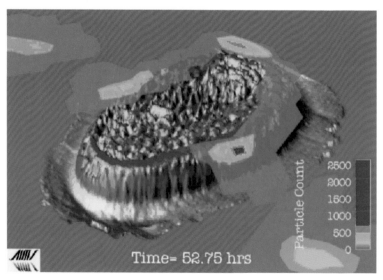

Figure BX9.5. A snapshot from a computer animation showing observed and predicted distribution of damselfish larvae around Bowden Reef, Great Barrier Reef, taking into account water currents and fish larvae horizontal swimming behavior. Adapted from Wolanski et al. 1999.

complementary physical measurement is vertical velocity. There can be confounding of acoustical signals intended for deriving estimates of vertical velocities. These are derived from backscatter from organisms, which in some cases are swimmers or vertical migrators using buoyancy control. This is a topic requiring more study.

Fishermen have been using acoustics in the form of echosounders (fish finders) for several decades now. Advances in the use of acoustics for research have involved the development of multifrequency systems (see figure 9.5) (see Smith et al. 1992; Holliday et al. 1998). Models have been developed to use the multifrequency data not only to make estimates of total biomass of zooplankton, but also to determine size distributions (in some cases with vertical resolution of ten's of centimeters). Three-dimensional acoustic, optical, and video systems are being developed to collect data concurrently, optimizing the advantages of each method. Such hybrid sampling may provide information concerning predator-prey interactions in the future. Occasional net sampling is needed to calibrate and interpret these data (Benfield et al. 1996).

Zooplankton size distributions in space and time can be obtained by deploying acoustic systems from ships and moorings. Since the choice of the size ranges of targeted organisms is dictated by the choice of transducers (low frequencies for large organisms; higher frequencies for smaller organisms), a rather complete size spectrum of organisms can be obtained in principle. Some investigators have also used the backscatter acoustical signal data obtained from monofrequency ADCPs to estimate biomass (e.g., Smith et al. 1992; Roe et al. 1996; Benfield et al. 1998). The power of this approach is that ADCP data are being collected by commercial as well as research interests; however, interpretation (e.g., size classes observed) remains problematic without multiple frequencies.

TOWARD THE FUTURE

Interesting new tools for biological oceanography involve molecular genetic and species-specific molecular probes (Parrish 1999). Molecules such as ribosomal RNAs have already begun to be used for detection and determinations of abundance of some bacterioplankton species and to study evolutionary relationships among several species. Species-specific probes (DNA antibody) for HABs have been developed and offer great promise in helping to understand the context of these events (Scholin et al. 1997). Applications of genetic techniques to higher trophic levels (zooplankton to whales) are being pursued as well (Parrish 1999).

It should be emphasized that interpretation of video and acoustical data is challenging and will require considerable research effort. However, neural network approaches may be valuable. Several types of laboratory studies can be useful as well. For example, laboratory video studies of animal behavior and predator-prey interactions have already generated new insights. Meanwhile, acoustical studies of nontarget species such as siphonophores and pteropods indicate that they can produce larger echoes than targeted zooplankton. Some methodologies such as holography appear promising and amenable for in situ application, but also will require more research emphasis. Measurements of bioluminescent plankton and nekton may also prove to be effective proxy indicators of higher trophic level organisms and fish. Acoustics using different frequency ranges, preferably with some overlapping, will be needed. Resolution of size classes will dictate the number of different transducers or, alternatively, the selection of frequency band choices using chirp acoustical scanning over a broad frequency domain. New systems, which can scan across the frequency spectrum of interest, may provide exceptionally powerful data. Deployments of acoustic systems in side-looking mode suggest that distributions of migrating fish (e.g., salmon) can be observed in this fashion. Also, acoustic tomography may be useful for some applications (e.g., mapping fish parameters over large areas by using measures of absorption losses at resonance frequencies of key organisms).

Some of the overarching needs remain in the areas of:

- quantifying the gross numbers of organisms in the sea and eventually estimating their changes regionally and globally (census of life);
- species and larval stage identification; and
- rates of feeding/predation, swimming, reproduction, and mortality.

The challenges of sampling zooplankton and higher trophic levels are arguably greater than for any other ocean parameters. The use of in situ platforms appears to be the most feasible option at present. Fortunately, several new platforms described earlier could be used to deploy smaller versions of presently used and future video and acoustical systems. Novel platforms, which mimic organismal movement or simply travel with targeted organisms, have also been the subject of research and development. In principle, arrays of miniature sensors can be deployed from these "organismal" platforms. Clearly, the optimal use of platforms is of paramount importance for biological observations. Using satellite-based observations as a guide, "smart sampling" could be focused using AUVs and ships. Interdisciplinary models will be important not only for synthesis and prediction purposes, but also for optimizing sampling strategies (e.g., Robinson and Dickey 1997). Testbed studies using promising sampling methods (most in near real-time) and data assimilation models (e.g., Dickey 2002) are likely critical for various reasons. For example:

- Interpretation of many of the data derived from the new technologies and even traditional net sampling remains uncertain.
- Design of future sampling programs will require comprehensive "oversampled" data sets that will likely uncover important "missing" parameters as well.
- There is only limited experience in utilizing complex and diverse biological observations for model simulations and predictive data assimilations (Dickey 2002).

The scope of such an activity should encompass all trophic levels. International coordination will be especially important because of the long ambit ranges of many fish and inter-region teleconnections (e.g., ENSO, NAO, and PDO).

Marine Geology and Geophysics

PRESENT AND NEAR-FUTURE CAPABILITIES

Marine geology and geophysics encompass a broad range of diverse study areas, including structural geology, geochemistry, geophysics, paleontology, stratigraphy, sedimentology, and hydrology. It is not feasible to address the technologies of all of these here, so we will only consider a few specific technologies. Several of the chemical sensors and systems as well as tracer techniques described earlier are relevant to geochemistry in general and deep circulation and hydrothermal vents in particular. Also, sediment transport issues have been discussed previously in the physics and optics sections.

Accuracy of bathymetry remains a limitation for several important problems, including marine safety and circulation models. Side-scan sonar provides images, but no direct measurements of depth. However, improved quantitative maps can now be obtained using new technologies such as interferometric swath bathymetric survey sonars and laser line scanning cameras for obtaining high definition images over distances greater than conventional photography by an order of magnitude. Combinations of these various technologies have been effective in advancing our knowledge of ocean features such as mid-ocean rift valleys. The subsurface can now be probed and visualized using multichannel seismic and 3-D seismic systems.

TOWARD THE FUTURE

Nearshore processes are difficult to measure in terms of sediment problems as well as for physics, chemistry, and biology. New technologies involving acoustics and optics and novel methods for mounting instruments (e.g., special bottom tripods) or remote sensing of key variables (e.g., from shore-based and satellite systems) will need to be developed.

An area of continuing importance is the development of geological time series for determining climatological variability. Periodic forcing (e.g., orbital) is only part of the reason for observed climate change and apparently cannot account for sharp transitional shifts in climate states. It appears that complicated modes of internal oscillation and feedbacks may result in nonlinear interactions and thus a cascade of climate variability to both lower and higher frequencies. Clearly there is a substantial need for new methods, including geochemical (e.g., radiocarbon dating) and biological proxy indicators with high enough resolution to detect transitions between glacial and interglacial states that may have occurred over just a few decades. Use of multiple proxy indicators is likely to bear great dividends.

The Ocean Drilling Program has provided a critical means of obtaining ocean cores for many different scientific purposes. New piston coring devices will include critical solid earth probes. Alteration of geochemical signals in marine sediments represents a major obstacle in interpreting the climatic record. New schemes (and geochemical indicators) for diagnosing diagenesis need to be developed. Finally, it will be necessary to sample from a variety of geographic locations in order to understand global teleconnections. A promising new scientific approach is to establish spatial arrays of sensors and systems that can measure full suites of interdisciplinary variables with high temporal resolution to capture important short-lived, but intense, events. Several interesting examples follow:

- The U.S. Navy's underwater sound and surveillance system has been used since 1993 to monitor earthquakes and animal communications and to study ocean temperature change over long distances using acoustic thermography (Orcutt et al. 2000).
- The Hawaii Undersea Geo-Observatory was established to facilitate observations of undersea volcanism associated with the submarine volcano Loihi off the island of Hawaii (Caplan-Auerbach et al. 2001). A shore station supplies power and receives data via an electro-optical cable terminating at the summit of Loihi. A high data rate hydrophone and seismometers are the key recording instruments.
- Another deep ocean observatory (H_2O) has been established midway between Hawaii and California (5000 m depth) where a retired cable is used for power and data telemetry (Chave et al. 1999). The impetus for H_2O was originally

9. Oceanographic Instrumentation and Technologies in the 21st Century | 243

for seismology, in that most seismometers are located on continents or islands, leaving the vast ocean bottom devoid of seismic data. Opportunities for using H_2O also exist for geomagnetism, tsunami, biological, geochemical, and physical oceanographic studies as well.

• A final exciting observatory example is the planned NEPTUNE program (Delaney and Chave 1999; Delaney et al. 2000), which is described as a "telescope to inner space" (box 9.2). NEPTUNE would take advantage of many of the new technologies described in this chapter. In particular, high-speed submarine fiberoptic cable links would connect a network of instruments at the lithospheric plate scale in the Juan de Fuca area off the northwest United

---

Box 9.2. NEPTUNE: Telescope to Inner Space

John Delaney

Ocean sciences are on the threshold of a new era. We are going to have to enter the entire ocean environment and establish interactive networks for in situ, adaptive observations of, and experiments with, key phenomena in the global earth-ocean system. The complexity of earth–ocean–atmosphere interactions calls for a new research mode using recent advances in computer sciences, robotics, and telecommunications and in the power and sensor industries. These critical new sensor and robotic technologies will progressively displace ships as the dominant observational platform for sustained time-series investigations and experimentation in the oceans.

The goal of the NEPTUNE project is to establish, at tectonic plate scale, an undersea observatory based on electro-optical networking (figure BX9.2). Using the Internet, it will connect many remote, interactive natural laboratory nodes designed for real-time, four-dimensional experiments on, above, and below the seafloor (www.neptune.washington.edu). NEPTUNE will be located in the northeastern Pacific and spatially associated with the Juan de Fuca Plate. It will involve decadal studies of a broad suite of oceanographic and plate-tectonic processes. By combining fiberoptic/power networks and full water-column moorings, NEPTUNE's capabilities will make it possible to observe processes that operate within the ocean and below the seafloor at scales of up to 500 by 1000 km. The network will provide significant power to instruments, high bandwidth for real-time data transmission, and two-way command-control capabilities for interaction with fixed instruments and with robotic undersea vehicles operated from shore. Remote, interactive experimental sites will be connected with land-based research laboratories and classrooms (figure BX9.3).

Scientific opportunities and needs will largely drive the NEPTUNE project, but the ultimate measure of success will be the quality of the scientific innovation and the educational benefits that the system enables. Because NEPTUNE can offer

(continues)

Box 9.2. Continued

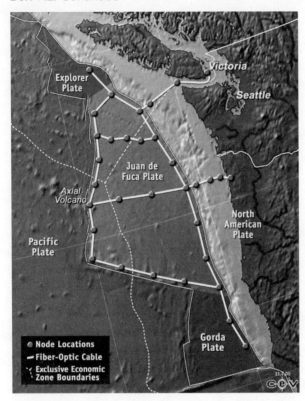

Figure BX9.2. An example of a generic NEPTUNE experimental site, draped over Axial Volcano and based on the New Millenium Observatory of the National Atmospheric and Oceanic Administration/ Pacific Marine Environmental Laboratory. Developed for NEPTUNE by the Center for Environmental Visualization, University of Washington.

great opportunities to many areas of marine science, a broad spectrum of scientific studies is currently envisioned as part of the observatory network. Science working groups have identified the major scientific issues that NEPTUNE could address:

SEISMOLOGY AND GEODYNAMICS

The study area includes all major types of oceanic plate boundaries, including the Cascadia Subduction Zone. There is interest in understanding earthquake behavior associated with these boundaries, which lie near the major population centers of Vancouver, Seattle, and Portland.

RIDGE-CREST PROCESSES

NEPTUNE will enable continuous, controllable experiments over periods of years, and seafloor-based robotic vehicles that can be rapidly deployed (within hours). These will be used to monitor, observe, and record seismic or eruptive events to establish the specific nature of links and variations between geological, physical, chemical, and biological processes at active mid-ocean ridges. Among other things, this will allow optimal sampling of high-temperature microbes expelled from the deepest portions of active volcanoes during eruptions.

Figure BX9.3. Land-based scientists and the public are linked in real time and interactively with sensors, sensor networks, and various mobile sensor platforms in, on, and above the seafloor. NEPTUNE's fiberoptic/power cable and associated technology provide the enabling network infrastructure. Developed for NEPTUNE by Paul Zibton, University of Washington.

SUBSURFACE HYDROGEOLOGY AND BIOGEOCHEMISTRY

Instrumented boreholes within the oceanic crust can serve as laboratories for studying the interdependence of tectonics, fluid and thermal flows, and biological activity. NEPTUNE's provision of continuous power, high bandwidth, and real-time control will allow interactive experiments, such as hole-to-hole pumping, and long-term observations, not easily accomplished by other means.

FLUID VENTING AND GAS HYDRATES IN SUBDUCTION ZONE PROCESSES

NEPTUNE will enable studies of tectonically induced release of fluids, and gases like methane, in zones of plate convergence. Methane is of major societal interest because of its suspected link to past episodes of climate warming, its role in chemosynthetically based benthic ecosystems, and its resource potential in the form of gas hydrates.

CROSS-MARGIN PARTICULATE FLUX STUDIES

Rugged terrain onshore, high coastal rainfall, and a narrow continental shelf combine to generate large fluxes of sediment to the adjacent deep-sea floor, mostly during major storms. NEPTUNE's capabilities will permit measuring, sampling, and experi-

(continues)

Box 9.2. Continued

mentation to characterize the poorly known cross-margin flux during such episodic events.

WATER-COLUMN PROCESSES

NEPTUNE will build long-term data series to improve current physical-chemical-biological models of globally significant surface and bottom water processes. These will include deep surface mixing during winter storms and energy and mass transfer from the atmosphere to the ocean via coupled physical and biological processes.

FISHERIES AND MARINE MAMMALS

NEPTUNE's acoustic sensors will allow pioneering work in fisheries and detailed studies of the migratory patterns and feeding behavior of whales and other large marine mammals.

DEEP-SEA ECOLOGY

By allowing extensive spatial and temporal sampling, measurements, and in situ experiments, the NEPTUNE network will develop a functional understanding of the ecology of the little studied deep-sea communities that cover about 60 percent of the planet's surface.

Because NEPTUNE will allow synchronous measurements across time scales of seconds to decades it will be possible to develop spectral and cross-spectral estimates of the variability of many water-column and seafloor processes, leading to the discovery of hitherto unexpected cause-and-effect relationships. This aspect of NEPTUNE will benefit from recent work involving interdisciplinary mooring-based time-series measurements collected on time scales from minutes to years, as described in the text section entitled "Moorings, Bottom Tripods, Offshore, and Shore-Based Platforms."

The NEPTUNE system must have the following characteristics to meet the scientific requirements:

- Plate scale (covering the full Juan de Fuca tectonic plate)
- Bandwidth (many gigabits/sec) to support HDTV and acoustic studies
- Power (tens of kilowatts) to operate AUVs and bottom rovers
- Very accurate timing signals to support dense seismometer arrays
- Real-time information on the status of instruments
- Real-time ability to change measurement parameters and download data
- Reliable measurements and user-friendly information management system
- Available for nominal twenty to thirty years.

A feasibility study demonstrated that a cabled system can deliver considerable power and communications bandwidth with reliability and flexibility at an affordable price. NEPTUNE will appear as a seamless extension of the global Internet, connecting users anywhere on shore to the sensors on the seafloor. NEPTUNE products will

9. Oceanographic Instrumentation and Technologies in the 21st Century | 247

---

offer rich educational opportunities for students of all ages and will be well suited for use in classrooms, laboratories, and even the living rooms of interested learners.
The project will be managed through the NEPTUNE Office at the University of Washington. Detailed planning and system design will take place in 2000–2003, with procurement and installation in 2004–2007 and operations from 2006–2036. Properly configured and executed, NEPTUNE may become the virtual equivalent of a gigantic ship that sails through time, not space. In this sense it represents an additional step toward conducting oceanographic research and education in the twenty-first century.

States and southwest Canada. This would make it possible to carry out studies of tectonic plate activity, erupting volcanoes, and the deep hot microbial biosphere. The framework of such a system could be used for water column studies related to hydrothermal vents, biogeochemical cycling, ocean circulation, and sediment transport, while providing telepresence access to many oceanographers, the general public, and educational institutions. Long-term commitment is a crucial aspect of this innovative approach.

Interdisciplinary Systems

PRESENT AND NEAR-FUTURE CAPABILITIES

The previous sections have indicated that many recently developed sampling systems contain suites of physical, chemical, biological, geophysical, optical, and acoustical sensors. Scientific hypotheses commonly concern physical versus biological versus chemical versus light controls. The power of measuring a variety of interdisciplinary variables has already been demonstrated by interesting correlations in some cases and a confounding lack of correlation in others. Sampling of biology, chemistry, geophysics, optics, and acoustics on virtually the same time and space scales as physics is certainly one of the major achievements of the oceanographic community during the past two decades.

TOWARD THE FUTURE

The past success and economy of this approach will motivate new platforms to carry even more sensors and systems in the future. Challenges will lie in developing smaller, lighter, and less power-demanding sensors and systems. It is likely that two approaches will evolve. The first may follow the philosophy of sampling large numbers of variables using relatively large vehicles or using highly rugged and capable deep-sea moorings. A second may use large numbers of smaller platforms or vehicles with fewer sensors, sacrificing the number of different sensors per platform and simultaneity of measurements. Each concept has advantages, and both could be used in complementary modes.

Telemetry: Shore-Based and Satellite-Based

PRESENT AND NEAR-FUTURE CAPABILITIES

Shore-based or satellite-based systems—and coaxial cable or fiberoptic cable links—can be used to telemeter data from open ocean moorings or other fixed location platforms (e.g., Dickey et al. 1993, 1998a; Detrick et al. 1999; Frye et al. 2000). Power can also be supplied via the coaxial cable method while both modes allow for high bandwidth transmissions. These cabling methods have great advantages, but can be quite expensive if cable networks are not already in place. Where direct link transfer is not feasible, data can be telemetered from moorings. Two steps are involved: (1) data can be sent from instruments at depth to a surface buoy and then (2) the data can be sent from the buoy to land stations using radio or satellite modes. Inductive modems have been used for subsurface transmission of data to the surface from moored instrumentation, with the mooring wire acting as the transmission element and by using subsurface acoustic modems. Telemetry of these data from surface buoys can be accomplished using line-of-sight radio frequency, cellular phone, or communication satellite transmissions (e.g., Dickey et al. 1993, 1998a). The radio frequency method works well for near-coastal applications, and relays can also be used to extend the range farther offshore. Open ocean telemetry requires dedicated communication satellites. Surface hardware for the communication of data is often commercially available. An excellent example of the power of data telemetry is the use of data transmitted from the TOGA-TAO array in the equatorial Pacific for ENSO predictions (e.g., McPhaden 1995; McPhaden et al. 2001).

TOWARD THE FUTURE

The amount of data that can presently be transmitted is quite limited because of bandwidth availability. Major programs are underway to expand bandwidth using large numbers of low earth orbit satellites. Cost per transmission is an important factor and will likely be prohibitive for some applications. For nearshore applications, data transmission may be accomplished using fiberoptics (great bandwidth), and in particular locations in the open ocean (e.g., near islands or existing cable nodes), submarine communication cables can be used. Virtually all platforms reviewed here need improved telemetry both for data transmission and instrument control. Power also remains as a constraining factor for some applications.

DATA UTILIZATION

A recurring theme of this review is that the oceans are vastly undersampled. New sensors, systems, and platforms will certainly improve our ability to monitor and study the ocean as well as to make predictions (e.g., Curtin et al. 1993). However, since limited resources will be available, judicious deployment of ocean assets is of utmost importance. End-to-end data management systems will be required for data synthesis, quality control, multiuse, and modeling (figure 9.6). Some regions (e.g., coastal and equatorial regions) clearly need higher spatial and temporal resolution than others. Strong interactions among observers and modelers will be needed to select key sampling locations and to design sampling arrays. The concept of nesting of platforms and model computational grids will be necessary (Dickey 1991). An important, though ambitious, goal is to obtain global, 3-D time-series of key ocean parameters. Means for achieving this objective are described in this chapter's boxes.

One of the challenging issues will revolve around the idea of more intensive sampling of features or locations, which are perceived to have extraordinarily great influence globally. Since much of the world ocean is still very poorly sampled (e.g., especially the South Pacific and South Atlantic, Indian Ocean, Southern Ocean), mobile and mooring platforms will be necessary, even in areas that may presently be thought to be of lesser importance on a large scale. A major community effort will be required to design a global ocean observing system that will have to satisfy a broad set of interdisciplinary needs and applications in coastal and open ocean environments.

It will be necessary to develop strong national and international collaborations for optimal sampling and exchange of data sets. The Internet is already being used effectively for information transfer, yet the Web (or some future counterpart) will need to play an even greater role in the future. Coordinated sampling networks present an important means to this end. Already, some nations are forming alliances to collect and share data obtained from buoys and floats (e.g., in coastal regions of Europe). Future comprehensive networks must be carefully designed to take fullest advantage of the various platforms.

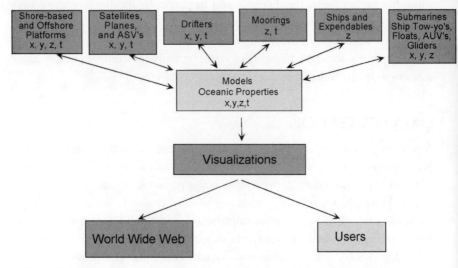

Figure 9.6. Schematic illustrating synthesis of various data types using ocean models. Visualizations and transmittal of data to the Web and specific users such as managers are also depicted. The double arrows at the top connote feedbacks of model predictions to enable adaptive sampling by the various platforms. AUV, autonomous underwater vehicle. ASV, autonomous surface vehicle.

This will probably require the formation of cooperative partnerships involving governmental agencies at various levels (e.g., local, regional, national, and international), private industry, private foundations, and academia.

One of the longstanding problems is sustainability of ocean sampling. An important factor in sustaining ocean observing networks is public interest and demonstration of their use. Again, a good example is the use of the TOGA-TAO array (e.g., McPhaden et al. 2001) and complementary satellite observations along with model predictions, which are contributing greatly to the public's knowledge of ENSO and its societal and economic impacts. Clearly, the synthesis and visualization of these data sets and model simulations are critical elements. The oceanographic community can build on experiences of the atmospheric community and the growing field of information communication (e.g., volume visualization using geographic information systems). It is possible to use television and the Internet to provide the public with easily understood oceanographic information (e.g., beach reports concerning red tide or HAB outbreaks, bacterial and viral counts, and surf and current conditions; public safety notices about tsunamis, storm surges, and rogue waves). Examples of data visualization applications are discussed in box 9.3.

Box 9.3. Information Technology in Marine Science
Eric Wolanski

Physical, biological, and chemical data sets exhibit great spatial and temporal variability, yet are linked by underlying processes. For effective marine environmental management, scientists have to explain to engineers and marine resource managers the processes hidden in such vast, complex data sets. Computerized visualization provides scientists with a powerful tool for presenting such complex information in readily understandable ways to themselves and to the public. Visualizations are likely to become more and more important as a way to facilitate communication about marine scientific processes and to help manage the sustainable exploitation of marine resources (Wolanski et al. 1999). Some examples of the practical applications of computer visualization to practical marine problems follow.

GULF OF THAILAND

Using advanced computing techniques, scientists have merged, visualized, and correlated the previously unexplored SEAWATCH oceanographic data from the National Research Council of Thailand and the data on dissolved and dispersed petroleum hydrocarbon concentration from Chulalongkorn University (see figure BX9.4 in the color section) (Wattayakorn et al., 1998). The computer animations of these data show that water currents control the fate of hydrocarbons in the Gulf.

During the Southwest Monsoon, hydrocarbon-contaminated water from the Inner Gulf is slowly advected toward the industrialized eastern seaboard, where more hydrocarbons are added. During the Northeast Monsoon, this contaminated water returns to the Inner Gulf where more hydrocarbons are discharged. Because these coastal waters mix but do not flush, much of the Inner Gulf and eastern seaboard water is chronically polluted by hydrocarbons. The data on currents and hydrocarbons were used to drive an oil spill model, which predicted that acute contamination occurs at least once a year anywhere in the Inner Gulf. The computer animation was a pivotal demonstration helping to ensure the passing of stronger oil pollution control laws.

CORAL REEF FISHERIES

Computer visualization was used to explore the large and complex database on oceanography and coral and fish larvae around selected coral reefs of the Great Barrier Reef (Wolanski et al. 1997). Computer animation of the field data showed that the fish larvae were not initially resident around these reefs, but were washed in by currents. On arrival, they aggregated in hydrodynamic shadow zones around the reefs. The aggregation patterns were successfully predicted by oceanographic models that incorporated the swimming behavior of larval fish (see figure BX9.5 in the color section). The model and its analysis suggest that the reef is characterized by high rates of

(continues)

> **Box 9.3. Continued**
>
> recruitment and so may be more sensitive to overfishing than previously believed. This has important management implications. For instance, aggregating fish stocks during mass spawning need protection. Legislation has recently been passed to close fishing on particular reefs for particular fish during such periods. This technology could be readily applied to the coral reef fisheries in other areas.
>
> **CONTRASTING REAL AND MODELED SEAS**
>
> Visualization provides a powerful tool to explain model output and compare it with field data. Such explanations are difficult when both model output and field data are patchy or variable in time and space. Computer animations of the data used to visually compare observations with model output suggest that in coastal systems, models tend to underestimate natural variability and patchiness (see, e.g., Schwab et al. 2000; Signell et al. 2000; He and Hamblin 2000).
>
> **APPLICATION TO SAFETY OF LIFE, AND OIL SPILLS**
>
> Computer visualization technology will also increasingly be used in human emergencies. One clear application is its use with dispersion models to predict the likely drift of people lost at sea and to guide search planes and ships. In oil spill emergencies, managers routinely use the same visualization packages. These applications demand the use of oceanographic data in real-time for assimilation into numerical models.
>
> The support of AIMS, IBM, Simon Spagnol, and Richard Brinkman is gratefully acknowledged.

CHALLENGES AND RECOMMENDATIONS FOR DEVELOPING AND USING NEW TECHNOLOGIES

Advances in our understanding of ocean conditions and our capabilities to observe and predict them will not come easily, because several important challenges remain. For example:

- Most of the systems and platforms described here are not designed for sampling within 30 m of the shoreline, which is the most important ocean zone for direct human interaction.
- Data telemetry is bandwidth limited at present. More capable communication satellite and cabled network systems are essential.
- Propelled AUVs are still power limited, and fuel cell advances are needed.
- Biofouling of many of the in situ sensors is still a problem for longer duration sampling.
- Satellite observation of higher trophic organisms remains a major research goal.

- Resources for developing and implementing global sampling networks must come from a variety of sources and be internationally based.
- Many of the new sensors, systems, and platforms described above are presently in developmental phases or in the hands of only a few researchers.
- Interpretation of many of the signals remains an issue, and intensive cross-sampling and intercalibrations are critically needed.
- New instruments and platforms must be transferred to the commercial sector for widespread use.
- The technology is moving so rapidly that many of the measurement systems will become obsolete quite quickly. However, continuity of standardized and well-calibrated measurements must be hallmarks of ocean observing systems intended to quantify long-term change.
- One possible benefit of advanced technologies may well be simpler sampling systems (e.g., "chip-based" or nanotechnology sensors with automated data processing), which would benefit oceanographic programs regardless of present technical capabilities and skill levels.
- More effective use of data will require clever manipulations of data sets. For example, many of the processes of interest are nonlinear in nature, so newly emerging methods of analysis will need to be used.
- Oceanographers are already collecting large volumes of data. Thus, more ocean scientists and analysts will be needed worldwide. In fact, nations not otherwise involved in collecting data may be able to play important roles in data processing and analyses of in situ and remotely sensed data sets and modeling.
- Flexibility and creativity in all aspects will be important, as will information from diverse communities, including environmental groups. Already sampling of water for bacterial, viral, and DNA analyses in coastal waters by volunteers is well developed in some areas. Environmental groups have also collected flotsam and jetsam from the North Pacific gyre. Analyses of these samples suggest that microscopic plastic may be entering the food chain at even the lower trophic levels.

Despite the myriad challenges described here, the oceanographic community has never been in such a strong position to make major advances, largely because of the burgeoning technologies, which can be brought to bear on problems of such great societal relevance and interest.

ACKNOWLEDGMENTS

The present work has benefited greatly from a series of interactions with numerous international collaborators and colleagues who have shared their knowledge through joint research activities, publications, and workshops.

Particularly valuable recent input for this paper has come from Jim Bellingham, Paul Bissett, Peter Brewer, Robert Byrne, Francisco Chavez, Peter Coles, Laura Dobeck, Paul Falkowski, Scott Glenn, Gwyn Griffiths, Julie Hall, Van Holliday, Ben Holt, Charles Kiedman, Chet Koblinsky, Swam Krishnaswami, Derek Manov, Chuck McClain, Casey Moore, Eduardo Marone, David Porter, Chris Scholin, David Sigurdson, Colin Summerhayes, and John Tokar. John Delaney, Gwyn Griffiths, and Eric Wolanski are thanked for contributing highlight boxes. Liz Gross is thanked for her many efforts to promote oceanography and this chapter in particular. Support of my research has been provided by the U.S. Office of Naval Research, National Science Foundation, National Aeronautics and Space Agency, National Oceanic and Atmospheric Administration, Minerals Management Service, the National Ocean Partnership Program, and the University of California, Santa Barbara.

ENDNOTES FOR FURTHER READING

Dickey, T., and P. Falkowski. 2002. Solar energy and its biological-physical interactions in the sea. *The Sea, Volume 12.* Chichester: Wiley. A. R. Robinson, J. J. McCarthy, and B. J. Rothschild, chapter 10, pp. 401–440.

Jaffe, J. S. 1999. Technology workshop for a census of marine life. *Oceanography* 12(3):8–11.

Koblinsky, C. J., and N. R. Smith (eds.). 2001. *Observing the Oceans in the Twentieth Century.* Melbourne, Australia: GODAE, Bureau of Meteorology, Australia.

Smith, S. L., R. Pieper, M. Moore, L. Rudstam, C. Gren, J. Zamon, C. Flagg, and C. Williamson. 1992. Acoustic techniques for in situ observations of zooplankton. *Archiv für Hydrobiologie, Beiheft 36, Ergebnisse der Limnologie* 35:23–43.

Varney, M. 2000. *Chemical Sensors in Oceanography.* Amsterdam: Gordon and Breach Scientific Publishers.

PART III
SOCIAL AND INSTITUTIONAL CHALLENGES

Chapter 10

Framework of Cooperation

Geoffrey Holland and Patricio Bernal

Ocean science operates and interacts within a social and institutional context that needs to be better understood. The study of the oceans yields answers that, directly or indirectly, influence most of society and conversely the demands from society dictate the funding and application of ocean science. In addition, the very nature and complexity of the oceans has led the evolution of marine sciences to a systems approach, encompassing all the natural sciences and, more recently, the social sciences as well. It is accepted that any discussion of the future direction for ocean science must also address societal demands and examine the structure within which the cooperation and coordination of the ocean research effort takes place.

The twentieth century saw ocean science grow from an exploratory phase into one demanding increasing application. The limitations for observing ocean phenomena that have challenged ocean researchers from the outset have also been responsible for the early use and development of mathematical and systems modeling in oceanography. These have also contributed directly to the vanguard of scientific progress.

For most of the century, the research vessel has been the essential tool of marine sciences, and every oceanographer is familiar with the task of extrapolating the tantalizing quasi-synoptic picture obtained from a cruise into regional or global descriptions using classical theories and models. The understanding of fundamental ocean properties depends on the availability of high-quality observations conducted on a basin-wide or global scale. It is this feature that imposes some unique organizational demands on the ocean sciences. The expense of constructing and operating research vessels has been one of the challenges facing ocean observations over the years. Today, the economic value of large-scale synoptic observations and the increased effectiveness of the new observational technologies represent a significant opportunity for advances. Nevertheless, to develop the framework so that society benefits is a major challenge.

As for society's demands upon the science, it must be recognized that the oceans are a global commons. Local actions will ultimately impact on regional and global ocean environments and the resources they contain; they will not only influence the ability of others to share in and benefit from the management of such resources, but possibly reduce their quality of life if the environment is impaired. In the past, the impact of anthropogenic actions on the ocean has been relatively small, and with little measurable effect outside the local environment, but this is no longer the case. For example, one need only consider the impact of persistent organic pollutants (POPs) in the Arctic, where the buildup of residual chemicals from pesticides and herbicides from farms in the tropics are accumulating in the traditional foods of northern peoples.

Governments are becoming more sensitive to opportunities and conflicts in their respective coastal waters and how these impact on the well-being of their respective economies and societies. They have been slower to recognize their collective responsibility toward the sustainable management of the global ocean commons. Ocean sciences must respond to both global and regional problems, although this is often difficult to accomplish when funding is from national sources and must address the more obvious national priorities. For the most part, governmental action will be driven by the immediate political, legal, social, and economic considerations that are occupying the electorate. These marine considerations are many and varied, ranging from the development of new industries to the wholesomeness of sea products, from the provision of food to the threat of tsunamis, and from the benefits of tourism to the impact of sewage disposal. The more pressing national priorities will tend to take precedence over regional and global environmental issues that may possibly be more serious in terms of their ultimate impact on society. The broader issues include the ocean's role in climate change, seasonal weather patterns, the water and carbon cycles, biodiversity, and the health of the ocean. Addressing global issues remains difficult and expensive, but if sustainable development is to be achieved, national and regional management decisions must also accommodate the global need.

Intergovernmental negotiations and agreements provide the most common basis for cooperative action. Once drafted and signed, agreements still need to be implemented, and governments must ratify and pursue the agreed objectives with national legislation and programs. Governments will not respond without compelling data and scientific argument, including the assessment of social and economic benefits and the risks involved. Both public opinion and political conviction are needed, and science will be called upon to play a role at every level.

THE INSTITUTIONAL FRAMEWORK

Science as a Social Institution

From the point of view of societal decision making, scientists may exert a disproportionate influence, despite being a conspicuous minority. The impact and social prestige of science in the past two centuries comes from the very strong influence it has exerted on our everyday life through technological innovation. However, this influence is based on the delivery of goods and services, not so much from the increased understanding of nature, life, or our origins that science provides. Recent studies in anthropology show that our schools do not yet provide curricula to help students grasp the meaning and sense of concepts such as environment and planetary life-support systems. The challenge to scientific education is enormous.

Guilds and the Invisible Colleges: The Commitment to Peer Review Sciences

Scientific work follows its own rules of validation, based on the free exchange of scientific results and the acceptance of the collective judgment of one's own work by peers. The peer review process for proposals and for publication secures the integrity of the system. Scientists form guilds, that is, a group of people sharing a methodological approach and a body of knowledge that through time has been codified into a self-perpetuating practice. This embodies the ever-present threat of developing its own professional jargon and codes, potentially distancing its results and benefits from the very clients that it must deal with.

Guilds provide for collegiality, the willingness of the community to work together for the benefit of all concerned. The guild system and collegiality ritualize competition for resources. All these are positive features of the system, and most scientists feel comfortable working under those rules. However, in a system with finite resources, major oceanographic programs can be seen as a threat by those researchers outside the proposals. A recent study conducted by the U.S. National Research Council (NRC 1999) (box 10.1) concluded that global programs provide an important service to science and to the ocean community,

Box 10.1. The Needs for Organization and Cooperation in Ocean Science
Rana A. Fine

A U.S. National Research Council report (Global Ocean Science: Toward an Integrated Approach) (NRC 1999) concluded that major ocean research programs have had an important impact on ocean science. Many breakthroughs and discoveries in ocean processes that operate on large spatial scales and over a range of time frames have been made that could not have been expected without the concentrated effort of a variety of specialists working together on major challenges. Examples include increased understanding of the causes of mass extinction, the role of ocean circulation in climate (for example, through El Niño) and in the decline in fisheries, and the ability of ocean and marine organisms to buffer changes in concentrations of so-called greenhouse gases.

The programs also leave behind a legacy of high-quality, high-resolution, multiparameter data sets, offering ample opportunity for data syntheses like those carried out at the conclusion of the World Ocean Circulation Experiment and the Joint Global Ocean Flux Study. The programs also have also created important new time series, new and improved facilities and technical developments, and a large number of trained technicians and young scientists. Many program discoveries and data are already being used in the classroom, or by the wider community, and in training graduate students. The discoveries, data, and facilities will continue to be used to increase the understanding of fundamental Earth system processes well after the current generation of programs has ended, which will further enlarge the impact of these programs.

The report recognized that the global scale and multifaceted nature of some of the present scientific challenges will require major integrative ocean research programs in the future. The report foresaw a need for an increase in interdisciplinary research through multi-investigator projects, reflecting the emerging national emphasis on global environmental and climate issues.

The report concluded that large-scale interdisciplinary projects, to ensure that they are fully effective, should ideally be managed separately from other programs by the National Science Foundation and in partnership between different government agencies to ensure a focused effort. New procedures are needed for initiating and selecting interdisciplinary programs, for example, beginning with planning workshops and encouraging broad community participation. The foundation should support a broad spectrum of interdisciplinary research activities, varying in size from collaboration between a few scientists to programs perhaps even larger in scope than the present major oceanographic programs.

The foundation should also ensure a healthy investment in smaller scale research conducted by individual investigators in the core disciplines.

> The report identified a number of generic gaps that should be considered in planning future ocean science. Mechanisms are needed to deal with contingencies and to establish priorities for moving long time series and other observations into an operational mode. Modelers and observers should work together during all stages of program design and implementation to enhance modeling, data assimilation, data synthesis capabilities, and the funding of dedicated computers. Federal agencies in partnership with the National Oceanographic Data Center should support data synthesis.
>
> As is done already for the research fleet, thorough periodic reviews should be made of other major facilities and the procedures for establishing and maintaining them, as the basis for prioritizing support. Strategic planning for ship and nonship facilities should be coordinated across agencies with long-range science plans, with help from the ocean sciences community.

even though some individuals feel threatened by the existence of large interdisciplinary programs, feeling that the funding of such programs limits their ability to obtain funds for their individual research. Here we see reflected the battle lines between Big Science and Little Science (de Solla Price 1963).

Market and Societal Demands

For market and societal demands, unsolicited proposals generated by the research community do not provide sufficient direction to science for it to deliver appropriate answers. Compounding this failure has been a tendency in the past for science to prefer the pursuit of its own research-oriented goals. The situation seems to be changing rapidly. Today, the pharmaceutical industry is extremely close to bioengineering research units. Corporations devoted to the development of instrumentation, in close contact with innovative research groups, are establishing relationships of mutual benefit. Additionally, although markets help to identify commercial opportunities, they do not ensure strategic foresight. To ensure the adoption of research direction for wider benefits to society, the ocean community must also work with governments to address national policy for the medium and long term. An example is the existing Marine Foresight Process initiative by the U.K. government, outlined in box 10.2. Another example is manifest in the potential shown in box 10.3 for a possible industry/intergovernmental/government alliance in the marine transportation industry.

Box 10.2. Fostering a Coherent Approach to the Development of the Marine Information Business: The Marine Foresight Process

Ralph Rayner

Foresight is a U.K. initiative that brings together representatives of industry, academia, and government with the objective of determining a strategy for the development of specific sectoral activities. One such activity is the marine information business. Although it is a significant area of academic and commercial activity with far-reaching policy implications and considerable importance in wealth creation, the U.K. marine information sector suffers from a high degree of fragmentation. To seek to bring a greater degree of coherence to this sector, a Foresight Task Force was asked to look at the changing needs for marine information over a twenty-year time scale and to establish a strategy for the development of the U.K.'s marine information sector. This recently published strategy addresses:

- the issues of forming more effective partnerships between academic researchers and industry;
- improving ocean monitoring and forecasting capability;
- improving access to data;
- improving relevant education and training; and
- underpinning the formation of a representative industry association for the many small- to medium-sized companies active in the manufacture of oceanographic instrumentation and the provision of marine information services.

To implement the strategy, a Marine Information Council has been formed, funded by subscription, and managed under the auspices of the Society for Underwater Technology. The council comprises representatives of government departments and agencies with marine interests, academic research organizations, and industry associations, including a recently formed Oceanographic Industry Association. The council provides a single national focal point for the implementation and further development of the strategy outlined in the task force report and brings much needed coherence to the U.K. marine information sector.

Box 10.3. An International Forum for Intergovernmental and Industrial Cooperation in the Application of Science and Engineering to the Maritime Transportation Industry

David P. Rogers

There is a need for a common agenda to improve the safety and efficiency of marine transportation. The dearth of adequate observations of the ocean environment that could improve ship operations is a critical problem. As a global commons, the ocean has no single body that is responsible for its management. Each user generally operates within a framework established by a specific community: industry, academia, and government. The boundaries between each tend to be sharp and inhibit cooperation. Providing ocean data for climate application is primarily the purview of governments. Marine shipping provides in situ marine meteorological and oceanographic (METOC) data to weather services, but receives relatively little direct benefit from this information beyond the application to ocean routing. A new framework is needed that recognizes the benefit of METOC data to commercial marine operations, establishes priorities for data acquisition based on industry requirements, and encourages global standardization of information systems and data quality.

An international forum, such as suggested here, is required to bring together leading representatives of the marine industry, marine science and engineering experts, and national and international entities with responsibility for marine services. This group should establish requirements for marine shipping information systems that are of direct benefit to the industry. Working with the Intergovernmental Oceanographic Commission (IOC) of UNESCO, in partnership with the World Meteorological Organization and the International Maritime Organisation, these requirements can be promulgated throughout the maritime transportation community to help establish international standards to promote the global application of marine environmental information.

The forum should address the following:

- The establishment of a nonregulatory framework to encourage the open exchange of ideas that will serve the needs of the maritime transportation industry and other bodies.
- Identification of the needs of the maritime industry that can be at least partly solved by the availability of improved environmental information (for example, continuous measurement of water depth in restricted seaways and better information on currents that affect ship handling).
- Improvement of information systems that involve the participation of merchant vessels for data delivered to weather services around the world, and, perhaps more importantly, to ensure that the industry benefits directly from this information, thereby improving both the safety and efficiency of marine transportation.
- Identifying the costs associated with the exchange of information and the economic returns.
- Are there insurance benefits associated with improved use of environmental data for ship safety?
- Addressing the usefulness of international standards for such environmental information. When there is a single information provider, such as the Global Argos System, it is relatively easy to establish international protocols for the use of the data. Is it is possible to do something similar for other more diffuse data sources?
- Debating the technical difficulties that limit the availability of ocean data and who should be responsible for the development of these systems. For example, what would be the role of industry-government partnerships, national entities, and international organizations? And how should information be exchanged between providers and users?

Specialization versus Cooperation among Disciplines

Guilds tend to be specialized; therefore, very often they have been conservative about multidisciplinary and interdisciplinary exercises. There are few incentives in place to risk crossing the guild's borders. Multidisciplinarity is where several disciplines contribute to researching a science question. Interdisciplinarity brings the interaction between scientific disciplines, including social sciences. Transdisciplinarity brings the decision makers and stakeholders into the picture, making it the most intransigent of the three to accomplish successfully. One of the major achievements of International Geosphere–Biosphere Program: A Study of Global Change (IGBP) has been to generate a global cohort of practitioners of multidisciplinary and interdisciplinary studies, by accumulating critical mass, often through networking, and providing simple incentives, such as communication, joint planning, and assistance with travel expenses.

Opportunity to Join

One the most attractive features of the development of science, according to Derek de Solla Price (1963), is the possibility for latecomers (i.e., nations with little traditional opportunity in research) to be able to adopt and contribute to the science culture built up over extended periods by nations that have enjoyed ample capacity and large scientific establishments. Many different countries, and the global community itself, have benefited from this robust property of science, which is a consequence of the highly codified nature of scientific knowledge and of the open scientific practices of the scientific guilds. This promise, i.e., the availability of science to all countries that are willing to participate and cooperate, is a property to be preserved and cherished.

THE EVOLVING INSTITUTIONS IN SCIENCE

Science is one of the most internationalized activities in the world today, not only because of its intrinsic international character, but also of the developing international need to share the distribution of labor and cost. One of the main trends in science, most notably in physics, is the regionalization and internationalization of research activities (Chang et al. 1997). The facilities needed for top-level research are becoming larger and more costly. As a result major laboratories and academic groups in each country have attempted to concentrate on niche areas where their special expertise enables them to play a meaningful role in international teams. At the same time, national governments have recognized that greater concentration of investment funds is necessary to support such international specialization. Expressed in those terms, the analogy in economics would be the theory of comparative advantages.

International Facilities

The joint running of large international facilities demonstrates the benefit of close international cooperation. In ocean sciences, the Ocean Drilling Program, with its internationally shared funding of the operation of the deep sea drilling vessel *JOIDES Resolution,* is an example, as is the Research Ship Sharing Scheme of the European Union, outlined in box 10.4.

Box 10.4. Research Ship Sharing Scheme of the European Union

Colin P. Summerhayes

To make more efficient and cost-effective use of research ships, some countries have developed schemes that enable access by any funded principal investigator (PI) to the most appropriate research vessel or facilities, regardless of the institution to which the PI belongs. In the United States, the UNOLS (University-National Oceanographic Laboratory System) scheme has provided scientists with greater access to U.S. research ships since its inception in 1972. A similar scheme has since evolved in Europe (RVS 1999). In 1996 a tripartite agreement was signed between the United Kingdom (NERC [National Environmental Research Council]), France (IFREMER [French Research Institute for Exploration of the Sea]), and Germany (BMBF [German Ministry of Education and Research]) for the mutual cooperation of scientific interests and activities. Under this agreement, the managers and planners of the respective fleets of scientific research ships and major marine facilities meet annually with the objective of bartering ship time and exchanging major marine equipment without the need to charter or to exchange money.

This arrangement has two significant advantages. First, it allows scientists access to a wider range of facilities and equipment than would otherwise be available. These include fourteen research ships plus other facilities such as manned submersibles, towed arrays, and shipboard surveying systems that are so expensive that it would make little sense for all three countries to purchase their own. Second, it reduces wasted time (and therefore cost) spent on long passages between marine areas of scientific interest and allows scientists access to a wider range of geographic areas than they would normally get in any given year.

(continues)

Box 10.4. Continued

Although no money changes hands, no country provides free ship time. For each cruise on a foreign ship, the benefiting country must mount a full cruise on one of its own ships in return, and to an equivalent value. The operating costs fall to the ship owners, and each country has an appropriate scheme of banking to support the process. An equivalence points system has been agreed for the value of each ship, to ensure like-for-like value. Points are allocated per ship day used.

Access to these facilities is automatically incorporated into the planning cycle, and the three countries have adapted their original planning cycle to a common one to facilitate the process.

This tripartite agreement on cooperative programming and exchange of facilities may be the beginning of a coordinated research fleet of European partners, underpinned by nationally owned and managed resources, and is a model for the rest of Europe and other regions to follow.

The large cooperative programs are the hallmark of ocean sciences (the Joint Global Ocean Flux Study, Tropical Ocean Global Atmosphere, and World Ocean Circulation Experiment are recent examples). Extending the concept, these programs are equivalent to spatially distributed large facilities. For oceanographers, they correspond to the accelerators or radiotelescopes of other disciplines. However, whereas such large instrumented facilities have reached an internationally acceptable level of funding, no such equivalent is readily available for large ocean programs, and each one has had to be negotiated on an ad hoc basis.

In order to be legitimized by the scientific community, large ocean programs require a strong and transparent building of consensus. The images taken from the moon in 1969 provided us with a unique vision of Earth: a blue planet. After a rather long lag time, we are now recognizing the need to study this uniqueness as a single system. Some may be afraid that "one planet" means "one program." The bottom line is that keeping a healthy balance between program science and individually driven science priorities is essential.

THE SOCIETAL CONTEXT

A global picture is needed to examine progress in ocean science and the development of marine research policy and priorities, from both the national governmental and regional view. This involves legal, governance, and management implications and the important dimension of capacity building.

The Legal Regime

The United Nations Convention on the Law of the Sea finally came into force in November 1994. It was, and remains, a tremendous achievement and will provide the basis for international marine law for years to come. It is comprehensive, covering inter alia living and marine resources, environment, transportation, legal regimes, and marine science. Despite the unprecedented success in achieving such a comprehensive legal agreement, it must be said that the convention represents thinking that is twenty-five years old and is therefore already in need of review. This is relevant in the present context, because any legal review must be based on accurate data and a knowledgeable understanding of the factors involved.

One of the most significant ocean management changes accepted by parties to the convention is the acceptance of national jurisdiction out to 200 nautical miles from the coast. Many countries had anticipated this extension of their national management and had already adopted legislation. The assumption of control of economic resources within a national economic zone and, for the case of seabed resources, out to the limits of the continental shelf and allowable margin, also brings a responsibility for management. For many small island countries, the size of their extended ocean properties are many times that of their respective land masses. Most developing coastal states are ill-prepared for the additional demands of jurisdiction and management.

When the present convention was written, the cumulative impact of society on the global ocean was not considered. Management regimes covered local resources, bilateral pollution issues, and some shared responsibilities for migratory fish. Wider national responsibilities for global ocean health were not anticipated, and the international ocean areas, outside national jurisdiction, were assumed to be unaffected by human impact.

Present global concerns include overfishing, loss of coastal habitat to development, deterioration of water quality along ocean margins, increased conflict between competing marine activities, sea-level rise, and temperature changes. Improved management at national levels can mitigate some of these, but most will require collective legislative action by governments working together.

The articles dealing with marine scientific research allow for reasonable access to waters under national jurisdiction. Coastal states are naturally protective of resources that have devolved to them in these extended areas, and this has often led to unnecessary denial of access to scientists from other countries, who are conducting research into ocean basin phenomena. It is difficult for a country with limited scientific resources to differentiate between scientific studies that will ultimately yield knowledge about the regional and global marine environment that is of value to them and studies that may yield results on offshore resources that could give a commercial advantage to competitors. For the researching state, the cost and administrative difficulties in meeting the requirements of coastal state participation also generate a reluctance to comply. These difficulties will be diminished with increased sharing of capacity and knowledge among nations. The transfer of knowledge and availability of data in marine science are important issues that need to be addressed.

Governance

Governance covers the management framework needed to address ocean issues in a timely and adequate manner, whether the situation is national, regional, or global. The framework needs to accommodate all levels in a seamless manner. Coastal problems will generally have regional and global consequences that make intergovernmental cooperation not only desirable but essential. Global issues will also impact on coastal activities and decisions. For example, sea-level rise is a global phenomenon, yet the need to rebuild sea defenses, wharves, and ship-handling facilities and relocate buildings and facilities, etc., will generate a tremendous local expense. Even for shorter-term issues, such as storm surges or tsunamis, mitigation action demands advance warning monitoring requirements on a regional or ocean basin scale.

There are many other examples of the need for interaction between all levels of government. The development of coastal facilities is completely under the jurisdiction of a coastal state, yet the collective loss of habitat, whether coral reefs, mangrove swamp, or salt marsh is of direct concern to the global ecology.

Pollutants generated on land are the major source of contamination in the oceans, and their effects can be felt at large distances from the source. For example, many of the persistent organic pollutants are already a global problem. Their characteristics of volatility and bioaccumulation have already created unacceptable levels in the Arctic ecosystem. The control of the manufacture and use of these and other pollutants is subject to national jurisdiction. But their impact is worldwide and, therefore, action to curtail these anthropogenic pollutants demands governance on a global scale.

Actions to date include an Intergovernmental Agreement on the Prevention of Marine Pollution from Land-Based Activities, negotiated at Washington in 1996, that builds on complementary actions by national governments and regional organizations. A related convention to control POPs is being negotiated.

Other transboundary pollution problems span short-term and continuing issues, such as oil spills, the release of nutrients and industrial wastes into coastal waters, the transfer of exotic species by ballast water, and the early detection of dangerous toxic blooms. Even the aquaculture industry generates environmental questions that transcend national jurisdiction, such as the farming of species from foreign ecosystems, the development and potential release of genetically altered species, and the transfer of diseases.

Regional seas conventions are mechanisms that can be used to manage

---

Box 10.5. The Regional Seas Programme

Stjepan Keckes

The Regional Seas Programme, initiated by the United Nations Environmental Programme (UNEP) in 1974, consists of a series of regional action plans designed to protect and develop the marine and coastal environment through coordinated joint efforts of participating countries. The action plans are adopted and periodically revised by high-level intergovernmental conferences. Most action plans are structured in a similar way, although the specific activities reflect the priorities as defined by the governments of each region. Typically they relate to

- environmental assessment, that is, identification of the sources and magnitude of environmental problems;
- environmental management focusing on control of land-based activities affecting the marine and coastal environment;
- environmental legislation through development of regional legal instruments and development of national legislative and normative acts in conformity with globally and regionally agreed norms;
- institutional arrangements for the coordinated implementation of activities agreed in the framework of the action plan; and
- financial arrangements supporting the agreed activities.

(continues)

> **Box 10.5. Continued**
>
> Nine actions plans have been developed under UNEP's auspices and adopted for the following regions: the Mediterranean (1975), Persian Arabian Gulf (1978), Wider Caribbean (1981), East Asian Seas (1981), West and Central Africa (1981), Southeast Pacific (1982), South Pacific (1982), Eastern Africa (1985), and Northwest Pacific (1996). Six of these action plans are associated with regional conventions that provide legally binding frameworks. The conventions were adopted in Barcelona (1976), Kuwait (1978), Abidjan (1981), Lima (1981), Cartagena (1983), Nairobi (1985), and Noumea (1986). All adopted conventions are in force.
>
> Initially, the action plans focused on the protection of the marine environment from pollution, but they gradually broadened their scope in order to emphasize issues related to integrated management of coastal and marine environment along the lines recommended by Agenda 21. Fisheries problems are the responsibility of the Fisheries and Agricultural Organization and the only major topic not covered by the action plans.
>
> More than 140 countries participate in the Regional Seas Programme through the involvement of about 400 national institutions. Regional programs adopted for the Red Sea and Gulf of Aden (1976), the Black Sea (1992), and the South Asian Seas (1995) without major involvement of UNEP are closely associated with and considered as part of the Regional Seas Programme.

transboundary waters, resolving conflicts on shared resources and developing common actions to address environmental needs. The United Nations Environmental Programme's Regional Seas Programme elaborated in box 10.5 was a successful early example of such regional cooperation.

Shared responsibilities in such matters as safety, water quality, forecasting systems, etc., are efficient and effective ways for neighboring states to reduce management costs. Even when issues are transparently local in character, their ubiquitous nature allows the costs of finding solutions to be shared among countries to mutual benefit. Data gathering, exchange, and management activities are essential but expensive factors in all decision-making activities. Exchanging observations as well as ensuring that data are compatible, and that the quality of products is adequate for management purposes, can be the subject of regional and global agreements.

Unfortunately, the attention devoted to ocean matters, including the support of ocean science by national governments, falls far below the relative importance of the oceans to society. There are presently many existing regional agreements that are languishing due to governmental indifference. There are many intergovernmental organizations dealing with regional and global marine issues, but their respective efforts are fragmented. There is a failure in the top structures of the United Nations to take appropriate account of the ocean environment, which, nevertheless, occupies three quarters of the surface of the planet.

The Commission on Sustainable Development (CSD), at its 1999 session on oceans and seas in New York, recommended that "the U.N. General Assembly establish an open-ended informal consultative process with the sole function of facilitating the effective and constructive consideration of ocean related matters within its existing mandate." The meeting also issued guidelines for this process that, if followed, would give more attention to ocean debates at the highest level of the U.N. system. It could be a start to addressing issues of governance.

Management

The success of ocean and coastal management depends largely on three basic elements, namely information, knowledge, and a structure for decision making. Information provides a description of the developing situation at a time and scale appropriate to the process or event taking place. Knowledge encompasses a sufficient understanding of the processes and complex interactions to make the necessary decisions and to give strategic directions to government. The decision-making structure should conform to the scale of management being applied and the processes involved. It may be as small as a community addressing local conflicts of marine activities or involve national legislation, regional agreements, or global conventions.

Hierarchal Ocean Management

At past scales of exploitation and population pressures, methods based on indigenous knowledge and management were sufficient for sustainable management of coastal resources, although living marine resources were hunted rather than farmed. Knowledge, methodologies, and techniques gained over the years were passed down from generation to generation. As their terrestrial counterparts dealt with climate vagaries of drought and flood, the coastal communities would learn to accept and assimilate environmental changes in the ocean. As populations grew and the competition for resources increased, anthropogenic changes to the marine environment became significant. Local resources were depleted and had to be foraged farther afield, competing with other communities. Trade and the exploitation of local resources for export exerted new influences outside local control.

Today's management requires an ability to work cooperatively with other communities to resolve issues transcending local resolution and for nations to negotiate with regional neighbors. Many ocean management problems require attention at the global level. There is an inherent requirement for each respective level of management to be aware of the decisions they should rightly address and of those that should be more appropriately referred to a higher or lower level.

The same concept applies to information and knowledge. Changes in global ocean processes affect regional seas, and regional changes in turn affect local conditions. Information about, and an understanding of, the larger scale changes allow local managers to prepare for future situations and to take mitigating action. Countries rich in indigenous knowledge of local conditions may have a real problem in their ability to understand and adapt to the rapid changes in modern society or to situations where the balance of populations and available resources have suddenly changed from traditional values. Indeed, all countries are trying to cope with the rapidity of change in their respective societies.

At the end of the twentieth century, the three main elements of management (information, knowledge, and decision making) are deficient at all scales in the ocean and coastal infrastructure, although progress has been made in some areas over the past few decades. A number of global conventions and agreements have been achieved, some developed countries have adopted legislation that has set new standards for ocean management, recent global studies have increased our understanding of ocean processes, and a few governments are taking action to put global and regional monitoring systems in place. The priority during the next twenty years will be to build on these developments and initiatives.

THE REGIONAL PERSPECTIVE

It is not easy to establish an ocean science capability where little infrastructure exists. The recognized academic centers of ocean science are relatively few and mostly found in developed countries, primarily because the practice of ocean science is expensive in terms of equipment and facilities. As a first step to counteract this situation, more attention could be paid to the establishment of regional centers, sharing the cost of education and learning among several countries. The same benefits of cost sharing would apply to cooperative regional programs for gathering ocean information and data.

The international community must assist in the generation of increased regional and national capacity and ensure that local priorities, enunciated by regional countries, are addressed adequately. Cooperation of all governments is an eventual goal for tackling global environmental issues, and this cannot be achieved without first addressing capacity-building requirements at national and regional levels.

Regional Capabilities

Much has been made of the continually widening gap between the "have" countries and the "have-not" countries. This is not the place to enter into an in-depth discussion of the failures of the world governments to address this problem. It is worth saying, however, that global environmental issues are on the increase and solutions will need the cooperation and collective ability of all nations. Saving part of the planet is not a sustainable solution. At a meeting of African countries in Cape Town, 30 November to 4 December 1998, dealing with coastal and ocean management, the ministers attending recognized the need for both recipient and donor countries to change their approach to capacity-building programs into one of true partnership. The recipient countries must make an effort to establish national and regional frameworks and develop priorities that will give themselves more strategic control over the application of aid programs. Donor countries must support the national and regional initiatives rather than dictating their own priorities. In this way, true partnerships can be forged and sustainable progress achieved.

Technological developments are helping to reduce the costs of ocean science. Satellite coverage and automated instrumentation are yielding more data at substantially reduced costs. The computer power needed to reduce and analyze the information and provide the necessary management products is now available on desktop size and portable units. Advanced communication ability allows assistance from distant centers of learning to be at an operator's fingertips anywhere in the world. Despite these encouraging signs, however, it will still require much effort to put the needed improvements into place.

Governments must recognize the benefits of cooperation and the importance of tackling ocean problems on all scales from local to global. The most outstanding and well-documented example of such cooperation is the monitoring and forecast of the El Niño phenomenon in the tropical Pacific. Latin American countries on the west coast now use the El Niño forecast to manage their farm crop economies, predicating the choice of crops on the expected rainfall forecast from the situation developing in the Pacific Ocean. These governments support the programs of international science effort because of the beneficial results accruing to their respective economies.

Present and Future Regional Priorities

Countries with small economies and large social problems cannot be expected to embrace the whole gamut of national, regional, and global issues. Each must be encouraged to expend scarce resources on problems closest to their own most urgent requirements and to recognize that such contributions will be helping to satisfy the overall need.

With many developing coastal states, tourism can be the major source of wealth and the protection of the marine environment a top priority. Such countries need assistance with the management of their own shoreline development and pollution sources, but they will also be interested in the adequate control of oil pollution, sewage regulations for ocean liners, and other wider issues that are outside their national control. With countries with low coastal regions, especially some of the smaller island countries, sea-level rise is of prime national concern, even though, in this case, the problem is global.

Many countries are finding that aquaculture is more profitable than traditional sustenance fisheries, and the intensive farming of marketable species is growing. Insufficient knowledge of the management procedures for these new crops often leads to the destruction of natural habitat and a loss of water quality that ultimately damage both the capacity of the environment to sustain farming of the new species and the ability to return to the indigenous fishery. The aquaculture industry must learn to manage problems of fish health, conflicts with other marine users, and environmental impact. The lessons learned will lead to increased production and the maintenance of biodiversity. Increased growth rates and disease resistance will lead to improved economics. More of the higher value products may be produced in shore facilities, where environmental conditions and quality can be easily controlled.

Pressures of the commercial fisheries are sending distant-water fishing fleets farther and farther afield in search of their catch. It is estimated that 80 percent of the world's commercial fisheries are presently at their limit or overfished. The richest and most productive fishing waters often occur off the coasts of poorer nations that have neither the skills nor abilities to enforce management in their extended jurisdiction in order to profit from the situation or to prevent the more unscrupulous distant-water fleets from plundering their resources.

This is not to suggest that the world has reached the limit in terms of the growth of food from the oceans or that the expenditure of further research and development on the fishing industry is a waste of time. Present fishing practices are crude, often destroying the habitat or wasting large percentages of the catch of nontarget fish (bycatch). Improved technologies and methodologies are required and the greatest advances may still be to come. We may still be able to discover how to change from a hunting role to one of stewardship in the ocean environment.

As technologies develop, the ability to exploit mineral and hydrocarbon resources from offshore and the greater demand for these resources will bring wealth to some coastal states, together with associated problems of environmental quality and the ability to manage the resources to the benefit of the total society. The potential for growth is already evident, but the evolution of mineral exploitation has not followed the path originally predicted. When the Law of the Sea was being drafted in the early seventies, seabed mining for polymetallic nodules was the center of attention. That industry is still awaiting its economic window. It will arrive eventually, but in the meantime, placer deposits (like cassiterite and diamonds), gravel and sand, and even coral reef material are more important minerals for many countries. In particular, small island developing states have a lack of common building materials that larger terrestrial countries would take for granted. Their extraction of sand and crushed coral is a necessity that must be managed in a way that does not interfere with the other uses of the marine environment. There can be no sustained development in the long term if the environment is destroyed in the process.

In contrast to the exploitation of marine mineral resources, an insatiable appetite for oil and gas has spurred an increasing offshore hydrocarbon industry that now supplies an appreciable percentage of the world supply. Many coastal states have already profited greatly from the economic returns from this resource, and there is good reason to believe that this industry will continue to grow for the foreseeable future. For many coastal states, the delineation of offshore resources has yet to be carried out and will be a high priority on many government agendas over the next decade or so.

The search for pharmaceuticals and genetic materials from the sea is attracting a great deal of attention, particularly in the deep benthic and vent ecosystems. It is recognized that a major portion of the world's species live in the oceans, but most are still unknown and unclassified. The biological richness of the ocean is likely to generate billions of dollars. Even now the question of ownership of genetic materials is still unclear for terrestrial species. For the ocean, the existence of both national jurisdictions and international areas compounds the question. Twenty-five years ago, the U.N. Convention on the Law of the Sea foresaw the need to share the benefits of mining the international seabed, but genetic wealth was not even on the agenda. The science involved is advanced and governments must take action to ensure that the benefits do not pass merely to those societies already possessing the greatest capacity.

Water quality is also paramount in the priorities of countries whose natural freshwater supplies are insufficient. An increasing number of countries are finding that distillation of seawater is becoming a necessary addition to their freshwater supply. It has been recognized that water scarcity will become one of the great global issues in the twenty-first century. With 98 percent of the world's water in the ocean, the use of this source is likely to increase. Still to be explored is the possibility of genetic manipulation to make some food-producing plants adaptable to salt or brackish water, relieving some of the irrigation pressures on the freshwater resource.

The capacity of the ocean to provide renewable energy will continue to exercise engineering minds for decades. Such uses will satisfy local rather than global requirements. But there could be interesting side benefits to those uses. Wave power can be used directly to generate freshwater from seawater, ocean thermal energy supplies a source of nutrient-rich ocean bottom water for aquaculture, and there would be a co-located source of geothermal energy if the mining of hot vent minerals ever became economical.

Coastal populations are growing even faster than the overall growth of population because of migration toward the ocean. We are already infringing on the ocean edges, filling in salt marshes, building over mangrove swamps, and reclaiming coastal lands. Governments must take action collectively to preserve marine habitats. A complete look at ocean space is necessary for management, factoring in the value of the ocean environment to the cultural and social well-being of the population on the one hand, and the economics of using the ocean resources, its capacity for sewage and waste disposal, for recreation and tourism, and for marine transportation and other activities on the other hand. In the twenty-first century, the use of ocean space will expand. Today, engineering structures such as artificial harbors, oil development platforms, and airport extensions are becoming commonplace, impacting on the fragile coastal habitat. However, as the search for resources proceeds farther offshore, and the technologies develop, the economics of placing communities offshore will become evident.

As the oceans become more directly important to society and as more people become dependent on the ocean for their life and livelihood, priorities for marine knowledge and information will grow. There will be a need for many more ocean observations; greater knowledge of physical, chemical, and biological ocean processes and their interactions; and a better interpretation of information into useful and timely operational services. In decision making, scientific knowledge and information are a valuable and critical resource, giving credibility to arguments and establishing the acceptability or vulnerability of policies. When scientific uncertainty exists, as is inevitably the case in the increasingly complex marine issues facing us today, science should continue to present the best available interpretations together with an assessment of the risk attributed to actions or inaction on the issue under consideration. When subjected to careful peer review and reasoned opinion, even scientific uncertainty can be handled in a manner useful to decision making.

HOW CAN THESE NEW GLOBAL EFFORTS BE ORGANIZED AROUND THE WORLD?

The implementation of the new scientific programs at a global scale needs the distribution or decentralization of high-level scientific capacity. This will mean focusing capacity-building efforts at the global, regional, and national levels. Globally, an advanced international center for global ocean science should be considered as an interesting option. Such a center, conducting target-oriented research and developing appropriate systems engineering for global observation, could be the hub of a global network of scientific capacity-building for ocean science. It should work under the demanding standards of peer review science, but still be targeted.

The organization of the observation/information system itself will be regional, through regional centers for modeling, assimilation, and distribution of products. Therefore, regional centers of capacity building could match these centers, as part of the same network structure.

Beyond cooperation, the new international commitments in which all societies have entered in the past decade of this century will require a corresponding increase in capacity to respond to these new demands. To become a legitimate partner, and to demonstrate having negotiated the new conventions in good faith, countries will have to organize themselves to contribute, to generate, and to make use of environmental information, especially within the ocean areas under their national jurisdiction.

A global observing system, without regular sources of data and information from the fringe 200-mile zone rim around every ocean basin, would be a very limited system by any standard. Most users, and therefore most impacts, are concentrated on the local, usually coastal scale. To bridge the gap between the large scale and the local scale in a successfully functional way is a major challenge for the design and implementation of the Global Ocean Observing System (GOOS). For example, in order to benefit from the products that will be generated by GOOS, each society will need to

---

Box 10.6. Partnership for Observation of the Global Oceans
Lisa Shaffer

The Partnership for Observation of the Global Oceans (POGO) is a forum recently created by directors and leaders of major oceanographic institutions around the world to promote global oceanography, particularly the implementation of an international and integrated global ocean observing system. POGO includes institutions performing ocean observations as well as representatives of existing international and regional programs and organizations, such as the Intergovernmental Oceanographic Commission and the Global Ocean Observing System. POGO is a partnership of institutions performing oceanographic observations, operating ships, building sensors, collecting and processing data, conducting scientific research, and in some cases providing operational services to the ocean and earth science communities. Many of these institutions are also involved in teaching and training. POGO will help focus attention on implementation issues, such as technical compatibility among observing networks and shared use of support infrastructure such as ships deploying or servicing automated platforms or telecommunications, and in public outreach and education.

POGO will not operate major systems of capabilities directly. This will be done by the individual institutions with government funding. Through POGO, however, advocacy can be organized and resources pursued to support needed integrative activities that link different projects that are often at the margins of traditional funding plans.

> Scripps Institution of Oceanography hosted the first formal meeting of POGO in early December 1999. This inaugural meeting included senior officials from seventeen institutions in twelve countries and representatives of seven international programs or groups. An initial work plan includes development of an advocacy plan for observing systems, participation in a "commitments process" to secure governmental commitments for high priority in situ ocean observing systems, a data interchange pilot project, and establishment of an information clearinghouse for POGO members and the broader community. The long-term goal of POGO is to be part of the creation and operation of an integrated global ocean observing strategy, addressing information needs of decision makers, researchers, service providers, and the general public. POGO's contribution to that goal is to provide an informal but effective forum for dialog among leaders of key oceanographic institutions. POGO can provide the "glue" to integrate the observational needs of different ocean disciplines (ocean circulation, biology, climate, etc.) and to reduce barriers between research and operational activities in this area.
>
> Through collaborative partnerships, POGO can help increase the capabilities of developing countries to participate in collecting and using timely ocean information for their own management needs. POGO can help make the case for extensive and continuing observations, along with research and modeling.

organize itself to reap the benefits of the new information made available and at the same time to contribute to its generation.

Leaders of many major oceanographic institutions around the world have joined in an initiative to promote ocean science and observation, as explained in box 10.6.

The way in which each society organizes itself is open to several options. However, the level and scale (global, regional, and subregional products) of processing the information will play a crucial role. Although publicly funded services or institutions will play an important role in making the basic information available locally, this should not preclude participation of the private sector in the development of a full range of added-value products targeted to different end users.

CONCLUSIONS

Ocean issues have not yet received the government attention that they deserve. This sentiment was expressed repeatedly during the debate at the

U.N. Commission for Sustainable Development when it addressed the subject of oceans and coastal seas in April 1999. Such attention is needed at all levels. It will demand a high level of cooperation within nations, as well as between nations at both the global and regional levels.

RECOMMENDATIONS AT THE GLOBAL, REGIONAL, AND NATIONAL LEVELS

Global Directions

- There is a need for governments, at a senior level, to discuss cooperative action to address global marine issues as well as regional and local issues that have global consequences.
- A high-level forum is necessary to coordinate the various related programs of the U.N. agencies.
- This increased priority must also be reflected within the programs and projects of the many U.N. agencies dealing with marine activities.

Regional Directions

- New directions at the global level must be accompanied by effective measures at the regional level. Currently, there is a remarkably underutilized potential for effective regional linkages. For example, to date only limited efforts have been undertaken in scientific cooperation or in education related to management of the oceans. Redundancy among individual efforts seems to be the rule rather than the exception, yet there are major avenues for collaboration at relatively little added cost.
- By the same token, with effective regional collaboration come possibilities for participating in the joint monitoring of ocean uses. If there is any truth in the maxim that "there is strength in numbers," then joint surveillance and monitoring are likely to be more credible and effective than individual efforts.
- At the core of any serious regional initiative must be attention to collaborative technology measures as well as joint capacity-building efforts. In this connection, the proverbial "stakeholder initiatives" so critical at the global level must be emphasized and reinforced at the regional levels.

National Directions

Finally, reinforcing the same new directions at the national level as at the global and regional levels means:

- effective national strategies for local capacity-building in education, information, and communication pertaining to environmental management generally and ocean affairs more specifically;
- reinforcing the efficacy of civil society by encouraging civic participation in local and national affairs;
- introducing "two-way" communication, that is, from national decision makers to various segments of civil society, and back from civil society to national decision makers;
- adopting marine science priorities that address practical needs of management and exploring potential opportunities to use the ocean in new and beneficial ways and in a sustainable manner;
- establishment of national institutional focal points of responsibility for maintaining linkages to regional networks, but ensuring national coordination across sectors;
- devoting more resources to marine research and to the application of that research to ocean management considerations; and
- increasing the technical transfer programs and educational support in marine science to allow for the adequate management of the vast coastal lands and resources now under the jurisdiction of coastal states and to encourage the participation of all states in addressing global issues related to the oceans.

ENDNOTES FOR FURTHER READING

Borgese, E. M. 1998. *The Oceanic Circle: Governing the Seas as a Global Resource.* Tokyo: U.N. University Press.

Irvine, J. (ed.). 1997. *Equipping Science for the 21st Century.* Cheltenham, U.K.: Edward Elgar.

Keckes, S. 1997. *Global Maritime Programmes and Organisations: An Overview.* Prepared for the Independent World Commission on the Oceans. Kuala Lumpur: Maritime Institute of Malaysia.

Kuhn, T. S. 1970. *The Structure of Scientific Revolutions.* Second ed. Chicago: University of Chicago Press.

NRC (U.S. National Research Council). 1999. *Global Ocean Science: Toward an Integrated Approach.* A report of the Committee on Major U.S. Oceanographic Research Programs. Washington, D.C.: National Academy Press.

The Ocean Our Future. Cambridge: Cambridge University Press, 1998. Produced for the Independent World Commission on the Oceans by John May.

Chapter 11

Capacity Building

Miguel D. Fortes and Gotthilf Hempel

All coastal states and all countries that care about the ocean need to make provision for marine research and monitoring. Much human and technical capacity is needed to carry out marine science and technology. Building that capacity means training marine scientists and technicians, developing and acquiring marine instrumentation and platforms, and creating networks for the exchange and storage of data as well as for communication between scientists. It also means establishing a modern institutional infrastructure at national and regional levels. In many countries this has already been part of a continuous process over many years. Others, particularly developing countries, lag behind and seek outside support.

In order to cope with the complex problems posed by the sustainable use of the oceans, new kinds of scientific capacity are needed worldwide. At the same time, the gap between countries and regions in terms of research and monitoring capacity has to be bridged through partnerships and mutual assistance. Capacity building has to be an integral part of global marine policy for sustainable development.

This chapter deals with three questions: What new kinds of capacity are needed to address complex ocean problems? What additional capacity is needed to enable all countries to participate fully in marine science research and development and in the sustainable use of the oceans? And what are the means available in terms of national efforts and international partnership to meet those needs?

WHAT NEW KINDS OF SCIENCE CAPACITY ARE NEEDED?

Two of the most significant driving forces on the marine environment are global warming and the overexploitation of the coastal zone. Both are largely the result of human activity, but have profound biophysical impacts. Addressing these activities and their impacts requires the integration of biological, physical, and social science disciplines. The building of interdisciplinary scientific capacity is thus critical. Climate-related research, in its broadest sense, has been a priority for oceanographers for the past two decades. This research has mostly been concerned with the nature and variability (in space and time) of processes and structures in the open ocean. In the future, the interaction between physical, chemical, and biological processes will become a focus of studies in ocean climate and will require a merging of those disciplines and the evolution of new kinds of ocean science capacity. Global programs have been established to address climate change concerns and to come up with scenarios and predictions of a variety of phenomena. These include ocean-wide interactions, like El Niño, which act on time scales varying from seasons to thousands of years and on spatial scales ranging from local to regional to global. Research and monitoring in these fields is enormously important for the economies and improvement of the quality of life in all coastal countries. But open-ocean research is expensive and can only be fully supported by a few governments in developed countries. These countries are prepared to advance their oceanographic capacity in terms of manpower recruited from different disciplines, instrumentation, and infrastructure, including facilities for ocean-wide research projects with climate-related goals in mind.

Developing countries are as affected by large-scale ocean and climate processes as the developed industrialized countries. The resources available to their local scientific communities, however, are too limited to enable them to take part in blue water (open ocean) research programs. Instead, governments of developing countries instruct their marine scientists to concentrate on the problems and resources of their coastal waters and exclusive economic zones (EEZs). Nevertheless, scientists from developing countries have contributed substantially to open-ocean studies. This trend should be fostered in the future. Equally, young scientists in developed countries should be encouraged and assisted to develop an interest in tropical coastal ecology and related fields, in close cooperation with local scientists. In this scenario, all participants benefit mutually from the exchange of knowledge and information. There are many opportunities for new kinds of collaboration. The newest trend in global change research programs (especially the International Geosphere–Biosphere Programme [IGBP]) is the incorporation of the human dimension. Attempts should be made to foster links not only between scientists in developed and developing countries, but also between new communities of expertise. Thus, natural scientists should be encouraged to work with their colleagues in the social sciences. In the shelf seas and coastal zones, many of the marine problems confronting developing countries are similar to those in industrialized countries, concerning fisheries, tourism, transport, waste disposal, pollution, oil, gas, sand and gravel extraction, etc.

In fact, both in developing and industrialized countries, most marine scientists work near the shore. Many of them become fascinated by the interdisciplinary and human aspects of their work. A number of large-scale international programs and cooperative research projects have grown out of this increased interest in coastal environmental problems, hence the demand for more information about coastal seas. Interest is also growing in comparative studies of shelf ecosystems and their reaction to climate change, sea-level rise, fishing, pollution, and other pressures caused by the expanding human population in coastal zones.

In the years to come, much of the ocean science carried out in support of sustainable development will be directed toward the coastal zones in tropical and subtropical areas, which are especially affected by global change and direct human activities. Sustainable exploitation of marine resources and protection of coastal zones are new challenges for the scientific community, requiring the formation of new, high-quality research capacity, again particularly in developing countries.

The concept of large marine ecosystems (LME) provides a regional framework, linking research, management, and governance with the goal of sustainable use of the shelf seas. Applying this concept will help to reduce the strong gradients in scientific potential within certain regions like southern Africa or the eastern Mediterranean. In addition, the collective need of nations to reduce the pollution of coastal seas is being addressed through international conventions and the action plans of the United Nations Environmental Programme's (UNEP) Regional Seas Programme. The fast developing Global Ocean Observing System (GOOS) provides an additional mechanism for meeting regional needs for operational oceanographic information in coastal seas (see www.ioc.unesco.org/goos).

In the long run, coastal zone management problems can only be solved by an integrated approach, bringing together disparate disciplines, including the natural and social sciences, as well as the various stakeholders. The gaps between these branches of science can only be bridged by a multiple approach, including new, modular teaching curricula and effective communication between natural and social scientists and managers. Dialogues are developing between scientists, politicians, and other stakeholders. There is an emerging trend, in which scientists learn with the local and regional communities, rather than only learning about them. Furthermore, the flow of experience and ideas on marine issues is no longer unidirectional, from developed to developing countries. It now goes in both directions.

The coastal sciences will become more and more multidisciplinary and interdisciplinary, often linking natural and social sciences. After decades of disciplinary research and teaching, we now work in a multidisciplinary mode. To resolve the complex environmental and developmental problems in many coastal regions and seas, a cooperative interdisciplinary approach is needed. This requires new kinds of teaching. Cross-disciplinary thinking by imaginative and innovative scientists will reveal new links between natural phenomena and human activities as well as producing blueprints for new scientific approaches. Scientific infrastructure and teaching in the marine sciences must be adapted to those needs. That includes multipurpose research vessels and platforms with interchangeable payloads, institutional networks, flexible funding, and—most important—modular teaching systems, to provide students with a broad spectrum of knowledge about the ocean, its resources and protection. Students trained in this way will be prepared for new careers in the sustainable management of marine resources and the environment on local, regional, and global scales.

More than ever, marine science cannot work in isolation. It has to communicate with its "customers," that is, the public, stakeholders, and government administrators. Communication and marketing ability must be built to meet the need for more public awareness about the ocean in all sectors and levels of society, ranging from user groups like fishers to policymakers and legislators. Marine science should make full use of the new opportunities provided by the global multimedia information revolution. Various new communication technologies, including the Internet, should be applied and exploited by new kinds of experts.

These developments toward more integration and communication will occur in most coastal countries, but the ability to respond to them will vary in different parts of the world, depending, in part, on the level of economic and academic development. What additional capacity is needed? An

Box 11.1. Capacity Building: A Thai Perspective
Manuwadi Hungspreugs

Marine science activities in Thailand are mostly carried out in the shallow waters of the Gulf of Thailand. Traditional small-scale fisheries used to provide the local people with abundant dietary protein until about forty years ago, when larger-scale commercial fisheries expanded its operations into neighboring waters and high seas. Thailand then became one of the world's top ten seafood-exporting countries. But the declaration of an exclusive economic zone made the Gulf of Thailand a virtually closed bay, depriving Thai fishermen of their traditional fishing grounds, leading to a decline in the industry. Coastal aquaculture, especially shrimp farming, was introduced to replace the capture fisheries, but it expanded so fast that the pristine coastal environment suffered considerable damage.

Interest in oceanography in Thailand gained momentum with the 1959–1961 NAGA Expedition, the Joint Thai–Vietnam–U.S. Survey of the Gulf of Thailand and the South China Sea. Soon after, Chulalongkorn University began to offer selected courses in marine biology. Then, in 1968, after a few years of staff preparation, two departments of marine science were set up simultaneously at Chulalongkorn University and Kasetsart University. The former concentrated on sciences, the latter on fisheries.

Because most students choose to major in biological sciences, there is a severe staff shortage in physical oceanography, geological oceanography, and chemical oceanography. A lack of strong leadership makes truly interdisciplinary research in ocean science at the national level a rarity. Short-term training for local researchers is available in advanced overseas institutions, but this is not enough to form good leaders. A regional graduate school of oceanography is needed, with international standards of excellence, offering programs in ocean and marine environmental sciences. Such a school would build on the capacity of existing institutions, enhanced by international faculties, libraries, adequate equipment, and seagoing research vessels.

The Phuket Marine Biological Research Centre started as a result of the International Indian Ocean Expedition in the early sixties, with aid from the Danish Government. The center's strategic location and nearby coral reefs and mangrove areas attract visiting overseas researchers, especially those from Denmark, Britain, and Japan. Several international training courses and research projects on coral reefs and mangroves are now based in Phuket. In addition, the Secretariat Office of the IOC Sub-Commission for the Western Pacific (WESTPAC) has been based at the National Research Council in Bangkok since 1994. However, this office has only a small workforce and needs to be upgraded in order to make a strong impact on the region.

example from Thailand (box 11.1) illustrates some of the approaches needed to meet these requirements.

Southeast Asian countries are making great efforts to review their coastal management programs in a thorough and competent way. This will lead to changes in practice and to policy reform. Topics for review include management techniques; program preparation and funding; institutional, legal, and political systems; and social and cultural attitudes and values. A great variety of experts and scientific generalists is needed for these reviews.

Priorities include:

- Adopting more appropriate technology. At the top of the list of priorities of several countries are innovative ways of producing energy from the oceans and new methods of growing and producing food for the region's burgeoning populations. The challenge is to develop and use ocean technology that is relatively cheap and easy to apply and that does not further burden the environment.
- Capturing the flows of knowledge and information in an increasingly globalized society. Asian nations should focus on ensuring that their research and development capability are internationally acceptable, requiring their scientists to participate in international activities in order to broaden their opportunities for continuing professional education.
- Ensuring regional cooperation in order to overcome isolation, to further strengthen science in the region, and to counter the tendency of many Asian researchers who are hesitant to cooperate regionally.

These approaches and their associated requirements for capacity building are primarily the responsibility of the individual coastal state and, collectively, of the states in the region concerned. There is, however, a vested global interest in the success of these approaches, both in terms of the marine science community and of global sustainability and welfare. The gap between developing and developed countries is growing in terms of human and material capacity and public awareness of marine science and marine affairs. Benefits of marine research will not accrue equally to all countries, and damage to the ocean environment will continue to take place. Assistance, whether it is bilateral, multilateral, or truly international, is the rational way to bridge the widening gap.

In general, marine management strategies look at the environment on three levels: the local level, where field operations are carried out; the central government level, where policies are made; and the international level of external support agencies. In order to be effective and sustainable, these strategies should consider what is needed in both the short term (e.g., coastal villagers rehabilitating a coral reef) and the long term (e.g., strengthening educational institutions), and they focus effort at the ecosystem level, instead of being concerned, for example, with problems relating to a single species.

These strategies are designed to fit into the current governmental structures in a manner that causes the least disruption of present institutional alignments. They and their usefulness are adapted to changing economic and social conditions.

These perspectives, in turn, are dependent on the following assumptions being satisfied:

- There will be a substantial positive change in national legislative agendas in favor of environmental concerns.
- National commitments will support and encourage the development of coastal marine science and management in the regions.
- A consensus will be reached on strategic approaches needed to ensure long-term sustainability of environment sector programs and projects.

International organizations and agencies with mandates involving scientific capacity building should develop a new role in which they will act as brokers to persuade donor agencies to spend their money in ways more consistent with regional needs. They will also continue to supply some capacity building themselves from their limited budgets.

NORTH/SOUTH AND SOUTH/SOUTH ASSISTANCE IN EDUCATION AND TRAINING

Achieving worldwide participation in marine science requires human and technical capacity building on a worldwide basis. Box 11.2 sets out many of the different ways in which capacity may be built.

Over the years, a number of developing countries have built up substantial research and teaching capacity in marine science. Several others have made great efforts to create the critical mass of marine scientists and infrastructure as part of their national policies to expand the use of coastal zones. Box 11.3 gives an example of capacity building approaches in Brazil.

Nevertheless, in the majority of developing countries, marine science capacity and public awareness of environmental problems are limited in comparison to their national needs and obligations for compliance with international conventions dealing with climate change and environmental protection.

The demand for scientific capacity in the universities, colleges, and

Box 11.2. Elements of Assistance in Capacity Building in Marine Science

Training
- overseas fellowships
- fellowships in the region
- advanced specialized training courses
- training on the job

Support
- advisors
- instruments
- research vessels (including maintenance and running costs)
- literature

Communication
- regional and global meetings and workshops
- electronic information and communication systems
- scientific journals

Institutional infrastructure
- universities
- research institutes
- national and regional organizations

Joint research projects
- joint cruises
- joint shore-based projects

Box 11.3. Capacity Building in Marine Sciences in South America: A Plea for Networking and General Education
Eduardo Marone and Paulo da Cunha Lana

South American countries have addressed the issue of marine science training and education in different ways, based on their own educational systems and capabilities. Although some countries have had marine science graduate and undergraduate programs since the 1960s and 1970s, there are still few undergraduate or graduate courses in marine science in the region. So, most countries simply send their young students and scientists abroad for training. Today, most countries in South America have a reasonable number of marine scientists in the usual fields of expertise, but they are unevenly distributed. The region would benefit from links being developed between existing marine science institutions. Shared research projects, regional meetings, visiting researchers, and professors would also help create synergy.

Key issues in building marine science for the future will be general education and the networking of existing capacities. General education in marine sciences should be made available at all levels, including the general public level. A basic education in marine affairs would encourage society to elect decision makers better prepared to address ocean issues. At undergraduate and graduate levels, a tilt is needed away from the current bias in favor of academic marine sciences toward more applied courses on the management of coastal resources. Networking will promote cooperation, by increasing national and regional interaction, making full use of existing capabilities and improving the so-called South-South and North-South flow of knowledge and resources. Laboratory facilities, data banks, modeling centers, instruments, ships, etc., are also needed. It is advisable to build this infrastructure by cooperative rather than competitive means.

A CASE STUDY IN EDUCATIONAL APPROACHES

The Centre for Marine Studies (CEM) of the Federal University of Paraná, Brazil, has started a new venture to provide general training in marine sciences at all levels. Its environmental educational unit, headed by a CEM staff member who is also one of the village's primary school teachers, provides education for young students and their families. For decision makers, the center acts as a training unit for the U.N.'s Train-SeaCoast program, which provides continuing education for employees from coastal municipalities, harbor authorities, governmental and nongovernmental organizations, environmental agencies, etc. Traditional undergraduate, graduate, and postgraduate courses complement these activities. Unlike other, similar initiatives, the undergraduate course at CEM has been designed to train professionals to respond to actual social demands. There is presently an enormous shortage of qualified professionals in coastal resource management in a country with more than 8000 km of coastline. So, besides a formal education that prepares students for research at the master's level, the undergraduate course in marine sciences also includes a major in coastal-resource management. Courses are organized as modules, each with a specific purpose. A master's degree course in structure and dynamic of coastal and marine systems is also being developed. Besides producing good young scientists in an academic context, this course will have the pragmatic objective of training professionals to find practical solutions to environmental issues in coastal areas, based in the "useful science" concept.

vocational schools in developing countries differs from that in the academic system of most developed countries. The management of highly heterogeneous coastal zones and their exploitation by small fishing units rather than by large, industrial-scale operations and by a multitude of other users requires large numbers of experts, advisors, and supervisors attached to local communities and national authorities. These people have to be trained in technical schools by instructors with university backgrounds in fields like tropical marine ecology and fisheries science.

Vocational training that makes use of multimedia and Internet technologies will become increasingly important for nonacademic technical staff, dealing with high-tech instrumentation in the field as well as in laboratories and computer centers. Research institutes and universities of industrialized countries have introduced various instruments to assist with capacity building. The financial support for those assistance programs comes mostly from national donor agencies and foundations. Those contributions were very small compared to the expenditures of the industrial countries for their own national marine activities. In many cases, the extra costs for capacity building in the framework of scientific cooperation with developing countries fall between two stools: aid programs for general development and scientific research projects. Furthermore, the communication and coordination among the donor organizations within and between the industrialized countries is far from being satisfactory.

The training of individual scientists is one of the most important challenges to be faced. In the past, privileged students from developing countries spent all their university careers in Europe or North America. On their return, some of them created the nucleus of science faculties at their local or national universities or joined national research centers, and others looked for more profitable occupations in commerce, government, or administration. Nowadays, many universities in developing countries offer at least a first degree in marine sciences, so the new generation of students can stay in their home regions until they are ready for master's and doctoral degree courses abroad.

A new trend is emerging in which postgraduate education abroad is often carried out in cooperation with the students' home universities. Course requirements are fulfilled at an academic institution in a developed country, with the thesis based on research topics and material especially relevant to the home country. Some programs permit graduate students to return home to conduct thesis research under the supervision of local scientists, returning to the overseas university to complete the thesis requirements. These programs ensure that research projects are relevant to the needs of the student's country and make it more likely that the new graduate will return home to seek employment.

Other innovative academic programs allow students who may already be employed as junior scientists in government laboratories to acquire a graduate degree while retaining their jobs, with time away from their home institutions for the course studies. The thesis research may then be conducted in the employee's institution, thereby reducing the amount of time spent away from a salaried position.

Most curricula of universities in Western Europe and North America are not tailored to the specific needs of postgraduate students from developing countries. But in recent years, several universities have developed international curricula in marine science that emphasize the problems of coastal management in tropical and subtropical countries. Ideally, these curricula are joint programs with tropical universities. Two examples are the master's courses in marine science fisheries and coastal zone management of the University of Bergen, Norway, and of the University of Bremen, Germany, in cooperation with universities in Costa Rica, Brazil, and Indonesia.

Another facet of capacity building is through the provision of opportunities to conduct research, for example, by the participation of students from developing countries on the research ships of developed nations. All developed nations should provide shipboard fellowships on research cruises whenever possible. These would be most effective if the student is partnered with a professional mentor who is willing to commit time and effort to this endeavor.

One of the most successful ways of disseminating information to a large number of people simultaneously is through distance learning. The Open University (United Kingdom) has for several years run a degree course in oceanography by using television as the medium for distance learning, supplemented by their widely used textbooks and supported through periodic tutorials where experts advise groups of students. Nowadays, the Internet offers very much the same potential, but at the global level. The development of an Internet-based degree-level course leading to a master's degree in marine science would make high-quality education readily available at relatively low cost to large numbers of students who currently have little or no access to the requisite staff or training materials. In addition, the Internet offers unprecedented access to a world of information hitherto largely unavailable to developing country scientists. A stumbling block at present is the lack of ready access to Internet in many developing countries, where the infrastructure is not in place and service provision is still exorbitantly expensive (see below). The Open University experience suggests that the Internet by itself may not be enough to achieve high standards through distance learning; at some point students must come into contact with local or regional experts.

Training networks like those developed under the Global Change System for Analysis Research and Training (START) of the International Geosphere–Biosphere Programme (IGBP), the World Climate Research Programme (WCRP), and the International Human Dimensions Programme (IHDP) are becoming increasingly important. Their regional synergy creates a critical mass of science capacity within individual regions and makes better use of external support. Various concepts are used to develop regional graduate education programs in different regions, by building on the human and material capacities of existing institutions. One model is the Oceanography School at the University of Concepción in Chile. It is supported by UNESCO, the German Academic exchange service (DAAD), and other donors. Its innovative aspect is that the local academic staff is supplemented by visiting academics (the "international faculty") who teach intensive three-week courses in their fields of expertise. It started with an undergraduate professional program in marine biology in 1974, followed by a master of science program in 1986, and a doctoral program in oceanography in 1992. It serves the interests of the network of five universities in temperate Latin American countries.

This experience highlights the importance of partnerships on one hand, and regional centers of excellence on the other hand. Partnerships with visiting mentors from overseas would help to enhance educational and research opportunities at home and hence to raise the quality of their graduate degrees to internationally attractive standards. The establishment of regional centers of excellence in teaching and research would also help to encourage students and young scientists to return home after studying abroad and to remain in their home countries. Here they can take the lead in the research needed to find solutions to local and regional problems. At present, a Regional Graduate School of Oceanography in Southeast Asia is being developed by IOC with the University of Concepción (Chile) as a model and template.

There is a growing interest in advanced training courses and workshops dealing with modern methodologies and new approaches in marine sciences. Other courses are directed to administrators who need a grounding in the application of marine science for marine management. In all courses, participants contribute their own data and specific problems and experiences from different parts of the world.

ASSISTANCE IN BUILDING TECHNICAL AND ORGANIZATIONAL INFRASTRUCTURE

Equipment

From the 1960s to the 1980s the provision of scientific equipment and research vessels played a major role in capacity building. Not all of these donations were of great practical value because of a lack of funds for operating costs and maintenance and/or a lack of skilled technical staff. The situation has improved over the years with the growing technical skill and infrastructure in the recipient countries and institutions. But donors of equipment should continue to assist in the training of the technicians and scientists who will use them.

Much of the recent effort in capacity building has to do with communication. Free access to the modern Internet service helps to overcome some of the financial and logistic constraints on access to scientific literature and teaching tools.

The acquisition of the necessary hardware and software may be an obstacle for many institutions in developing countries. Donors should consider this need, especially as the cost of computer equipment continues to decrease. It is true that in some areas, especially in Africa, the local telephone systems are inadequate to support electronic communication. But this situation is changing rapidly as many countries have begun to recognize that this is one of the most severe impediments to their development.

Data Access and Analysis

Provision of a sound information base for local and regional planning in coastal zones requires, among other things, creating a network of specialists trained in the use of data acquired by remote sensing from space satellites. In addition, means of access to space-derived data need to be readily available to developing countries. Access to the data, and the means to interpret them, would provide rich opportunities for scientific understanding and sustainable exploitation in regions that currently lack such opportunities because of their limited network of in situ sampling devices. The provision of computer hardware, software, and some training courses will also assist in removing the problems of access to databases.

In addition, there is a need to encourage developing countries to get their largely analog data records out of widely distributed files, digitized, and onto the web in forms that are suitable for analysis using modern methods and for exchange with other local or regional scientists.

As emphasized in chapter 8, once data have been acquired, numerical models can be used to extract information from them. By applying numerical models to data a broad range of products can be created, each tailored to the needs of different users (fisheries, tourism, port and harbor management). Modern desktop computing power is now so great that it has brought numerical modeling within easy reach. Data are commonly available from ships, buoys, tide gauges, or satellites. The missing ingredients are people trained in modeling.

Universities

A viable academic infrastructure is essential for capacity building efforts. The creation of universities in addition to research institutes played a major role in some traditional development programs. Now those systems are in place, and long-term advisors may be replaced by short-term specific experts. Continued strengthening of universities is most important for sustainable scientific capacity building. Not every university, and not even every nation among the developing countries, can run a high-quality university curriculum in marine science. Therefore, nationwide and regional agreements should be reached on the formation of national and regional centers of excellence in marine science teaching.

Libraries

Libraries suffer all over the world, but particularly in developing countries, from the high price of scientific literature. Donor organizations should be prepared to earmark substantial funds for the libraries of their partner institutions. The shift toward electronic publication helps developing countries to catch up, wherever those means become easily accessible. But it is not the complete answer to the needs of libraries in developing countries. Back issues of key journals and many books are not available electronically.

JOURNALS AND THE LANGUAGE BARRIER

For any scientist, international participation and recognition depend first of all on publishing in international journals. Scientists from non-English-speaking countries and less advanced laboratories find it very hard to get their papers published internationally and thus to have their work acknowledged. It is not only a language problem but also a question of how to present data and to write a good scientific paper.

Language is one of the most important obstacles to the full participation of many students and scientists in international scientific communication. English is the lingua franca of marine science and technology, but it is not the mother tongue of the majority of marine scientists and students. In developing countries, some scientists are quite competent in scientific English and are able to publish in the international literature. Many others may have a basic knowledge of several languages, but their command of English is limited. An enormous amount of information lies hidden in local journals, reports, and archives. The scientists who produced this information will never receive international recognition if they are not able to publish it in the more widely read international journals.

Solving this problem will be a slow process and will require a strong will on the side of scientists in developing countries as well as patience and assistance on the part of editors and referees of international journals. Many journal editors are not presently receptive to the manuscripts they may receive from authors whose English skills and experience in writing scientific papers are not well developed. Editors must be willing to exert the effort needed to recognize good science that is worthy of publication, but may be obscured by poor English. We need reviewers who are willing to commit time and effort to assist these scientists in gaining the skills they need to communicate the results of their research to a much broader audience. Another solution is for scientists in developed countries to actively coauthor papers with their colleagues in developing countries. Some regional and national journals of substantial general interest are increasingly recognized by the international science community because they are published mostly in English and supported by a strong international editorial board. Learning by doing in the framework of cooperative projects and joint publications is probably the best way to break the language and publication barrier.

Box 11.4. A Pacific Regional Perspective

Russell Howorth

The Pacific Ocean represents over half of the world's ocean space and covers about one-third of the Earth. Most of its sparse land space comprises the widely dispersed, small islands of the mostly independent and self-governing Pacific Small Island Developing States (Pacific SIDS). These have come together as a political regional group, the South Pacific Forum. Ironically, in 1998, the Year of the Ocean, no mention of the ocean appeared on the Forum's annual meeting agenda. Few states have established ocean-focused statutory bodies. The ocean is not getting the attention it deserves. There is also a lack of active management initiatives, irrespective of whether they be scientific or resource focused.

The physical and geological makeup differs markedly between the islands. It ranges from large, mountainous and continental in New Guinea, to volcanic in the Solomon Islands, to tiny coral atolls like those in the Marshall Islands. Populations also differ greatly from one island to another—ranging from 4 million in Papua New Guinea down to less than 20,000 in the Cook Islands. Not surprisingly, the issues and opportunities as far as the ocean is concerned also vary from one country to another.

The ocean has a massive effect on the predominantly small populations of these widespread, mostly tiny islands. It controls their weather, their climate, and, especially in the case of atolls, their exposure to changes in sea level. It is a source of wealth principally through fisheries. These are of two types: deep-water tuna and nearshore, the latter being usually small scale using low technology and often for subsistence. Aquaculture is important in a few countries only. Coral reefs are common, providing a source of food and a magnet for tourists. Flourishing reefs require a healthy environment and appropriate conservation measures. Sand, gravel, and, in some areas, coral are mined for local construction. Hydrocarbons are found in the coastal zone in Papua New Guinea. There is considerable potential to derive energy from waves and ocean thermal energy conversion (OTEC), but there are no pilot plants (except in Hawaii).

Maritime transport is vital to the economy. Coastal marine resources are under great pressure, and increased efforts are required to make development sustainable, not least to sustain the tourism that supplies about 5 percent of the gross domestic product. Many Pacific SIDS are expected to manage exclusive economic zone (EEZ) areas that are huge by comparison with their land areas. There are substantial concentrations of presently uneconomic deep-sea minerals in and near these EEZs.

To fully exploit the ocean and coastal environment in a sustainable way, there is a need to enhance the capacity of the region in marine science and technology. This need was addressed, for example, through the capacity-building program for the Global Ocean Observing System (GOOS). In 1998, the regional GOOS body, PacificGOOS, recommended the following priorities for regional capacity building:

- Capacity building needs to be defined more precisely in accordance with a country-by-country analysis of where the gains will be greatest from improved monitoring and its application.
- Greater advantage needs to be taken of marine science courses available in and near the region, for example, at the University of the South Pacific and universities in Papua New Guinea, Australia, New Zealand, New Caledonia, French Polynesia, Guam, and Hawaii.
- A regional assessment of communication needs to be undertaken in terms of hardware and of local skills and knowledge.
- Focal points should be designated in each country of the region, and any training needs associated with these countries included in the Pacific GOOS Strategic Plan.
- Awareness-building training programs need to be established to support Pacific GOOS. These should cover elements like monitoring, development and use of end products, data management, and modeling and should use existing arrangements where possible.

This is not just about physics. Mapping and monitoring critical habitats, like mangroves, seagrasses, and corals should become routine, as should the monitoring of critical pollutants in runoff from land.

The development of an observing and forecasting system requires the application of marine science. This exposed the need for fundamental training in the marine and environmental sciences to produce the people needed to apply operational methods to address key environmental and economic issues involving the ocean and its influence on the islands.

REGIONAL MEETINGS

Regional meetings are essential for the development of cooperation in research and monitoring, and also for the exchange of information and data. They are primarily a means of communication between scientists of the region, but they might profit from the attendance of some specifically experienced scientists from far away. Projects like the LME projects also provide tools for the improvement of scientific communication through regional meetings, resulting in joint field activities and publications. The overall importance of regional cooperation in the South Pacific is discussed in box 11.4. Such cooperation is particularly demanded for the development of ocean monitoring in support of social goals.

Equally important to the building of capacity are the development of an appropriate regional marine strategy and infrastructure supported politically by governments and appropriate marine and environmental agencies. Policies and needs have to be communicated to the people and to donors. As recognized elsewhere, such a strategy should be intersectoral and holistic and involve all stakeholders.

All Pacific islands would benefit from national oceans policies and from working together to create a regional policy. Ideally, these policies should promote an adaptive management strategy addressing the issue of resource-use conflicts. They should also make provisions to control the impacts of human intervention on the physical environment, while coordinating the initiatives of all public and private sectors and ensuring that common goals are attained through a unified approach.

REGIONAL AND GLOBAL NETWORKS

Without more regional communication, coordination, and cooperative marine science research and teaching activities, developing countries will always lag behind. The IOC, World Bank/GEF, UNEP, UNDP (United Nations Development Programme), European Union, and others are now working for the establishment of networks of regional cooperation and communication. Participation in the work of international oceanographic organizations of global scope like IOC and SCOR as well as in regional bodies is important for the broadening of the perspectives of marine scientists. National coordination bodies should bridge the gaps between the various scientific, industrial, and military user groups interested in the sea and its resources, not least to get scientists in different ministries cooperating with one another. They should also foster public awareness of marine science and marine affairs. Coherent national programs should be tailored to the overall national needs and make the best use of international cooperative activities. In Eastern Europe and Northern Asia, the marine science institutions of the former Soviet Union are in a very difficult situation. In order to maintain the existing human and material potential in the face of their weak economies, the strong support of the international marine science community is called for. Without external support, much of the research capabilities, skills, and information will erode and only a very small fraction of the young generation will decide to enter marine science. Retaining and reforming scientific infrastructure and manpower in those regions so that they remain viable is of similar importance to the formation of new science capacity in other parts of the world.

The loss of students and researchers in the marine sciences to other fields and to other countries is not limited to particular states but is a very widespread problem. It can only be solved by creating attractive career prospects and acceptable living and working conditions. Only when such favorable conditions exist for the pursuit of a professional career in marine science can we expect the practitioners of that science to excel. The challenge to the community is to ensure that such conditions are available to everyone, regardless of their nationality or gender.

This, in turn, is partly a matter of public awareness of the importance of marine science for society. Potential students and the public, including media, nongovernmental organizations, industry, and policymakers, should be told what marine science is about, not in an overly romanticized way, but in terms that make it relevant to local and regional concerns. The general public in industrialized and developing countries should learn about the importance of marine science, its benefits to society, and its needs for continuous support for research and training.

From Assistance to Partnership

As interest in coastal zone management grows, scientific interests converge and might lead to new partnerships. From a pragmatic point of view, the political arguments related to permits for research in coastal waters and the collection of specimens and data appeal for greater collaboration between scientists in developed and developing countries. Scientists must recognize that they have no inherent right to conduct research wherever they please without an obligation to contribute to the advancement of scientific understanding of locally important issues. They must learn to take into account the needs of the area in which they wish to work and to make a commitment to share the results of their research if they expect to receive the necessary permits for it. Unexpected and rewarding new collaborations may develop when an individual scientist (or a research team) is open to the possibilities presented by research activities based in developing countries (box 11.5).

For the success of these kinds of projects, we list a number of criteria for partnership projects as tools for capacity building in marine science:

- Generated in the region with mutually beneficial objectives to be formulated by all partners.

Box 11.5. MADAM: Research and Capacity Building Through an Integrated Long-Term Project

Ulrich Saint-Paul

MADAM (Mangrove Dynamics and Management) is a jointly financed Brazilian-German research and training project established in northern Brazil. The research area at the peninsula of Bragança is located approximately 150 km southeast of the mouth of the Amazon River. Here, between a coastal plateau and the Atlantic Ocean, is a mangrove biome of 110 km^2 that in places is up to 20 km wide. The overall objective of MADAM is to investigate basic questions regarding the ecology and socioeconomy of mangrove ecosystems. It also aims to suggest possible solutions to the problems of sustainable utilization and environmental protection based on ecological knowledge and models of integrated coastal zone management. Teams of Brazilian and German botanists, marine chemists, fishery biologists, and socioeconomists come from the State University of Paraná and from the Center for Tropical Marine Ecology, Bremen. There are more than 100 scientists and students engaged. German masters and doctoral students work in Belém and Bragança under the local supervision of Brazilian professors, and Brazilian students go to Bremen to write up their results under the guidance of German professors. The MADAM project is planned to have a ten-year lifetime. The research is organized in three coordinated components. A synthesis group integrates the results.

The biogeochemistry component describes the flow of inorganic nutrients, nitrogen, and dissolved and particulate organic carbon in the system. Chemical marker analysis traces the sources of the organic matter in the mud and water. The hydrographic and meteorological parameters are recorded.

The biological component analyzes the community structure in the mangrove and in the river system. Biomass and primary production and population dynamics of the most important plants and microbial communities are studied on land and on the littoral mud banks. The land crabs and fish are studied in detail as the economically most important groups, along with trophic structure and the role of the different habitats in the various life history stages of fish in the rainy and dry seasons. The socioeconomic component deals with the economic, demographic, and sociocultural aspects of the mangrove system around Bragança. How does the local population use the mangrove, and what are the effects of population pressure, better roads, and growing tourism on the mangrove? What can be recommended for the sustainable development of the region, including fishery regulations, habitat protection, and recolonization of impacted areas?

The modeling group analyzes the results of all modules. Using trophic models of food and energy transfer and dynamic models of the main processes, they provide basic information about the dynamics of the mangrove system. Based on this holistic approach, the modeling group develops ideas for future management. No good maps were available for the region, but cartography is now underway, as is a geo-information system and database, including new and old data. Brazilian and a few foreign scientists in the study area have already carried out considerable work, but the results are mostly hidden in the gray literature or exist in the form of raw data. These old data sets will be reprocessed to be made available for science and management.

The program builds up scientific capacity in both countries. It strengthens the research and teaching capacity at the university in Belém while creating a core group of young German scientists who learn to work in tropical countries, with their colleagues there, on scientific problems of direct practical relevance. Brazilians and Germans will write their papers together—in English, which is a language foreign to all of them.

- Scientifically attractive; mostly multidisciplinary, preferably including social sciences.
- Relevant to the development of the host country in the framework of its national obligations.
- Able to strengthen science capacity in the host country and the region, particularly by training on the job and other means of advanced and specialized training.
- Training activities built around jointly planned research projects.
- Cooperative project planning and implementation involving all partners, making full use of the existing expertise in the host country.
- Long-term commitments with financial and material contributions by all partners.
- Full integration in the scientific structure of the host country and its universities and research and development organizations, connections to regional centers of excellence and training centers and networks.
- Able to strengthen public awareness in marine affairs.
- Generate links to regional and global programs.
- Unrestricted data exchange and joint publications, mostly in international journals.

These criteria have been adopted by German donor agencies funding overseas marine science projects.

One question remains: How can we encourage more scientists from developed countries to devote their time and efforts to partnership projects with developing countries? If they are committed to this effort, they may have to work under more difficult conditions than in their home laboratories. Money alone is not enough. Individual and community commitment is needed. Projects must be scientifically appealing to all partners. And this is not only in terms of the potential scientific results. The partnership itself should be rewarding to everyone involved.

Research and training are well integrated on a number of international research cruises. The expeditions of the Norwegian research vessel *Dr. Fridtjof Nansen* in many areas of the world provide an excellent example. Research vessels and the research groups aboard them must recognize that they are privileged guests while they are operating in foreign EEZs. The provisions of the Law of the Sea may profess to protect the rights of all scientists to conduct research in the ocean without restriction, but these provisions have often overridden by the strong wish of developing coastal states to ensure that the results of that research are made available to them. Accordingly, developing countries may place stringent conditions upon permits to conduct research within their EEZs. As guests in these areas, scientists from developed countries must adhere to these formal obligations in terms of data sharing. But this is also an opportunity to offer training to local scientists as part of a more equitable exchange.

Research vessels operating in such waters should always be big enough to carry a number of local trainees in addition to the regular research staff. Regional centers of excellence should include research vessels capable of modern interdisciplinary research with the capacity to accommodate research teams from the region and from abroad.

Truly multinational research and development projects may offer the best opportunities for training on the job, sharing of facilities, and creation of infrastructure. The LME projects in the Gulf of Mexico, Gulf of Guinea, Benguela Region, Yellow Sea, South China Sea, and Gulf of Bengal are opening a new era of regional cooperation with strong elements of capacity building and creation of public awareness. Recent successful examples include the multinational Red Sea Program and the Benguela region BENEFIT program of Angola, Namibia, and South Africa, with financial support from various European countries and international agencies.

Training obligations should be an integral part of all international research and monitoring programs and agreements, from the earliest planning stages and not a mere afterthought or add-on. For example, regional GOOS programs should mobilize scientific and technical manpower as well as providing monitoring installations (see box 11.4). Classes of activities in GOOS capacity building include:

- On-the-job training of individuals in both their home institutions and elsewhere.
- Provision of initial or refresher training to specialists.
- Fellowship to individuals for scientific, technical, and engineering training/formal education.
- Regional cooperative development projects directed at limited attainable objectives.
- Assistance in securing resources needed for developing/enhancing infrastructure needed for specific GOOS-related activities, either for institutions or nations.
- Short-term accredited (or diploma) residential courses/workshops dealing with specialized subjects with regional or global attendance.
- Courses taught by distance learning.
- Including strong capacity-building components in global and regional research programs.
- Creating awareness of the importance of GOOS activities and of the need for capacity building.

In a recent GOOS document, the following conclusions were drawn about marine capacity building. They go well beyond the needs for ocean observing systems and deserve emphasis here.

- Marine capacity building is a long-term process.
- The involvement of national governments is crucial.
- Approaches should be tailored to specific country or regional needs.
- For building indigenous capacity, the active involvement of the community in the developing countries is necessary. Partners in these countries are the most effective and persistent advocates for marine science and technology.
- Capacity-building activities can vary from a training course to the implementation of a complete environmental monitoring system.
- The best instruments are activities in which scientists, technicians, and users work closely together (learning by doing, teaching the teachers) in the execution of projects, programs, and partnership.
- Governments, scientific and international organizations, the private sector, and donors should join forces. In this regard, substantial awareness raising is also needed within science foundations and donor organizations (even in industrialized countries). Most donor organizations are unsure of marine issues.
- All participants must recognize the need to maintain marine science capacity

once it has been brought to a certain level. Awareness of both public and policymakers is essential to raise national and international support.

CONCLUSIONS AND RECOMMENDATIONS

The culture of our scientific enterprise is on the brink of a "sea change." The environmental problems and resource demands of one nation rapidly become the problems of all countries. The scientific community is beginning to recognize the value of partnerships of all kinds in the quest for understanding and solving not only to problems of the marine environment and its sustainable development. These new partnerships involve research collaboration between scientists in all countries, whatever their state of development. Only when opportunities are equal can all members of a partnership benefit equally from their collaboration. The privileged have an inherent obligation to assist those who do not have access to the same resources. In this respect:

- Financial commitments have to be more substantial than hitherto. They will have rewarding feedback by enhancing global and regional programs and bilateral projects.
- This assistance, whether at an individual, community, or national level, involves more truly joint research projects, commitments to teaching, and graduate student research in academic institutions in developing countries.
- This assistance also implies help in overcoming the obstacles to publication in internationally recognized journals, better access to electronic communication technologies, enhancing the research capabilities of institutions in developing countries and, above all, ensuring that the scientific communities of developing countries are nurtured rather than drained of their best talent.
- Nations must develop their own clearly stated priorities and strategies in marine research. Once this first step has been achieved, bilateral, multilateral, regional, and international cooperation becomes much more feasible because it is grounded in well-defined objectives for each of the partners.
- This requires the development of public and political awareness of marine environmental issues by appropriate means (teaching, media) and an active public interest in such issues.
- Regional conditions and local specific problems must be recognized and play an important part in capacity building.
- Skills in both fundamental and applied science need to be developed. These may require different training.
- Regional centers of excellence need support and development. It is important

to ensure continuity of support for such centers, both from national sources and via funding from partner countries.
- Partnerships (networks) between different national and international agencies, universities, and countries must be developed in order to focus effort for greater effectiveness.
- Optimal use of electronically based methods can help raise standards of training and teaching.
- Traditional classroom approaches should be supported and sustained by international web-based tuition networks, etc., to help students on a long-term basis.

ACKNOWLEDGMENTS

The authors benefited greatly from the discussions of the Potsdam workshop and their summary by Erlich Desa. A wealth of information and comments were received from Eduardo Marone, with particular emphasis on the dichotomy of blue-ocean versus coastal research and on the importance of the language barrier. Colin Summerhayes and Geoffrey Holland contributed information on the role of GOOS in global capacity building. Elizabeth Gross put much effort in editing the chapter and making it more homogeneous after input by various colleagues. We thank them all for their support. Karin Lochte made a number of valuable comments and provided a summary of recommendations.

Chapter 12

The Vision to 2020

John G. Field, Colin P. Summerhayes, and Gotthilf Hempel

Making predictions is very difficult, particularly about the future.
—Nils Bohr

A world where all societies respect and care for the oceans, being well-informed of the contribution of the oceans to the unique and sustainable life-support characteristics of Planet Earth.

ADVANCING THE VISION

In contributing to the vision, marine scientists and technologists recognize that major changes will occur in areas outside marine science and technology that may cause surprises in the way we do things in the oceans in future. We must therefore keep in close contact with those who lead social, economic, and political change.

Change is likely to come about in response to the growing appreciation that exploitation should be sustainable. It has to be said that the international response to Agenda 21—the blueprint for action for global sustainable development into the twenty-first century—has been rather limited to date. The Independent World Commission on the Oceans noted that "the world community has so far failed to recognize the seriousness of the deterioration that has taken place." It proposed the formation of a World Ocean Affairs Observatory to monitor the system of ocean governance and to keep a watch on ocean affairs. It also proposed an Independent World Ocean Forum to articulate concerns and to express hopes and aspirations (IWCO 1998). The advent of such new institutions would at least improve information and understanding, which, in turn, should lead to improvements in ocean policies. Recognizing the need for some kind of world forum, the General Assembly of the U.N. decided, in late 1999, to create an informal consultative process that will allow key questions regarding the oceans to be aired at a high level.

Regarding fisheries, views differ widely on what is needed in future. The overall goal of maximum sustainable use of the living resources of the ocean is in conflict with the immediate needs of coastal people and with global market forces. Furthermore, the protection of marine biodiversity has to be reconciled with fisheries. We will have to find new ways of management, using novel scientific approaches. Although there are undoubtedly long-term, large-scale environmental influences on fisheries, we should not blame natural variability in ecosystems for population collapses caused by fisheries. Such collapses can commonly be the result of continued heavy fishing after a series of recruitment failures caused by adverse environmental conditions (Larkin 1996), hence the need for a broad ecological approach to fisheries management.

Some scientists see the future of sustainable fisheries in small-scale fisheries, and others assume that market forces will demand that large-scale fishing fleets continue to be the major players. Most people believe in the further rapid development of aquaculture, but much has to be done to avoid negative effects on the environment and to protect the farms against diseases and other hazards. Ecosystem research is needed to provide blueprints for the further development of fisheries and aquaculture as well as for marine conservation. It is important that regional solutions be developed for aquaculture and fisheries development, taking into account the needs of local people in addition to market forces and environmental protection. In this context, it is important that there be better cooperation among science, administration, and industry as well as more data in the public domain.

In our prediction of technological trends, we follow the extrapolations by Kaku (1998). The computer revolution is not just transforming business, communication, and lifestyles, but also science. Computer power doubles roughly every two years. By 2020, the price of microchips will be so low it will allow us to place intelligent systems wherever they are needed. Future applications of global positioning systems (GPS) will become virtually unlimited, enabling us to locate remote instruments precisely. These various developments will greatly enhance our ability to monitor and simulate the weather and the ocean.

Miniaturization is another important trend that is already affecting marine science and technology. Between now and 2020, the first generation of micromachines (miniature sensors and motors about the size of dust particles) are expected to find widespread applications. The idea of seeding the ocean with many small, cheap, observing devices is coming closer to reality.

The communications revolution is radically changing the way people work. Within a few years the Internet will have connected millions of computer networks. The ability of anyone anywhere to tap into all of recorded science will radically change local marine science capacity, especially in developing countries. By 2020, advances in artificial intelligence and robotics will lead to increased automation of data collection and analysis in the marine world. Robotic devices should, in the future, become as useful to the exploration of "inner space" (i.e., the oceans) as they have in outer space.

Research vessels will continue to be the backbone of marine science. However, interdisciplinary and international teamwork on board already requires a new generation of relatively large vessels with the capacity to accommodate more than twenty scientists at a time. These vessels should be equipped to handle heavy, yet delicate, tools. Smaller vessels will also be needed for work in coastal waters. Many of the improvements in technology mentioned in this book will not only change marine science, they will also increase the ability of nations to monitor, police, and control the use of the seas.

The above vision is of a marine world in which remote sensing, information technology, and communication radically change the ways in which we gather information about the oceans and the problems they pose and feed the analyzed information to society. The following paragraphs spell out particular areas that we believe deserve priority in meeting the needs of society in the first two decades of the twenty-first century.

LOOKING TO THE FUTURE

In looking to the future of marine science to the year 2020, it is clearly impossible to predict completely new scientific paradigms. The best we can do is to look back on developments that have shaped marine science during the most recent decades and extrapolate forward from those. We have selected twelve out of many possible areas for brief comment. The examples provided here differ widely in scope and scale.

1. Remote Sensing

Satellite remote sensing has revolutionized ocean science in many ways, allowing synoptic views of ocean surface temperature, ocean color, and sea level, with estimates of ocean plant pigments, currents, heat storage, and many other derived products. One of the major problems is coping with the vast quantities of spatial data that are generated. We can expect coordinated developments in areas that presently pose a problem for remote sensing, such as fields of salinity, primary production, and improved resolution of coastal margins.

There have been remarkable developments in technology for sensing physical, chemical, and even biological properties of seawater from buoys, drifters, floats, autonomous underwater vehicles, and remotely operated vehicles. All of these dramatically increase the potential for sampling the ocean interior, providing information to complement observations from space satellites. Further developments are imminent, with many new applications of these technologies. Similarly, the use of vertical profiling floats and buoys with new sensors for a variety of chemical species such as CO_2, molecular biological probes, and others will allow routine time-series of observations on a much larger scale and in remote regions of the ocean. The development of bio-optical and bioacoustic methods, together with pattern recognition algorithms and neural network learning, are likely to revolutionize biological oceanography and fisheries biology (see chapter 9).

2. The Information Revolution and Ocean Science

Use of the Internet has already dramatically changed the way we communicate data and ideas. And these developments are continuing to grow exponentially. Information is widely available to scientists around the world, such as the compact disks of data produced by the World Ocean Circulation Experiment and the Joint Global Ocean Flux Study (JGOFS). Such widespread data sharing was unthinkable in the seventies. Another example is the availability on the Web of a global database of information on fish (www.fishbase.org), ranging from taxonomy to population dynamics. There is a need for more widespread use of animated graphics to present scientific results (see box 9.3). This will help improve communication between scientists and the public, and especially between scientists and decision makers.

3. The Globalization of Modeling Capacity

There is a trend toward making local, regional, and global models more widely available, so that researchers can spend more time developing their ideas and analyzing data using existing code and less time developing new models. Intercomparison and calibration of models and forecasts developed in different countries around the world are essential as a test of their robustness. Equally important are tests of the ability of models to assimilate and integrate data from different sources, like the in situ data from ocean instruments and the remotely sensed ocean data from satellites. The Global Ocean Data Assimilation Experiment is a large-scale experiment along these lines, and it is a critical next step in developing and testing the credibility of a Global Ocean Observing System to underpin ocean, weather, and climate forecasting.

This trend toward globalization of models and the assimilation of data into models is likely to continue, with increasing benefits to forecasting and understanding the oceans in relation to climate change. Just as chaos theory predicts that there are limits to weather forecasting, so, too, there are limits to predicting complex interconnections in the ocean. The limits to predictability in the ocean are likely to become clearer, and techniques need to be developed to model complex physical-biological coupling in a simplified way. For example, we need to develop ways to predict fish production from remotely sensing the chlorophyll in the ocean surface layers using empirical relationships that do not attempt to model all the complex links in the food web from physics to fish.

4. Discovering Functional Biodiversity

The rapid development of molecular biology and informatics over the past decade has revealed our ignorance of prokaryotic organisms in the oceans. Perhaps only 1 percent of the species are recognized. However, the same disciplines have also provided tools for rapid phylogenetic identification of populations of individual cells. In the future, molecular probes will be developed to detect the expression of functional genes in marine microorganisms and thereby understand their ecological role. Combination with silicon technology (DNA chips) will make it possible to achieve sufficient capacity to screen marine systems rapidly for their response to environmental change, for example, organic pollution, toxins, and global warming. With increasing sequencing capacity, as has been provided through the human genome project, it will become routine to sequence whole genomes of microorganisms by 2020. This opens the possibility of discovering novel genes and catalytic functions of key organisms in the ocean, with important environmental or biotechnological implications.

5. Global Climate Change

After a decade of the International Geosphere–Biosphere Global Change Research Program (IGBP), several "big questions" still need answers. These include, How can natural variability be separated from anthropogenic change? and How will global warming affect ocean ecology? There will undoubtedly be changes in community structure; the functioning and the rate of the biological CO_2 pump will change; there may be switches and changes in the thermohaline circulation, causing rapid climate changes over a short period, when presently unknown thresholds are reached.

In order to make these advances, we will need to improve our sampling strategies, scaling up from conventional biological process studies conducted from one or two ships over weeks to months, to study and predict on the decade to century scale. For this, we will need time-series observations of key parameters, such as physical conditions, nutrients, dissolved organic matter, phytoplankton biomass, and community structure. Key sites, such as those off Bermuda and Hawaii and in the Southern Ocean, North Pacific, and North Atlantic, have already provided invaluable data sets and knowledge about shifts and cyclic behavior in ecosystems. It is of great importance to maintain such time-series observations at critical sites (see chapter 9).

6. Waste Disposal

The thorny question of how to dispose of the increasing quantities of waste generated by human activity and industry will not go away. Clearly, we cannot allow the oceans to be used as a dumping ground. But there may be ways of disposing of waste in a manner that is harmless and safe for both humans and marine life. Before any more waste is disposed of in the oceans, there must be adequate research to satisfy the precautionary principle. There is thus a need to study the potential and hazards of the deep ocean for disposal. Such studies should include the submarine earth crust and its movements as well as fine-scale deep-ocean water currents. Both pose potential hazards for waste disposal at particular sites.

7. Deep Sea Floor Biosphere

Plate tectonics drive the geological processes of mid-ocean ridge formation and ocean margin subduction. These are associated with hot vents and cold seeps, respectively, both of which harbor a rich invertebrate fauna based on symbiotic, chemosynthetic bacteria and the chemical energy of hydrogen sulfide or methane. This has been one of the major discoveries of the century in marine science, changing our perspective on the evolution and limits of life on Earth. It is likely that early life evolved under the unique chemical conditions associated with hydrothermal activity.

Microbial processes may also affect major geological processes, for example, through methanogenesis, which restructures the ocean floor by mud volcanoes, carbonate mounds, and gas hydrate formation. Deep-sea vents are a window to the deep biosphere, which was discovered through the Ocean Drilling Program in the last quarter of the last century. Communities of bacteria occur under the sea floor to a depth of at least a kilometer, in deposits 10 million years old. They may make up 10 percent or more of the total living biomass on Earth. Their importance for short- and long-term ocean processes has yet to be recognized.

8. The Land and Sea Interface

Conflicting demands on the use of the world's coastal zones will call for a much higher profile of political, social, and scientific attention. Their importance for most fisheries and much of biodiversity, as well as for recreation, conflicts with their proximity to the largest cities and exploitation for development, transport, waste disposal, and pollution. Clearly, coastal zones have to be managed as sustainable ecosystems, yet we have little management science for such ecosystems. Both theory and applications need to be developed for conserving the ecosystem life-support systems.

Marine protected areas (MPAs) have been proposed as one of several tools for conserving particularly valuable parts of both inshore and offshore marine ecosystems. But what size should an MPA be to be effective? Clearly, each situation will be slightly different, and different factors will influence each ecosystem under different circumstances. But we need to have research on guidelines or starting points for estimating the size and use of MPAs. A number of types of coastal zone are in special need of attention because of particular threats to their functioning as natural systems. These include coral reefs, mangroves, and many estuaries. Science-based integrated coastal management is needed to allow development that preserves the resources for future generations.

9. Growth of Interdisciplinarity

There was such a marked advance in collaboration between previously separate disciplines in the 1980s and 1990s that new fields have developed in marine science and technology. One such new area is paleoceanography, which allows us to judge from fossil records in sediments and ice cores how climate varied over the decade-to-century time scale. Similarly, by bringing together marine chemists, geochemists, and biologists, the JGOFS gave birth to marine biogeochemistry (see box 2.1). In the technology area, new disciplines of bio-optics and bioacoustics have created new windows into the ocean, with potential for studying fish and plankton on the appropriate scales for understanding global change (see chapter 9). New developments are likely at other interdisciplinary interfaces. Thus, a new generation of young oceanographers should be able to think and work across the boundaries of disciplines, while making use of advances in modern physics, molecular biology, information science, etc.

10. Involvement of Society

There is a clear need to involve society in managing the ocean's limited resources. Local management decisions are unlikely to be successful if they do not involve local communities with a long-term stake in the resource. Such decisions cannot be left just to scientists or technocrats. When it comes to issues of global scale, where local communities may be less effective, what is needed is increased awareness by the public and the media. To enable people to deal with both local and global issues there is a significant need for widespread education on scientific issues and the dissemination of scientific knowledge to the public. There are already several examples of successful trends in this direction, one being co-management of the Australian Great Barrier Reef coastal zone (see box 3.5). Another is the development of fisheries operational management procedures, negotiated between government authorities, scientists, and fishers, as in the North Sea Herring fishery (see box 5.3) and in some fisheries in South Africa, New Zealand, and Australia, among others. These set agreed-upon rules on quotas under different likely scenarios for three to five years at a time, after which they are revised.

11. Fisheries

It has long been known that there are few really big fish in a heavily fished fishery. What occurs in most fisheries is that the number of year-classes is reduced by heavy fishing, eliminating the older fish that are also the most fecund. What has only recently been realized is that in most cool and temperate waters, strong fish year-classes often occur at decadal-scale intervals, and long-lived fish have evolved longevity in order to survive long enough to ride over the many years of poor conditions when most spawnings have little success. The consequences of this fact are now much clearer and are not yet modeled by most traditional fisheries models. New management strategies need to be developed urgently to allow sufficient fish to survive to become large, old, and fecund. This may even require revising traditional dogma that young fish should be left in the sea to grow longer, protected by minimum size restrictions. It may turn out to be better to catch some of the young fish (which usually have a heavy natural mortality anyway), provided that fishing pressure on the large fish is reduced. This would allow their escape and return to the natural age-distribution that has evolved to enable the stock to reproduce in good years on decadal time scales. Traditional maximum sustainable yield levels are likely to be revised to reduce total catches, particularly those of older fish. For further reading about the ecosystem approach to fisheries, the reader is directed to ICES 2000 and to a web-based report on responsible fisheries in the marine environment (www.refisheries2001.org).

12. Capacity Building

There is a widespread need to build marine science and technology capacity in all parts of the world, especially in developing countries. This may involve two-way flows of funds and ideas between countries. People in all parts of the world need to be able to use constantly evolving technology to solve local and general problems and to design appropriate surveys and experiments to answer the questions posed. And they need to be able to analyze and interpret the data collected and to model (in the broadest sense) the problem at hand. These are general needs, not confined to developing countries. Optimal use should be made of the new information technologies in order to spread marine knowledge globally. With appropriate partnerships between developed and developing countries, and increased educational, technological, institutional, and/or financial support from developed countries, great strides can and should be made, for the future of marine science and of humankind in general.

CONCLUSION

The first decades of the twenty-first century pose many challenging problems in studying, harvesting, and managing the oceans. We offer a vision in which remote sensing, information technology, and improved communications radically change the ways in which information is gathered about the oceans and the problems they pose and in the way they are presented to society. The twelve areas discussed above are those in which exciting developments are likely that will both advance the science and aid sustainable development in the first two decades of the twenty-first century. There will be significant increases in the use and application of remote-sensing technologies to measure the properties of the ocean and its contents both in outer space and in the body of the ocean, which will be useful to many user communities, especially in biological oceanography and fisheries biology. There will be significant increases in the use of the Internet to communicate and exchange data and ideas, not least in the use of visualization to present scientific results to decision makers and the public. Numerical modeling will become more widespread, cheaper, and easier to carry out, facilitating the assimilation of real-time data into maps of the present and future state of the oceans and their contents and of climate. Major improvements are expected, especially in biological and ecological modeling, which will have a spinoff for fisheries. Significant enhancements are likely in molecular biology and genetics, vastly improving our understanding of marine microorganisms and their ecological role. Expanded observations of physical, chemical, and biological ocean properties on long time scales at critical sites will greatly improve understanding about, and the ability to forecast, shifts and cyclic behavior in ecosystems. Continued studies will be needed of the potential for, and hazards of, using the ocean for the disposal of wastes of various kinds, especially of carbon dioxide. More research is needed on the newly discovered deep earth biosphere of microbes buried deep in ocean sediments and active at hydrothermal vents, to improve our perspective on the evolution and limits of life on Earth. Vast improvements are needed in our understanding of the scientific processes that govern the development of the coastal zone and its ecosystems, so as to improve the management of this much used but inherently fragile environment. A considerable increase is needed in studies that cross disciplinary boundaries to improve understanding of complex ocean processes. Scientists must become much more engaged with the wider community to provide a sound scientific basis for decision making. New and ecosystem-based management strategies are needed to enable fish to be harvested in an increasingly sustainable way. Finally, greatly improved mechanisms are needed to build the scientific and technical capacity of developing countries to understand their marine environments as the basis of sustainable exploitation. To cope with all these developments is going to require huge efforts in communication and capacity building within the scientific community, and especially between scientists and society. We live in exciting times!

ACKNOWLEDGMENTS

Discussions with the many attendees at the Potsdam workshop in October 1999 helped to refine these ideas, which were further elaborated with the particular help of J. Baker, B. Barker Joergensen, and K. Lochte.

REFERENCES

Abbott, M. R., K. H. Brink, C. R. Booth, D. Blasco, L. A. Codispotti, P. P. Niiler, and S. R. Ramp. 1990. Observations of phytoplankton and nutrients from a Lagrangian drifter off northern California. *Journal of Geophysical Research* 95:9393–9409.
Abbott, M. R., J. G. Richman, R. M. Letelier, and J. S. Bartlett. 2000. The spring bloom in the Antarctic Polar Frontal Zone as observed from a mesoscale array of bio-optical sensors. *Deep-Sea Research II* 47(15–16):3285–3314.
Adger, W. N. 1996. Working paper GEC 96-05, Center for Social and Economic Research on the Global Environment, University of East Anglia, Norwich, U.K.
Aebischer, N. J., J. C. Coulson, and S. M. Colebrook. 1990. Parallel long-term trends across four trophic levels and weather. *Nature* 347:735–55.
Afshan, A., and J. A. Curry. 1997. Determination of surface turbulent fluxes over leads in Arctic sea ice. *Journal of Geophysical Research* 102:3331–43.
Agrawal, Y. C., and H. C. Pottsmith. 1994. Laser diffraction particle sizing in STRESS. *Continental Shelf Research* 14:1101–1121.
Alverson, D. L., M. Freeberg, J. Pope, and S. Murawski. 1994. A global assessment of fisheries by-catch and discards: A summary overview. *FAO Fisheries Technical Paper*, no. 339, Rome.
Anderson, D. T., E. S. Sarachik, P. J. Webster, and L. M. Rothstein. 1998. The TOGA decade: Reviewing the progress of El Niño research and prediction. *Journal of Geophysical Research* 103:14, 167.
Andreassen, C. 1999. U.S. defence nautical charting. *Sea Technology*, March, 10–15.
Angel, M. V. 1996. Waste disposal in the deep ocean. In C. P. Summerhayes and S. A. Thorpe (eds.), *Oceanography: An Illustrated Guide*, 338–45. London: Manson.
Angel, M. V., and A. L. Rice. 1996. The ecology of the deep ocean and its relevance to global waste management. *Journal of Applied Ecology* 33:915–26.
Anon. 1999a. Thrust into the limelight. *Offshore Engineer*, March, 28–29.
Anon. 1999b. Race is on for autonomous survey vehicles. *Offshore Engineer*, May, 26–30.

Apt, J., M. Helfert, and J. Wilkinson. 1996. *Orbit*. Washington, D.C.: National Geographic Society.
Argo Science Team. 2001. *Argo: The global array of profiling floats*. In C. J. Koblinksy and N. R. Smith (eds.), *Observing the Oceans in the 21st Century*, pp. 248–258. Melbourne, Australia: GODAE Project Office, Bureau of Meteorology.
Atkinson, A., M. J. Whitehouse, J. Priddle, G. C. Cripps, P. Ward, and M. A. Brandon. 2001. South Georgia, Antarctica: A productive, cold water, pelagic ecosystem. *Marine Ecology Progress Series* 216:279–308.
Bakun, A. 1996. *Patterns of the Ocean*. California Sea Grant College System.
Baranov, F. I. 1918. On the question of the biological basis of fisheries. *Nauchnyi Issledovatelskii Ikhtiologicheskii Insitut Isvestia* 1:81–128.
Barnett T. P., K. Hasselmann, M. Chelliah, T. Delworth, G. Hegerl, P. Jones, E. Rasmusson, E. Roeckner, C. Ropelewski, B. Santer, and S. Tett. 1999. Detection and attribution of recent climate change: A status report. *Bulletin of the American Meteorological Society* 80:2631–59.
Bates, N. R., A. F. Michaels, and A. H. Knap. 1996. Seasonal and interannual variability of oceanic carbon dioxide species at the U.S. JGOFS Bermuda Atlantic Time-series Study Site. *Deep-Sea Research II* 43:347–84.
Bell, N., and J. Smith. 1999. Coral growing on North Sea oil rigs. *Nature* 402:601.
Bellingham, J. G., K. Streitlien, J. Overland, S. Rajan, P. Stein, J. Stannard, W. Kirkwood, and D. Yoerger. 2000. An arctic basin observational capability using AUVs. *Oceanography* 13(2):64–70.
Benfield, M. C., C. S. Davis, P. H. Wiebe, S. M. Gallager, R. G. Lough, and N. J. Copley. 1996. Video Plankton Recorder estimates of copepod, pteropod, and larvacean distributions from a stratified region of Georges Bank with comparative measurements from a MOCNESS sampler. *Deep-Sea Research II* 43(7–8):1925–45.
Benfield, M. C., P. H. Wiebe, T. K. Stanton, C. S. Davis, S. M. Gallager, and C. H. Greene. 1998. Estimating the spatial distribution of zooplankton biomass by combining Video Plankton Recorder and single-frequency acoustic data. *Deep-Sea Research II* 45(7):1175–99.
Bergman, M. J. N., and H. J. Lindeboom. 1999. Natural variability and the effects of fisheries in the North Sea: Towards an integrated fisheries and ecosystem management. In J. S. Gray et al. (eds.), *Biogeochemical Cycling and Sediment Ecology*, 173–84. Amsterdam: Kluwer Academic Publishers.
Beverton, R. J. H., and S. J. Holt. 1957. *On the Dynamics of Exploited Fish Populations: Fisheries Investigations*. London: Ministry of Agriculture, Fisheries and Food, ser. II 19.
Bishop, J. K. B. 1999. Transmissometer measurement of POC. *Deep-Sea Research I* 46:353–69.
Bluth, G. J. S., S. D. Doiron, C. C. Schnetzler, A. J. Krueger, and L. S. Walter. 1992. Global tracking of the SO_2 clouds from the June 1991 Mount Pinatubo eruptions. *Geophysical Research Laboratories* 19:151–54.

Bogucki, D., T. Dickey, and L. Redekopp. 1997. Sediment resuspension and mixing through resonantly-generated internal solitary waves. *Journal of Physical Oceanography* 27:1181–1196.

Bohnsack, J. A. 1993. Marine reserves: They enhance fisheries, reduce conflicts, and protect resources. *Oceanus* 36:63–71.

Bond, G. C., H. Heinrich, W. Broecker, L. Labeyrie, J. McManus, J. Andrews, S. Huon, R. Jantschik, S. Clasen, C. Simet, K. Tedesco, M. Klas, G. Bonani, and S. Ivy. 1992. Evidence for massive discharges of icebergs into the North Atlantic Ocean during the last glacial period. *Nature* 360:245–9.

Bond, G. C., and R. Lotti. 1995. Iceberg discharges into the North Atlantic on millenial time scales during the last glaciation. *Science* 267:1006–10.

Brewer, P. G. 2000. Contemplating action: Storing carbon dioxide in the ocean. *Oceanography* 13(3):84–92.

Brewer, P. G., K. W. Bruland, R. W. Eppley, and J. J. McCarthy. 1986. The Global Ocean Flux Study (GOFS): Status of the U.S. GOFS program. *Eos* 44:827–32, 835–37.

Brewer, P. G., G. Friederich, E. T. Peltzer, and F. M. Orr. 1999. Direct experiments on the ocean disposal of fossil fuel CO_2. *Science* 284:943–45.

Brink, K. H., and A. R. Robinson, eds. 1998. *The Global Coastal Ocean: Processes and Methods.* The Sea, vol. 10. Chichester: Wiley.

Broecker, W. S. 1991. The great ocean conveyor. *Oceanography* 4(2):79–89.

Broecker, W. S. 1997. Thermohaline circulation, the Achilles heel of our climate system: Will man-made CO_2 upset the current balance? *Science* 278:1582–88.

Brown, M. G. 1994. An assessment of the feasibility of an integrated OTEC system for Grand Cayman. *Oceanology International* London: Spearhead.

Bückmann, A. 1938. Über den Höchstertrag der Fischerei und die Gesetze organischen Wachstums. *Ber. dt. wiss. Komm. f. Meeresf.* 9:16–48.

Byrne, R. H., W. S. Yao, E. Kaltenbacher, and R. D. Waterbury. 2000. Construction of a compact spectrofluorometer/spectrophotometer system using a flexible liquid core waveguide. *Talanta* 50(6):1307–1312.

Camphuysen, C. J., and S. Garthe. 1999. Seabirds and commercial fisheries: Population trends of piscivorous seabirds explained? In M. J. Kaiser and S. J. de Groot (eds.), *Effects of Fishing on Non-Target Species and Habitats: Biological, Conservation, and Socio-Economic Issues,* 163–84. Oxford: Blackwell Science.

Campos, E., A. Buslacchi, S. Garzoli, J. Lutjeharms, R. Matano, P. Nobre, D. Olson, A. Piola, C. Tanajura, and I. Wainer. 2001. Important Aspects of the South Atlantic to the Understanding of Global Climate. In C. J. Koblinsky and N. R. Smith (eds.), *Observing the Oceans in the 21st Century: A Strategy for Global Ocean Observations.* GODAE Project Office, Bureau of Meteorology, Melbourne, pp 453–472.

Cane, M. A. 1992. Tropical Pacific ENSO models: ENSO as a mode of the coupled system. In K. Trenberth (ed.),*Climate System Modeling,* 583–614. Cambridge: Cambridge University Press.

Caplan-Auerbach, J., C. G. Fox, and F. Duennebier. 2001. Hydroacoustic detection of submarine landslides on Kilauea volcano. *Geophysical Research Letters* 28:1811–13.

Carder, K. L., and R. G. Steward. 1985. A remote-sensing reflectance model of a red-tide dinoflagellate off west Florida. *Limnology and Oceanography* 30:286–298.

Carpenter, S. R., S. W. Chisholm, C. J. Krebs, D. W. Schindler, and R. F. Wright. 1995. Ecosystem experiments. *Science* 269:324–27.

Cavalieri, D. J., P. Gloersen, C. L. Parkinson, J. C. Comiso, and H. J. Zwally. 1997. Observed hemispheric asymmetry in global sea ice changes. *Science* 278:1104–1106.

Chang, G. C., and T. D. Dickey. 1999. Partitioning in situ total spectral absorption by use of moored spectral absorption and attenuation meters. *Applied Optics* 38:3876–87.

Chang, G. C., and T. D. Dickey. 2001. Optical and physical variability on time scales from minutes to the seasonal cycle on the New England continental shelf: July 1996–June 1997. *Journal of Geophysical Research* 106:9435–53.

Chang, G. C., T. D. Dickey, and A. J. Williams III. 2001. Sediment resuspension on the Middle Atlantic Bight continental shelf during Hurricanes Edouard and Hortense: September 1996. *Journal of Geophysical Research* 106:9517–31.

Chang, H., F. W. Saris, F. Van der Veen, and M. Van der Wiel. 1997. The need for skilled technicians and well equipped technical workshops in leading-edge experimental research. In J. Irvine (ed.), *Equipping Science for the 21st Century*, 370–82. Proceedings of the International Workshop on Equipping Science for the 21st Century, Amsterdam, 7–9 April 1992. Cheltenham, U.K.: Edward Elgar.

Charlson, R. J., J. E. Lovelock, M. O. Andreae, and S. G. Warren. 1987. Oceanic phytoplankton, atmospheric sulfur, cloud albedo, and climate. *Nature* 326:655–51.

Charnock, H. 1996. The atmosphere and the ocean. In C. P. Summerhayes and S. A. Thorpe (eds.), *Oceanography: An Illustrated Guide*, 24–40. London: Manson.

Chave, A. D., F. K. Dunnebier, and R. Butler. 1999. Putting H_2O in the ocean. *Oceanus* 42(1):6–9.

Chavez, F. P., J. T. Pennington, R. Herlien, H. Jannasch, G. Thurmond, and G. E. Friederich. 1997. Moorings and drifters for real-time interdisciplinary oceanography. *Journal of Atmospheric and Oceanic Technology* 14:1199–1211.

Chavez, F. P., P. G. Strutton, C. E. Friederich, R. A. Feely, G. C. Feldman, D. C. Foley, and M. J. McPhaden. 1999. Biological and chemical response of the equatorial Pacific Ocean to the 1997–98 El Niño. *Science* 286:2126–31.

Chavez, F. P., D. Wright, R. Herlien, M. Kelley, F. Shane, and P. G. Strutton. 2000. A device for protecting moored spectroradiometers from fouling. *Journal of Atmospheric and Oceanic Technology* 17: 215–19.

Clark, J. R. 1996. *Coastal Zone Management Handbook*. CRC Marine Science Series. Boca Raton: CRC Press.

Cloern, J. E. 1996. Phytoplankton bloom dynamics in coastal ecosystems: A review with some general lessons from sustained investigations of San Francisco Bay. California. *Reviews of Geophysics* 34:127–168.

Coale, K. H., C. S. Chin, G. J. Massoth, K. S. Johnson, and E. T. Baker. 1991. In situ chemical mapping of dissolved iron and manganese in hydrothermal plumes. *Nature* 352:325–28.

Colwell, R. R. 1996. Global climate and infectious disease: The cholera paradigm, *Science* 274:2025–31.

Colwell, R. R. 1999. Marine biotechnology. *Sea Technology,* January, 50–51.

Cook, P. J. 1996. Societal trends and their impact on the coastal zone and adjacent seas. *British Geological Survey Technical Reports* WQ/96/3:12 pp.

Coultier, M. C. 1984. Seabird conservation in the Galápagos Islands, Ecuador. In J. P. Croxall, P. G. H. Evans, and R. W. Schreiber (eds.), *Status and Conservation of the World's Seabirds,* 237–44. Technical Publication no. 2. Cambridge: IGBP.

Cruickshank, M. J. 1996. Manganese nodules = Kitchen fixtures. *Sea Technology,* October, 76.

Cruickshank, M. J. 1999. International Seabed Authority workshop convened in Kingston, Jamaica. *Sea Technology,* August, 101.

Cruickshank, M. J., and Masutani, S. M. 1999. Methane hydrate research and development. *Sea Technology,* August, 69–74.

Csirke, J. 1987. *Los recursos pesqueros patagónicos y las pesquerías de altura en el Atlántico Sud-Occidental. FAO Documentos Técnicos de Pesca,* 286. Rome: FAO.

Curtin, T. B., J. B. Bellingham, J. Catipovic, and D. Webb. 1993. Autonomous oceanographic sampling networks. *Oceanography* 6(3):86–94.

Cushing, D. H. 1988. *The Provident Sea.* Cambridge: Cambridge University Press.

Daan, N. 1980. A review of replacement of depleted stocks by other species and the mechanics underlying such replacement. *Rapports et Procés-Verbaux des Réunions.* 177:405–21.

Davis, C. S., S. M. Gallager, M. Marra, and W. K. Stewart. 1996. Rapid visualization of plankton abundance and taxonomic composition using the Video Plankton Recorder. *Deep-Sea Research II* 43(7–8):1947–70.

Davis, C. O., M. Kappus, B.-C. Gao, W. P. Bissett and W. Snyder. 1998. The Naval Earth Map Observer (NEMO) science and naval products. *Proceedings of the SPIE* 3437:11–19. The International Society for Optical Engineering (SPIE), Bellingham, WA (USA).

Dawes, T. C., D. S. Stakes, P. McGill, S. Etchemendy, and J. P. Barry. 1998. Sub-sea instrument deployment: Methodologies and techniques using a work-class remotely operated vehicle (ROV). In *Proceedings of the Oceans 1998 Conference, Nice, France.* New York: IEEE, Oceanic Engineering Society.

DeGrandpré, M. D., T. R. Hammar, D. W. R. Wallace, and C. D. Wirick. 1997. Simultaneous mooring-based measurements of seawater CO_2 and O_2 off Cape Hatteras, North Carolina. *Limnology and Oceanography* 42:21–28.

DeGrandpré, M. D., M. M. Baehr, and T. R. Hammer. 2000. Development of an optical chemical sensor for oceanographic applications: The submersible autonomous moored instrument for seawater CO_2. In M. S. Varney (ed.), *Chemical Sensors in Oceanography*, 123–142. Reading, U.K.: Gordon and Breach Science Publishers.

Delaney, J., and A. Chave. 1999. NEPTUNE. *Oceanus* 42(1):10–11.

Delaney, J. R., G. R. Heath, B. Howe, A. D. Chave, and H. Kirkham. 2000. NEPTUNE: Real-time ocean and earth sciences at the scale of a tectonic plate. *Oceanography* 13(2):71–79.

Desiderio, R. A., C. Moore, C. Lantz, and T. J. Cowles. 1997. Multiple excitation fluorometer for in situ oceanographic applications. *Applied Optics* 36:1289–96.

de Solla Price, D. J. 1963. *Little Science, Big Science*. New York: Columbia University Press.

Detrick, R. S., J. A. Collins, and D. E. Frye. 1999. Seafloor to surface to satellite to shore. *Oceanus* 42(1):12–13.

Dickey, T. 1991. Concurrent high resolution physical and bio-optical measurements in the upper ocean and their applications. *Reviews of Geophysics* 29:383–413.

Dickey, T. (ed.). 1993. Sampling and observing systems. In *GLOBEC Report Series 3*. Solomons, Md.: GLOBEC International, Chesapeake Biological Laboratory.

Dickey, T. 1997a. Emerging interdisciplinary technologies in biological, chemical, optical, and physical sampling of the oceans. *NOAA Technical Report NESDIS* 87:115–24.

Dickey, T. 1997b. *Uses of Offshore Platforms: Role of Offshore Platforms in Environmental and Coastal Research*. Norway: Nansen Environmental & Remote Sensing Center and the University of Bergen.

Dickey, T. 2001. Sensors: Inherent and apparent optical properties. In J. H. Steele, S. A. Thorpe, and K. K. Turekian (eds.), *Encyclopedia of Ocean Science*, 1313–23. New York: Academic Press.

Dickey, T. 2002. Future ocean observations for interdisciplinary data assimilation models. *Journal of Marine Sciences*. In review.

Dickey, T., and P. Falkowski. 2002. Solar Energy and Its Biological–Physical Interactions in the Sea. In A. R. Robinson, J. J. McCarthy, and B. J. Rothchild (eds.), *Biological–Physical Interaction in the Sea*. The Sea, vol. 12, chapter 10, pp. 401–440. Chichester: Wiley.

Dickey, T. D., and A. J. Williams III. 2001. Interdisciplinary ocean process studies on the New England shelf. *Journal of Geophysical Research* 106:9427–34.

Dickey, T. D., R. H. Douglass, D. Manov, and D. Bogucki. 1993. An experiment in duplex communication with a multi-variable moored system in coastal waters. *Journal of Atmospheric and Oceanic Technology* 10:637–44.

Dickey, T., J. Marra, M. Stramska, C, Langdon, T. Granata, R. Weller, A. Plueddemann, and J. Yoder. 1994. Bio-optical and physical variability in the sub-arctic

North Atlantic Ocean during the spring of 1989. *Journal of Geophysical Research* 99: 22541–56.

Dickey, T., D. Frye, H. Jannasch, E. Boyle, D. Manov, D. Sigurdson, J. McNeil, M. Stramska, A. Michaels, N. Nelson, D. Siegel, G. Chang, J. Wu, and A. Knap. 1998a. Initial results from the Bermuda Testbed Mooring Program. *Deep-Sea Research I* 45:771–94.

Dickey, T., J. Marra, D. E. Sigurdson, R. A. Weller, C. S. Kinkade, E. Zedler, J. D. Wiggert, and C. Langdon. 1998b. Seasonal variability of bio-optical and physical properties in the Arabian Sea: October 1994–October 1995. *Deep-Sea Research II* 45:2001–25.

Dickey, T., A. Plueddemann, and R. Weller. 1998c. Current and water property measurements in the coastal ocean. In K. H. Brink and A. R. Robinson (eds.), *The Sea, Volume 10*, 367–98. Chichester: Wiley.

Dickey, T. D., G. C. Chang, Y. C. Agrawal, A. J. Williams III, and P. S Hill. 1998d. Sediment resuspension in the wakes of Hurricanes Edouard and Hortense. *Geophysical Research Laboratories* 25:3533–36.

Dickey, T., D. Frye, J. McNeil, D. Manov, N. Nelson, D. Sigurdson, H. Jannasch, D. Siegel, T. Michaels, and R. Johnson. 1998e. Upper-ocean temperature response to Hurricane Felix as measured by the Bermuda Testbed Mooring. *Monthly Weather Review* 126:1195–1201.

Dickey, T., S. Zedler, D. Frye, H. Jannasch, D. Manov, D. Sigurdson, J. D. McNeil, L. Dobeck, X. Yu, T. Gilboy, C. Bravo, S. C. Doney, D. A. Siegel, and N. Nelson. 2001a. Physical and biogeochemical variability from hours to years at the Bermuda Testbed Mooring site: June 1994–March 1998. *Deep-Sea Research II* 48:2105–31.

Dickey, T., N. Bates, R. Byrne, F. Chavez, R. Feely, C. Moore, and R. Waninkhof. 2001b. A review of the NOPP Ocean-System for Chemical, Optical, and Physical Experiments (O-SCOPE) project. In *Proceedings of the Fifth Symposium on Integrated Observing Systems*. 14–19 January, Albuquerque, New Mexico. Boston: American Meteorology Society.

DOT. 1998. *Transportation Statistics Annual Report 1998*. BTS98-S-01. Washington, D.C.: Bureau of Transportation Statistics.

DTI. 1997. *Foresight Report 16: The Report of the Marine Foresight Panel*. London: Office of Science and Technology, Department of Trade and Industry.

DTI. 1999a. *The Greenwich Project: A Marine Information Strategy for the United Kingdom. Foresight Task Force*. London: Office of Science and Technology, Department of Trade and Industry.

DTI. 1999b. *Energies from the Sea—Towards 2000*. London: HMSO Stationery Office.

DuRand, M. D., and R. J. Olson. 1996. Contributions of phytoplankton light scattering and cell concentration changes to diel variations in beam attenuation in the equatorial Pacific from flow cytometric measurements of pico-, ultra-, and nanoplankton. *Deep-Sea Research II* 43:891–906.

Dushaw, B., G. Bold, C.-S. Chiu, J. Colosi, B. Cornuelle, Y. Desaubies, M. Dzieciuch, A. Forbes, F. Gaillard, A. Garrilov, J. Gould, B. Howe, M. Lawrence, J. Lynch, D. Menmenlis, J. Mercer, P. M. Mikhalevsky, W. Munk, I. Nakano, F. Schott, U. Send, R. Spindel, T. Terre, P. Worcester, and C. Wunsch. 2001. Observing the oceans in the 2000s: A strategy for the role of acoustic tomography in ocean climate observation. In C. J. Koblinksy and N. R. Smith (eds.), *Observing the Oceans in the 21st Century*, pp. 391–418. Melbourne, Australia: GUDAE Project Office, Bureau of Meteorology.

Earle, S. A. 1995. *Sea Change: A Message of the Oceans.* New York: Ballantine Books.

Earle, S. A., and W. Henry. 1999. *Wild Ocean: America's Park under the Sea.* Washington, D.C.: National Geographic Society.

Ebert, E. E., and J. A. Curry. 1993. An intermediate one-dimensional thermodynamic sea ice model for investigating ice-atmosphere interactions. *Journal of Geophysical Research* 98:10085–109.

Falkowski, P. G., R. T. Barber, and V. Smetacek. 1998. Biogeochemical controls and feedbacks on Ocean Primary Production. *Science* 281:200–205.

Fichefet, T., and M. A. Morales Maqueda. 1997. Sensitivity of a global sea ice model to the treatment of ice thermodynamics and dynamics. *Journal of Geophysical Research* 102:12609–46.

Field, C. B., M. J. Behrenfeld, J. T. Randerson, and P. Falkowski. 1998. Primary production of the geosphere: Integrating terrestrial and oceanic components. *Science* 281:237–40.

Fine, R. A., L. Merlivat, W. Roether, W. M. Smethie Jr., and R. Wanninkhof. 2001. Observing tracers and the carbon cycle. In C. J. Koblinsky, and N. R. Smith (eds.), *Observing the Oceans in the 21st Century: A Strategy for Global Ocean Observations.* GODAE Project Office, Bureau of Meteorology, Melbourne, pp 361–375.

Finlayson, A. C. 1994. *Fishing for Truth: A Sociological Analysis of Northern Cod Assessments from 1977 to 1990.* St. John's, Newfoundland: ISER, Memorial University.

Fischer, J., and N. C. Flemming. 1999. *Operational Oceanography: Data Requirements Survey.* EuroGOOS Pub. no. 12 (EG99.04). Southampton: Southampton Oceanography Centre.

Flather, R. A. 2000. Existing operational oceanography. *Coastal Engineering* 41:13–40.

Flato, G. M. 1995. Spatial and temporal variability of Arctic ice thickness. Annals of Glaciology 21:323–29.

Flemming, N. C. 1994. Oceanography, part III, analytical report. In *OECD Megascience Forum Expert Meeting on Oceanography*, pp. 71–164. Paris: Organization for Economic Cooperation and Development.

Flemming, N. C. 2001. Dividends from investing in ocean observations: A European perspective. In C. J. Koblinsky, and N. R. Smith (eds.), *Observing the Oceans in the 21st Century: A Strategy for Global Ocean Observations*. GODAE Project Office, Bureau of Meteorology, Melbourne, pp. 66–86.

Foley, D., T. Dickey, M. McPhaden, R. Bidigare, M. Lewis, R. Barber, C. Garside, and D. Manov. 1997. Time-series of physical, bio-optical, and geochemical properties in the central equatorial Pacific at 0°, 140° W February 1992–March 1993. *Deep-Sea Research II* 44:1801–26.

Fosså, J. H., and P. B. Mortensen. 1998. The diversity of associated species on *Lophelia* coral reefs and methods for mapping and monitoring. *Fisken og Havet* 17:1–95.

Fosså, J. H, P. B. Mortensen, and D. M. Furevik. 2002. The deep-water coral *Lophelia pertusa* in Norwegian waters: Distribution and fishery impacts. *Hydrobiologia* 471:1–12.

FRAM Group. 1991. An eddy-resolving model of the Southern Ocean. *Eos* 72(15):169, 174–75.

Friederich, G. E., P. G. Brewer, R. Herlien, and F. Chavez. 1995. Measurement of sea surface partial pressure of CO_2 from a moored buoy. *Deep-Sea Research I* 42:1175–86.

Froese, R., and D. Pauly (eds.). 2000. *FishBase 2000: Concepts, Design and Data Sources*. Manila: ICLARM. Distributed with four CD-ROMs; see www.fishbase.org for updates.

Fromentin, J.-M., and B. Planque. 1996. *Calanus* and environment in the eastern North Atlantic. II. Influence of the North Atlantic Oscilation on *C finmarchicus* and *C helgolandicus*. *Marine Ecology Progress Series* 134:111–18.

Frye, D., B. Butman, M. Johnson, K. von der Heydt, and S. Lerner. 2000. Portable coastal observatories. *Oceanography* 13(2):24–31.

Garzoli, S., D. Olson, E. Chassignet, R. Matano, H. Berbery, E. Campos, J. Miller, A. Piola, G. Podesta, R. Fine, and R. Molinari. 1996. South Atlantic and Climate Change—SACC. Draft document. Available at http://www.oce.orst.edu/po/research/matano2/sacc_2.html.

Gattuso, J.-P., M. Frankignoulle, I. Bourge, S. Romaine, and R. W. Buddemeier. 1998. Effect of calcium carbonate saturation of seawater on coral calcification. *Global and Planetary Change* 18:37–46.

GESAMP. 2001a. Protecting the oceans from land-based activities. *GESAMP Reports and Studies*, no. 71. UNEP, Nairobi.

GESAMP. 2001b. A sea of troubles. *GESAMP Reports and Studies*, no. 70. UNEP, Nairobi.

Gilman, B. C. 1999. Future ROV requirements in deep-water. *Offshore Magazine*, December.

Glenn, S. M., T. D. Dickey, B. Parker, and W. Boicourt. 2000. Long-term real-time coastal ocean observation networks. *Oceanography* 13:24–34.

Gloersen, P., W. J. Campbell, D. J. Cavalieri, J. C. Comiso, C. L. Parkinson, and H. J. Zwally. 1992. Arctic and Antarctic sea ice, 1978–1987: Satellite passive-microwave observations and analysis. *NASA Special Publications,* SP-511.

Goldenfeld, N., and L. P. Kadanoff. 1999. Simple lessons from complexity. *Science* 284:87–92.

Goolsby, D. A. 2000. Mississippi Basin nitrogen flux believed to cause Gulf hypoxia. *Eos* 81:325–27.

Gordon, A. L. 1986. Interocean exchange of thermocline water. *Journal of Geophysical Research* 91:5037–46.

Gordon, A. L. 1988. Temporal variations in the separation of Brazil and Malvinas currents. Deep-Sea Research 35:1971–90.

Gordon, H. B., and S. P. O'Farrell. 1997. Transient climate change in the CSIRO coupled model with dynamic sea ice *Monthly Weather Review* 125:875–907.

Graham, M. 1943. *The Fish Gate.* London: Faber and Faber.

Granata, T., J. Wiggert, and T. Dickey. 1995. Trapped near-inertial waves and enhanced chlorophyll distributions. *Journal of Geophysical Research* 100:20793–804.

Granéli, E., G. A. Codd, B. Dale, E. Lipiatou, S. Y. Maestrini, and H. Rosenthal. 1999. *Harmful Algal Blooms in European Marine and Brackish Waters.* Luxembourg: Office for Official Publications of the European Communities.

Gray, J. S., and H. Christie. 1983. Predicting long-term changes in marine benthic communities. *Marine Ecology Progress Series* 13:87–94.

Gray, J. S., M. Delpledge, and A. Knap. 1996. Global climate controversy. *Journal of the American Medical Association* 276(5):372–373.

Griffiths, G., R. Davis, C. Eriksen, D. Frye, P. Marchand, and T. Dickey. 2001. Towards new platform technology for sustained observations. In C. J. Koblinsky and N. R. Smith (eds.), *Observing the Oceans in the 21st Century: A Strategy for Global Ocean Observations.* GODAE Project Office, Bureau of Meteorology, Melbourne, pp. 324–338.

Griffiths, G., A. Knap, and T. Dickey. 2000. Autonomous vehicle experiment. *Sea Technology* 41(2):35–45.

Hallegraeff, G. M. 1993. A review of harmful algal blooms and their apparent increase. *Phycologia* 32:79–99.

Hallegraeff, G. M., and C. J. Bolch. 1991. Transport of toxic dinoflagellate cysts via ships' ballast water. *Marine Pollution Bulletin* 22(1):27–30.

Hallegraeff, G. M., and C. J. Bolch. 1992. Transport of diatom and dinoflagellate resting spores in ships' ballast water: Implications for plankton biogeography and aquaculture. *Journal of Plankton Research* 14(8):1067–84.

Hansen, J., M. Sato, A. Lacis, R. Ruedy, I. Tegen, and E. Mathews. 1998. Climate forcings in the industrial era. *Proceedings of the National Academy of Science USA* 95(22):12753–58.

Harbison, G. R., and S. P. Volovick. 1994. The ctenophore, *Mnemiopsis leidyi,* in the

Black Sea: A holoplanktonic organism transported in the ballast water of ships. In D. Cottingham (ed.), *Nonindigenous and Introduced Marine Species.* U.S. Department of Commerce, NOAA Technical Report. Washington, D. C.: Government Printing Office.

Harvey, N. 1999. Australian integrated coastal management: A case study of the Great Barrier Reef. In W. Salomons, R. K. Turner, L. D. de Lacerda, and S. Ramachandran (eds.), *Perspectives on Integrated Coastal Zone Management.* Heidelberg: Springer-Verlag.

Haury, L., and J. A. McGowan. 1998. Time-space scales in marine biogeography. In A. C. Pierrot-Bults and S. van der Spoel (eds.), *Pelagic Biogeography ICOPB II. Proceedings of the Second International Conference,* 163–70. IOC Workshop report no. 142. Paris: Intergovernmental Oceanographic Commission, UNESCO.

Hawken, P. 1993. *The Ecology of Commerce: A Declaration of Sustainability.* New York: Harper Collins.

Hawkes, G. S. 1997. Microsubs go to sea. *Scientific American* 277(4):132–35.

He, C., and P. F. Hamblin. 2000. Visualisation of model results and field observations in irregular coastal regions. *Estuarine, Coastal, and Shelf Science* 50(1):73–80.

Hibler, W. D. III. 1997. On modeling sea ice fracture and flow in numerical investigations of climate. *Ann. Glaciol.* 25:26–32.

Holliday, D. V., R. E. Pieper, C. F. Greenlaw, and J. K. Dawson. 1998. Acoustical sensing of small-scale vertical structures in zooplankton assemblages. *Oceanography* 11(1):18–23.

Holt, M. W. 2002. Real time forecast modelling for the NW European shelf seas. In N. C. Flemming, S. Vallerga, N. Pinardi, H. W. A. Behrens, G. Manzella, D. Prandle, and J. H. Stel (eds.), *Operational Oceanography: Implementation at the European and Regional Scales.* Proceedings of the Second EuroGOOS Conference, 11–13 March 1999, Rome. Amsterdam: Elsevier Oceanography Series, no. 66, pp. 69–76.

Hopkins, E. H., K. S. Sell, A. L. Soli, and R. H. Byrne. 2000. In-situ spectrophotometric pH measurements: The effect of pressure on thymol blue protonation and absorbance characteristics. *Marine Chemistry* 71:103–109.

Houghton, J. T., L. G. Meira Filho, J. Bruce, H. Lee, B. A. Calander, E. Haites, N. Harris, and K. Maskell. 1994. *Climate Change 1994: Radiative Forcing of Climate Change and an Evaluation of the IPCC IS92 Emission Scenarios.* Cambridge: Cambridge University Press.

Howarth, R. W. (ed.). 1996. Nitrogen cycling in the North Atlantic Ocean and its watersheds: Report of the International SCOPE Nitrogen Project. *Biogeochemistry* 35:1–304.

Hoyt, D. V., and R. Schatten. 1993. A discussion of plausible solar irradiance variations: 1700–1992. *Journal of Geophysical Research* 98:18895–906.

Hubold, G. 1999. Perspektiven für eine nachhaltige Befischung der Meere. *Meer und Museum* 15:21–24.

Hyde, W. T., K. Y. Kim, T. J. Crowley, and G. R. North. 1990. On the relation between polar continentality and climate: Studies with a nonlinear seasonal energy balance model. *Journal of Geophysical Research* 95:18653–9668.

ICES. 1999a. *Extract of the Report of the Advisory Committee on Fisheries Management, May 1999: Herring in Sub-area IV, Division VIId, and Division IIIa (Autumn Spawners)*. Copenhagen: International Council for the Exploration of the Seas.

ICES. 1999b. *Report of the Herring Assessment Working Group for the Area South of 62°N*. ICES CM 1999/ACFM: 12. Copenhagen: International Council for the Exploration of the Seas.

ICES. 2000. Ecosystem effects of fishing. Proceedings of an ICES/SCOR Symposium held in Montpellier, France, 16–19 March 1999. *ICES Journal of Marine Science* 57(3):1–791.

INIDEP. 1986. *Impacto ecológico y económico de las capturas alrededor de las Malvinas después de 1982*. Informe preparado en el Instituto Nacional de Investigación y Desarrollo Pesquero (INIDEP), Mar del Plata. Argentina, March 1986.

IOC. 1984. *Ocean Science for the Year 2000*. Paris: Intergovernmental Oceanographic Commission, UNESCO.

IOC. 1996. *A Strategic Plan for the Assessment and Prediction of the Health of the Ocean: A Module of the Global Ocean Observing System*. IOC/INF 1044. Paris: Intergovernmental Oceanographic Commission, UNESCO.

IOC. 1998a. *Strategic Plan and Principles for the Global Ocean Observing System (GOOS), Version 1.0*. GOOS Report 41. Paris: Intergovernmental Oceanographic Commission, UNESCO.

IOC. 1998b. *The Global Ocean Observing System 1998 Prospectus, GOOS 1998*. GOOS Report 42. Paris: Intergovernmental Oceanographic Commission, UNESCO.

IOC. 1998c. *IOC/WESTPAC Coordinating Committee for the North-East Asian Regional–Global Ocean Observing System (NEAR-GOOS), Third Session*. GOOS Report 60. Paris: Intergovernmental Oceanographic Commission, UNESCO.

IOC. 1999. *Global Physical Ocean Observations for GOOS/GCOS: An Action Plan for Existing Bodies and Mechanisms*. GOOS Report 66. Paris: Intergovernmental Oceanographic Commission, UNESCO.

IOC. 2000a. *Strategic Design Plan for the Coastal Component of the Global Ocean Observing System (GOOS)*. GOOS Report 90. Also printed as IOC/INF 1146. Paris: Intergovernmental Oceanographic Commission, UNESCO.

IOC. 2000b. *Strategic Design Plan for the Living Marine Resources Panel of the Global Ocean Observing System (GOOS)*. GOOS Report 94. Also printed as IOC/INF 1150. Paris: Intergovernmental Oceanographic Commission, UNESCO.

IOCCG. 1999. *Status and Plans for Satellite Ocean-Colour Missions: Considerations for Complementary Missions*. IOCCG Report 2. Dartmouth, Nova Scotia: International Ocean Colour Coordinating Group.

IPCC. 1998. *The Regional Impacts of Climate Change: An Assessment of the Vulnera-*

bility. A Special Report of IPCC Working Group II. Cambridge: Cambridge University Press.
IWCO. 1998. *The Ocean, Our Future: The Report of the Independent World Commission on the Oceans.* Cambridge: Cambridge University Press.
Jackson, J. B. C., M. X. Kirby, W. H. Berger, K. A. Bjorndal, L. W. Rotsford, B. J. Bourque, R. Cooke, J. A. Estes, T. P. Hughes, S. Kidwell, C. B. Lange, H. S. Lenihan, J. M. Pandolfi, C. H. Peterson, R. S. Steneck, M. J. Tegner, and R. R. Warner. 2001. Historical overfishing and the recent collapse of coastal ecosystems. *Science* 293:629–38.
Jannasch, H. W., K. S. Johnson, and S. M. Sakamoto. 1994. Submersible, osmotically pumped analyzers for continuous determination of nitrate in situ. *Analytical Chemistry* 66:3352–61.
JGR. 1986. Swath bathymetric mapping. *Journal of Geophysical Research* 19(B3). Special issue.
Johannessen, J., H. Rebhan, C. Le Provost, H. Drange, M. Srokosz, P. Woodworth, P. Schlussel, P. Le Grand, Y. Kerr, and D. Wingham. 1999. Emerging new earth observation capabilities in the context of the ocean observing system for climate. In *Proceedings of the International Conference on the Ocean Observing System for Climate, Volume 1.* OceanObs99, San Raphael, France, 18–22 October 1999. Toulouse, France: CNES.
Johannessen, O. M., E. Korsbakken, P. Samuel, A. D. Jenkins and H. A. Espedal. 1997a. COAST WATCH: Using SAR imagery in an operational system for monitoring coastal currents, wind, surfactants, and oil spills. In J. H. Stel, H. W. A. Behrens, J. C. Borst, L. J. Droppert, J. v. d. Meulen (eds.), *Operational Oceanography: The Challenge for European Co-operation,* 234–42. Amsterdam: Elsevier Science.
Johannessen, O. M., A. M. Volkov, V. D. Grischenko, L. P. Bobylev, S. Sandven, K. Kloster, T. Hamre, V. Asmus, V. G. Smirnov, V. V. Melentyev, and L. Zaitsev. 1997b. ICEWATCH: Ice SAR monitoring of the Northern Sea Route. In J. H. Stel, H. W. A. Behrens, J. C. Borst, L. J. Droppert, J. v. d. Meulen (eds.), *Operational Oceanography: The Challenge for European Co-operation,* 224–33. Amsterdam: Elsevier Science.
Johannessen, O. M., E. V. Shalina, M. W. Miles. 1999: Satellite evidence for an Arctic sea ice cover in transformation. *Science* 286:1937–39.
Johnston, A. K. 1856. *The Geographical Distribution of the Currents of Air, Showing the Regions of the Trade Winds, Variable Winds and Hurricanes with Their Effects on Determining the Different Tracks of Navigation.* Edinburgh: W. and A. K. Johnston.
Jones, P. D., T. M. L. Wigley, C. K. Folland, D. E. Parker, J. K. Angell, S. Lebedeff, and J. E. Hansen. 1988. Evidence for global warming in the past decade. *Nature* 332(6167):790.
Kaku, M. 1998. *Visions: How Science Will Revolutionize the Twenty-first Century.* New York: Oxford University Press.

Karl, T. R., R. W. Knight, and R. G. Quayle. 1995. Trends in high-frequency climate variability in the twentieth century. *Nature* 377:217–20.

Kennett, J. P., and Stott, L. D. 1991. Abrupt deep-sea warming, paleoceanographic changes, and benthic extinctions at the end of the Paleocene. *Nature* 353:225–29.

Kleypas, J. A., R. W. Buddemeier, D. Archer, J.-P. Gattuso, C. Langdon, and B. N. Opdyke. 1999. Geochemical consequences of increased atmospheric carbon dioxide on coral reefs. *Science* 284:118–20.

Knox, R., and J. Wallace. 1999. Future plans for research ships in global ocean observations. In *Proceedings of the International Conference on the Ocean Observing System for Climate, Volume 2*. OceanObs99, San Raphael, France, 18–22 October 1999. Toulouse, France: CNES.

Koblentz-Mishke, O. J., V. V. Volkevinsky, and J. C. Kabanova. 1970. Plankton primary production of the world ocean. In W. S. Wooster (ed.), *Scientific Exploration of the Southern Pacific*, chap. 4. Washington, D.C.: National Academy of Sciences.

Kolber, Z. S., O. Prasil, and P. Falkowski. 1998. Measurements of variable chlorophyll fluorescence using fast repetition rate techniques: Defining methodology and experimental protocols. *Biochimica et Biophysica Acta* 1376:88–106.

Koblinsky, C. J. and N. R. Smith (eds.). 2001. *Observing the Oceans in the 21st Century*. Melbourne, Australia: GODAE Project Office, Bureau of Meteorology.

Krueger, A. J., S. L. Walter, P. K Bhartia, C. C. Schnetzler, N. A. Krotkov, I. Sprod, and G. J. S. Bluth. 1995. Volcanic sulfur dioxide measurements from the total ozone mapping spectrometer instruments. *Journal of Geophysical Research* 100:14057–76.

Lagerloef, G., and T. Delcroix. 2001. Sea surface salinity: A regional case study for the tropical Pacific. In C. J. Koblinksy and N. R. Smith (eds.), *Observing the Oceans in the 21st Century*, pp. 137–148. Melbourne, Australia: GUDAE Project Office, Bureau of Meteorology.

Larkin, P. 1996. Concepts and issues in marine ecosystem management. *Reviews of Fish Biology and Fisheries* 6:139–64.

Le Traon, P. Y., M. Reinecker, N. Smith, P. Bahurel, M. Bell, H. Hurlburt, and P. Dandin. 2001. Operational oceanography and prediction: A GODAE perspective. In C. J. Koblinsky and N. R. Smith (eds.), *Observing the Oceans in the 21st Century: A Strategy for Global Ocean Observations*. GODAE Project Office, Bureau of Meteorology, Melbourne, pp. 529–545.

Lean, J. 1991. Variations in the sun's radiative output. *Reviews of Geophysics* 29:505–35.

Ledwell, J. R., E. T. Montgomery, K. L. Polzin, L. C. St. Laurent, R. W. Schmitt, and J. M. Toole. 2000. Evidence for enhanced mixing over rough topography in the abyssal ocean. *Nature* 403(6766):179–82.

Leetma, A., W. Higgins, D. Anderson, P. Delecluse, and M. Latif. 2001. Application of seasonal to interannual predictions: A northern hemisphere perspective. In C. J. Koblinsky and N. R. Smith (eds.), *Observing the Oceans in the 21st Century:*

A Strategy for Global Ocean Observations. GODAE Project Office, Bureau of Meteorology, Melbourne, pp. 29–38.

Lloyds Register. 1999a. *World Fleet Statistics 1998.* London: Lloyds Register of Shipping.

Lloyds Register. 1999b. *World Casualty Statistics 1998.* London: Lloyds Register of Shipping.

Longhurst, A. 1998. Cod: Perhaps if we all stood back a bit? *Fisheries Research* 38:101–108.

Lorenz, E. N. 1968. Climate determinism. *Meteorological Monographs* 8(30):1–3.

Lorenz, E. N. 1969. The predictability of a flow which possesses many scales of motion. *Tellus* 21:289–307.

Ludwig, D., R. Hilborn, and C. Walters. 1993. Uncertainty, resources exploitation, and conservation: Lessons from history. *Science* 260: 17, 36.

Mace, P. M. 1997. Developing and sustaining world fisheries resources: The state of fisheries and management. In D. H. Hancock, D. C. Smith, and J. Beumer (eds.), *Developing and Sustaining World Fisheries Resources,* 1–20. Proceedings of the Second World Fisheries Congress, July 1996, Brisbane. Collingwood, VIC, Australia: CSIRO Publishing.

Manabe, S., and R. J. Stouffer. 1993. Century-scale effects of increased atmospheric CO_2 on the ocean-atmosphere system. *Nature* 364:215–18.

Manabe, S., and R. Stouffer. 1994. Multiple century response of a coupled ocean-atmosphere model to an increase of atmospheric carbon dioxide. *Journal of Climate* 7:5–23.

Mann, K. H. 2000. *Ecology of Coastal Waters, with Implications for Management.* 2nd ed. Malden, Mass.: Blackwell Science.

Mann, K. H., and J. R. N. Lazier. 1996. *Dynamics of Marine Ecosystems: Biological-Physical Interactions in the Oceans.* Cambridge, Mass.: Blackwell Science.

Marchetti, C. 1977. On geoengineering and the CO_2 problem. *Climate Change* 1:59–68.

Marra, J., R. R. Bidigare, and T. D. Dickey. 1990. Nutrients and mixing, chlorophyll, and phytoplankton growth. *Deep-Sea Research I* 37:127–43.

Martinson, D. G., and R. A. Iannuzzi. 1998. Antarctic ocean-ice interaction: Implications from ocean bulk property distributions in the Weddell Gyre. In M. O. Jeffries (ed.), *Antarctic Sea Ice, Physical Processes, Interactions and Variability,* 243–71. Antarctic Research Series 74. Washington, D.C.: American Geophysical Union.

Martinson, D. G., K. Bryan, M. Ghil, M. Hall, T. R. Karl, E. S. Sarachik, S. Soroshin, L. D. Talley (eds.). 1995. *Natural Climate Variability on Decade to Century Time Scales.* Washington, D.C.: National Academy Press.

Masson, D. G., N. H. Kenyon, and P. P. E. Weaver. 1996. Slides, debris flows and turbidity currents. In C. P. Summerhayes and S. A. Thorpe (eds.), *Oceanography: An Illustrated Guide,* 136–51. London: Manson.

Masson, D. G., A. B. Watts, M. R. J. Gee, R. Urgeles, N. C. Mitchell, T. P. Le Bas, and M. Canals. 2002. Slope failures on the flanks of the western Canary Islands. *Earth-Science Reviews* 57, 1–35.

Matsuda, F., J. Szyper, P. Takahashi, and J. R. Vadus. 1999. The ultimate ocean ranch. *Sea Technology* August, 17–26.

Maury, M. F. 1860. *The Physical Geography of the Sea, and Its Meteorology.* London: Sampson Low, Son and Co.

McGillicuddy, D. J., A. R. Robinson, D. A. Siegel, H. W. Jannasch, R. Johnson, T. D. Dickey, J. D. McNeil, A. F. Michaels, and A. H. Knap. 1998. New evidence for the impact of mesoscale eddies on biogeochemical cycling in the Sargasso Sea. *Nature* 394:263–66.

McGinn, A. P. 1999. Charting a new course for oceans. In L. R. Brown (ed.), *State of the World.* New York: W. W. Norton.

McGlade, J. M. 1999. *Advances in Theoretical Ecology.* Oxford: Blackwell Science.

McGowan, J. A., D. B. Chelton, and A. Conversi. 1996. Plankton patterns, climate and change in the California Current. *California Cooperative Oceanic Fisheries Investigation* 37:45–68.

McKenzie, D. 1997. Report of the Marine Technology Foresight Panel Working Group on Energy. *Journal of the Society for Underwater Technology* 22(4):139–54.

McNeil, J. D., H. Jannasch, T. Dickey, D. McGillicuddy, M. Brzezinski, and C. M. Sakamoto. 1999. New chemical, bio-optical, and physical observations of upper ocean response to the passage of a mesoscale eddy. *Journal of Geophysical Research* 104:15537–48.

McPhaden, M. J. 1995. The Tropical Atmosphere-Ocean Array is completed. *Bulletin of the American Meteorological Society* 76:739–41.

McPhaden, M. J., A. J. Busalacchi, R. Chevvey, J. R. Donqury, K. S. Gage, D. Halpern, Ming Ji, P. Julian, G. Meyers, G. T. Mitchumm, P. Niiler, J. Pacaut, R. W. Reynolds, N. Smith, and K. Takauchi. 1998. The tropical ocean-global atmosphere observing system: A decade of progress. *Journal of Geophysical Research* 103(12):14169–240.

McPhaden, M. J., T. Delcroix, K. Hanawa, Y. Kuroda, G. Meyers, J. Picaut, and M. Swenson. 2001. The El-Niño-southern oscillation (ENSO) observing system. In C. J. Koblinsky and N. R. Smith (eds.), *Observing the Oceans in the 21st Century: A Strategy for Global Ocean Observations.* GODAE Project Office, Bureau of Meteorology, Melbourne, pp. 239–247.

McPhee, M. G., T. P. Stanton, J. H. Morison, and D. G. Martinson. 1998. Freshening of the upper ocean in the Central Arctic: Is perennial sea ice disappearing? *Geophysical Research Letters* 25:1729–33.

Meier, M. F. 1984. Contribution of small glaciers to global sea level. *Science* 226:1418–21.

Merlivat, L., and P. Brault. 1995. CARIOCA Buoy: Carbon dioxide monitor. *Sea Technology* 10:23–30.

Mitchell, J. F. B., R. A. Davis, W. J. Ingram, and C. A. Senior. 1995. On surface temperature, greenhouse gases, and aerosols: Models and observations. *Journal of Climate* 10:2364–86.

Monastersky, R. 1999. Good-bye to a greenhouse gas. *Science News* 155:392–94.

Moore, C. C., J. R. V. Zaneveld, and J. C. Kitchen. 1992. Preliminary results from in situ spectral absorption meter data. In *Proceedings of the Ocean Optics XI Conference.* Bellingham, Wash.: SPIE.

Moore, C. C., M. S. Twardowski, and J. R. V. Zaneveld. 2001. The EcoVSF: A sensor for determination of the volume scattering function. In *Proceedings of the Ocean Optics XV Conference.* Monaco, 16–20 October 2000. CD-ROM. Bellingham, Wash.: SPIE.

Mullin, M. M. 1993. *Webs and Scales: Physical and Ecological Processes in Marine Fish Recruitment.* Seattle: University of Washington Press.

Murphy, J. M., and J. F. B. Mitchell. 1995. Transient response of the Hadley Centre coupled ocean-atmosphere model to increasing carbon dioxide. Part II: Spatial and temporal structure of response. *Journal of Climate* 8:57–80.

Nakashima, T. 1995. Research activities at KOCHI artificial upwelling laboratory in Japan on the utilisation of deep water resources. *IOCA Newsletter* 6(4):1–5.

NAS. 2000. *Clean Coastal Waters: Understanding and Reducing the Effects of Nutrient Pollution.* Washington, D.C.: National Academy Press.

Needler, G., N. Smith, and A. Villwock. 1999. The action plan for GOOS/GCOS and sustained observations for CLIVAR. In *Proceedings of the International Conference on the Ocean Observing System for Climate, Volume 1.* OceanObs99, San Raphael, France, 18–22 October 1999. Toulouse, France: CNES.

Neeland, J. D., D. S. Battisi, A. C. Hirst, F. Jin, Y. Wakata, T. Yamagata, S. E. Zebliak. 1998. ENSO theory. *Journal of Geophysical Research* 103:14261–90.

Nerem, R. S., and G. T. Mitchum. 2000. Observations of sea level change from satellite altimetry. In B. C. Douglas, M. S. Kearney, and S. R. Leatherman (eds.), *Sea Level Rise,* 121–63. Academic Press International Geophysics Series, volume 75.

Nichols, F. H., J. K. Thompson, and L. E. Shemel. 1990. Remarkable invasion of San Francisco Bay (California USA) by the Asian clam *Potamocorbula amurensis.* II: Displacement of a former community. *Marine Ecology Progress Series* 66:95–101.

Niiler, P. 2000. The world ocean surface circulations. In G. Siedler, J. Church, and J. Gould (eds.), *Ocean Circulation and Climate: Observing and Modeling the Global Ocean.* New York: Academic Press.

Nixon, S. W. 1980. Between coastal marshes and coastal waters: A review of twenty years of speculation and research on the role of salt marshes and estuarine productivity. In P. Hamilton and K. B. MacDonald (eds.), *Estuarine Wetland Processes* 437–520. New York: Plenum Publishing.

Nixon, S. W. 1995. Coastal marine eutrophication: A definition, social causes, and future concerns. *Ophelia* 41:199–219.

Njoku, E., B. Wilson, S. Yueh, T. Liu, and G. Laegerloef. 1999. The OSIRIS concept for ocean salinity sensing, In *Proceedings of the International Conference on the Ocean Observing System for Climate, Volume 1*. OceanObs99, San Raphael, France, 18–22 October 1999. Toulouse, France: CNES.

Norris, R. D. 1997. Records of the apocalypse: ODP drills the K/T boundary. In *ODP's Greatest Hits*. Washington, D.C.: Joint Oceanographic Institutions.

North, G. R., and Q. Wu. 2001. Detecting climate signals using space-time EOFs. *Journal of Climate* 14:1839–63.

North, G. R., J. Schmandt, and J. Crawford (eds). 1995. *Impact of Global Warming on Texas*. Austin: University of Texas Press.

North, G. R., and M. Stevens. 1998. Detecting climate signals in the surface temperature record. *J. Climate* 11, 563–577.

Nowlin, W. D. 1999. A strategy for long-term ocean observations. *Bulletin of American Meterological Society* 80(4):621–27.

Nowlin, W. D., N. Smith, E. Harrison, C. Koblinsky, and G. Needler. 2001. An integrated, sustained ocean observing system. In C. J. Koblinsky and N. R. Smith (eds.), *Observing the Oceans in the 21st Century: A Strategy for Global Ocean Observations*. GODAE Project Office, Bureau of Meteorology, Melbourne, pp. 1–28.

NRC. 1999. *Sustaining Marine Fisheries*. Washington, D.C.: National Research Council, National Academic Press.

ODP. 1996. *Understanding our Dynamic Earth through Ocean Drilling: The Ocean Drilling Program Long Range Plan into the 21st Century*. Washington, D.C.: Ocean Drilling Program.

OECD. 1994. *Oceanography; Megascience: The OECD Forum*. Paris: Organisation for Economic Co-operation and Development.

Olsgard, F., and J. S. Gray. 1995. A comprehensive analysis of the effects of offshore oil and gas exploration and production on the benthic communities of the Norwegian continental shelf. *Marine Ecology Progress Series* 122:277–306.

Olson, D. B., G. P. Podesta, R. H. Evans, and O. Brown. 1988. Temporal variations in the separation of Brazil and Malvinas Currents. *Deep-Sea Research I* 35:1971–1990.

Ong, J. E. 1982. Aquaculture, forestry and conservation of Malaysian mangroves. *Ambio* 11:252–57.

OOSDP. 1995. *Scientific Design for the Common Module of the Global Ocean Observing System and the Global Climate Observing System: An Ocean Observing System for Climate*. College Station, Texas: Department of Oceanography at Texas A&M University.

Orcutt, J., C. deGroot-Hedlin, W. Hodgkiss, W. Kuperman, W. Munk, F. Vernon, P. Worcester, E. Bernard, R. Dziak, C. Fox, C. Chiu, C. Collins, J. Mercer, R. Odom, M. Park, D. Soukup, and R. Spindel. 2000. Long-term observations in acoustics: The ocean acoustic observatory federation. *Oceanography* 13(2):57–63.

Paduan, J., and L. K. Rosenfeld. 1996. Remotely sensed surface currents in Monterey Bay from shore–based HF radar (CODAR). *Journal of Geophysical Research* 101:20669–686.

Pap, J. M. 1997. Total solar irradiance variability: A review. International School of Physics "Enrico Fermi" Course CXXXIII. G. Cini Castagnoli and A. Provenzale (eds.), *Past and Present Variability of the Solar-Terrestrial System: Measurement, Data Analysis, and Theoretical Models*. Amsterdam: IOS Press.

Parkes, R. J., B. A. Cragg, S. J. Bale, J. M. Getliff, K. Goodman, P. A. Rochelle, J. C. Fry, A. J. Weightman, and S. M. Harvey. 1994. Deep bacterial biosphere in Pacific Ocean Sediments. *Nature* 371:410–13.

Parrish, J. S. 1999. Toward remote species indentification. *Oceanography* 12(3):30–32.

Pauly, D. 1997. Small-scale fisheries in the tropics: Marginality, marginalization, and some implication for fisheries management. In E. K. Pikitch, D. D. Huppert, and M. P. Sissenwine (eds.), *Global Trends: Fisheries Management*, 40–49. Symposium 20. Bethesda, Md.: American Fisheries Society.

Pauly, D., V. Christensen, J. Dalsgaard, R. Froese, and F. C. Torres Jr. 1998. Fishing down marine food webs. *Science* 279:860–63.

Pérez, M. A., A. Aubone, and M. A. Renzi. 1999. Comparación de los resultados de las campañas de evaluación del efectivo sur de 41°S, entre 1998 y 1999. Consideraciones sobre el estado del recurso y perspectivas. *Informe Técnico INIDEP,* 106. Mar del Plata, Argentina: Instituto Nacional de Investigación y Desarrollo Pesquero (INIDEP).

Petit, J. R., J. Jouzel, D. Raynaud, N.I. Barkov, J.-M. Barnola, I. Basile, M. Bender, J. Chappellaz, M. Davis, G. Delayque, M. Delmotte, V. M. Kotlyakov, M. Legrand, V. Y. Lipenkov, C. Lorius, L. Pépin, C. Ritz, E. Saltzman, and M. Stievenard. 1999. Climate and atmospheric history of the past 420,000 years from the Vostok ice core, Antarctica. *Nature* 399(6735):429–36.

Petrae, Gary (ed). 1995. *Barge Morris J. Berman Spill: NOAA's Scientific Response*. HAZMAT Report no. 95-10. Seattle: Hazardous Materials Response and Assessment Division, National Oceanic and Atmospheric Administration.

Petrenko, A. A., B. H. Jones, T. D. Dickey, M. LeHaitre, and C. Moore. 1997. Effects of a sewage plume on the biology, optical characteristics, and particle size distributions of coastal waters. *Journal of Geophysical Research* 102:25061–71.

Pike, J., and A. E. S. Kemp. 1997. Early Holocene decadal-scale ocean variability recorded in Gulf of California laminated sediments. *Paleoceanography* 12:227–38.

Pollard, B., and J. Martin. 1999. Wide-swath altimetry using radar interferometry. In *Proceedings of the International Conference on the Ocean Observing System for Climate, Volume 2*. OceanObs99, San Raphael, France, 18–22 October 1999. Toulouse, France: CNES.

Provost, C., A. Lourenco, K. Thoral, L. Egarnes, M. Du Chaffaut, and S. Genthon. 1999. YOYO 2001: A multi-sensor upper ocean profiler. In *Proceedings of the International Conference on the Ocean Observing System for Climate, Volume 2*.

OceanObs99, San Raphael, France, 18–22 October 1999. Toulouse, France: CNES.

Rabalais, N. N., R. E. Turner, D. Justic, Q. Dortch, and W. J. Wiseman Jr. 1999. *Characterization of Hypoxia: Topic 1 Report for the Integrated Assessment of Hypoxia in the Gulf of Mexico.* NOAA Coastal Ocean Program Decision Analysis Series, no. 15. Silver Springs, Md.: NOAA Coastal Ocean Program.

Rahmstorf, S. 1999. Shifting seas in the greenhouse? *Nature* 399:523–24.

Rahmstorf, S., and A. Ganopolski. 1999. Long-term global warming scenarios computed with an efficient coupled climate model. *Climatic Change* 43:353–67.

Rahmstorf, S., J. Marotzke, and J. Willebrand. 1996. Stability of the thermohaline circulation. In W. Krauss (ed.), *The Warm Water Sphere of the North Atlantic Ocean*, 129–58. Stuttgart: Borntraeger.

Raney, R. K., and D. L. Porter. 1999. WITTEX: A constellation of three small satellite radar altimeters. In *Proceedings of the International Conference on the Ocean Observing System for Climate, Volume 2.* OceanObs99, San Raphael, France, 18–22 October 1999. Toulouse, France: CNES.

Reid, P. C., B. Planque, and M. Edwards. 1998. Is observed variability in the long-term results of the CPR survey a response to climate change? *Fisheries Oceanography* 7(3/4):282–88.

Richardson, K., and B. B. Jørgensen. 1996. Eutrophication: Definition, history, and effects. In B. B. Jørgensen and K. Richardson (eds.), *Eutrophication in Coastal Marine Ecosystems*, 1–19. Coastal and Estuarine Studies, volume 52. Washington, D.C.: American Geophysical Union.

Rind, D. R., R. Healy, C. Parkinson, and D. Martinson. 1995. The role of sea ice in $2XCO_2$ climate model sensitivity. Part I: The total influence of sea ice thickness and extent. *Journal of Climate* 8:449–63.

Ring Group. 1981. Gulf Stream cold core rings: Their physics, chemistry, and biology. *Science* 212:1091–1100.

Roberts, C., W. J. Ballantine, C. D. Buxton, P. Dayton, L. B. Crowder, W. Milon, M. K. Orbach, D. Pauly, J. Trexler, and C. Walters. 1995. *Review of the Use of Marine Fishery Reserves in the U.S. Southeastern Atlantic.* NOAA Technical Memorandum, NMFS-SEFCS-376.

Robinson, A. R., and T. D. Dickey (eds.). 1997. *An Advanced Modeling/Observation System (AMOS) for Physical-Biological-Chemical Ecosystem Research and Monitoring.* GLOBEC International Special Contributions, no. 2. Plymouth Marine Laboratory, U.K.

Robinson, A. R., and K. H. Brink (eds.). 1998. *The Global Coastal Ocean: Regional Studies and Syntheses.* The Sea, vol. 11. Chichester: Wiley.

Roe, H. S. J., G. Griffiths, M. Hartman, and N. Crisp. 1996. Variability in biological distributions and hydrography from concurrent acoustic Doppler current profiler and Seasoar surveys. *ICES Journal of Marine Science* 53(2):131–38.

Roemmich, D. H., and C. Wunsch. 1985. Two transatlantic sections: Meridional circulation and heat flux in the subtropical North Atlantic Ocean. *Deep-Sea Research* 32: 619–64.

Roemmich, D., and J. A. McGowan. 1995. Climatic warming and the decline of zooplankton in the California current. *Science* 267:1324–26.

Roemmich, D., and W. B. Owens. 2000. The Argo Project: Global ocean observations for understanding and prediction of climate variability. *Oceanography* 13(2):45–50.

Roemmich, D., O. Boebel, Y. Desaubies, H. Freeland, B. King, P.-Y. LeTraon, R. Molinari, W. B. Owens, S. Riser, U. Send, K. Takeuchi, and S. Wijffels. 2001. Argo: The global array of profiling gloats. In C. J. Koblinsky and N. R. Smith (eds.), *Observing the Oceans in the 21st Century: A Strategy for Global Ocean Observations*. GODAE Project Office, Bureau of Meteorology, Melbourne, pp. 248–268.

Rogers, A. D. 1999. The biology of *Lophelia pertusa* (Linnaeus 1758) and other deep-water reef-forming corals and impacts from human activities. International Review of Hydrobiology 84:315–406.

Root, T. L., and S. H. Schneider. 1995. Ecology and climate: Research strategies and implications. *Science* 269:334–39.

Rosenthal, H. 1980. Implications of transplantations to aquaculture and ecosystems. *Marine Fisheries Review* 5:1–14.

Ruiz, G. M., J. T. Carlton, E. D. Grosholz, and A. H. Hines. 1997. Global invasions of marine and estuarine habitats by non-indigenous species: Mechanisms, extent, and consequences. *American Zoologist* 37:621–32.

Russell, E. S. 1931. Some theoretical considerations on the overfishing problem. *ICES Journal of Marine Science* 6(1):3–20.

RVS. 1999. *Marine Facilities Tripartite Group (Germany, France, U.K.): Ship's Particulars of Deep Ocean Fleet in Barter Scheme*. Swindon, U.K.: Research Vessel Services, Natural Environment Research Council (NERC).

Santer B. D., K. E. Taylor, T. M. L. Wigley, T. C. Johns, P. D. Jones, D. J. Karoly, J. F. B. Mitchell, A. H. Oort, J. E. Penner, V. Ramaswamy, M. D. Schwarzkopf, R. J. Stouffer, and S. Tett. 1996. A search for human influences on the thermal structure of the atmosphere. *Nature* 382:39–46.

Sarmiento, J. L. 1992. Biochemical ocean models. In K. Trenberth (ed.), *Climate Systems Modeling*, 519–551. Cambridge: Cambridge University Press.

Sarmiento, J. L. 1993. Ocean carbon cycle. *Chemical and Engineering News* 31:30–44.

Schaefer, M. B. 1954. Some aspects of the dynamics of populations important to the management of the commercial marine fisheries. *Inter-American Tropical Tuna Commission Bulletin* 1:27–56.

Schimel, D., I. G. Enting, H. Heimann, T. M. L. Wigley, D. Raynaud, D. Alves, U. Siegenthaler. 1994. CO_2 and the carbon cycle. In J. Houghton, L. G. Meira Filho, J. P. Bruce, H. Lee, B. A. Callander, E. F. Haites (eds.), *Climate Change 1994,*

Radiative Forcing of Climate Change and an Evaluation of the IPCC IS92 Emission Scenarios, 35–71. Cambridge: Cambridge University Press.

Schlesinger, W. H. 1991. *Biogeochemistry: An Analysis of Global Change.* New York: Academic Press.

Schlosser, P., and W. M. Smethie. 1996. Transient tracers as a tool to study variability of ocean circulation. In D. G. Martinson, K. Bryan, M. Ghil, M. M. Hall, T. R. Karl, E. S. Sarachik, S. Sorooshian, and L. D. Talley (eds.), *Natural Climate Variability on Decade to Century Time Scales,* 274–89. Washington, D.C.: National Academy Press.

Schmidt, G. A., and J. E. Hansen. 1999. Role of sea ice in global climate change pondered. *Eos* 80:317–19.

Schmitz, W. J. 1995. On the interbasin scale thermohaline circulation. *Reviews of Geophysics* 33:151–73.

Schmitz, W. J. 1996a. *On the World Ocean Circulation: Volume I. Some Global Features/North Atlantic Circulation.* Woods Hole Oceanographic Institution Technical Report WHOI-96-03.

Schmitz, W. J. 1996b. *On the World Ocean Circulation: Volume II. The Pacific and Indian Oceans: A Global Update.* Woods Hole Oceanographic Institution Technical Repprt WHOI-96-08.

Scholin, C., P. Miller, K. Buck, F. Chavez, P. Harris, P. Haydock, J. Howard, and G. Cangelosi. 1997. Detection and quantification of *Pseudo-nitzschia australis* in cultured and natural populations using LSU rRNA-targeted probes. *Limnology and Oceanography* 42:1265–72.

Schopf, P., and M. Suarez. 1990. Ocean wave dynamics and the timescale of ENSO. *Journal of Physical Oceanography* 20:629–45.

Schwab, D. J., D. Beletsky, and J. Lou. 2000. The 1998 coastal turbidity plume in Lake Michigan. *Estuarine, Coastal, and Shelf Science* 50(1):49–58. Special issue on visualization in marine science.

Scialabba, N. (ed.). 1998. *Integrated Coastal Area Management and Agriculture, Forestry, and Fisheries: FAO Guidelines.* Rome: Environment and Natural Resources Service, FAO.

SCOR. 1969. *Global Ocean Research.* Report of Working Group 30 on Scientific Aspects of International Ocean Research, Ponza and Rome (29 April to 7 May 1969). La Jolla, Calif.: Scripps Institution of Oceanography.

Send, U., E. Skarsoulis, M. Kalogerakis, F. Gaillard, C. Maillard, and D. Mauuary. 1999. Operational tools for ocean acoustic tomography (TOMOLAB). In *Proceedings of the International Conference on the Ocean Observing System for Climate, Volume 2.* OceanObs99, San Raphael, France, 18–22 October 1999. Toulouse, France: CNES.

Send, U., R. Weller, S. Cunningham, C. Eriksen, T. Dickey, M. Kawabe, R. Lukas, M. McCartney, and S. Osterhus. 2001. Oceanographic time series observatories. In C. J. Koblinsky and N. R. Smith (eds.), *Observing the Oceans in the 21st Cen-*

tury: A Strategy for Global Ocean Observations. GODAE Project Office, Bureau of Meteorology, Melbourne, pp. 376–390.
Severinghaus, J. P., and E. J. Brook. 1999. Abrupt climate change at the end of the last glacial period inferred from trapped air in polar ice. *Science* 286:930–34.
Sheldon, R. W., A. Prackash, and W. H. Suttcliffe Jr. 1972. The size distribution of particles in the ocean. *Limnology and Oceanography* 17:327–40.
Shepherd, J. G., J. R. Cann, H. Charnock, G. Eglinton, H. Elderfield, H. Ford, J. Gage, P. D. Killworth, A. Rice, J. Rullkotter, J.-C. Sibuet, H. Weikert, J. Widdows, W. J. Winkworth, and C. C. West. 1996. *First Report: Scientific Group on Decommissioning Offshore Structures.* Swindon, U.K.: Natural Environment Research Council (NERC).
Sholkovitz, E., G. Allsup, R. Arthur, and D. Hosom. 1998. Aerosol sampling from ocean buoys. *Eos* 79(3):29.
Signell, R. P., H. L. Jenter, and A. F. Blumberg. 2000. Predicting the physical effects of relocating Boston's sewage outfall. *Estuarine, Coastal, and Shelf Science* 50(1): 59–71. Special issue on visualization in marine science.
Skud, B. E. 1982. Dominance in fishes: The relation between environment and abundance. *Science* 216:144–49.
Smethie, W. M. 1993. Tracing the thermohaline circulation in the Western North Atlantic using chlorofluorocarbons. *Progress in Oceanography* 31:51–99.
Smethie, W. M., R. A. Fine, A. Putzka, E. P. Jones. 2000. Tracing the flow of North Atlantic Deep Water using chlorofluorocarbons. *Journal of Geophysical Research* 105:14297–14323.
Smith, N., R. Bailey, O. Alves, T. Delcroix, K. Hanawa, E. Harrison, R. Keeley, G. Meyers, R. Molinari, and D. Roemmich. 2001. The upper ocean thermal network. In C. J. Koblinsky, and N. R. Smith (eds.), *Observing the Oceans in the 21st Century: A Strategy for Global Ocean Observations.* GODAE Project Office, Bureau of Meteorology, Melbourne, pp. 269–284.
Stammer, D., E. P. Chassignet. 2000. Ocean state estimation and prediction in support of oceanographic research. *Oceanography* 13, 51–56.
Stel, J. H., H. W. A. Behrens, J. C. Borst, L. J. Droppert, and J. v.d. Meulen (eds.). 1997. *Operational Oceanography: The Challenge for European Co-operation.* Proceedings of the First International Conference on EuroGOOS. Elsevier Oceanography Series 62. Amsterdam: Elsevier.
Stewart, J. E. 1991. Introductions as factors in diseases of fish and aquatic invertebrates. *Canadian Journal of Fisheries and Aquatic Sciences* 48(1):110–17.
Stramska, M., and T. Dickey. 1992. Short-term variations of the bio-optical properties of the ocean in response to cloud-induced irradiance fluctuations. *Journal of Geophysical Research* 97:5713–21.
Summerhayes, C. P., R. Coles, B. Wheeler, M. Baker, D. S. Cronan, R. Burt, G. Griffiths, N. Veck, D. Anderson, H. Young, and M. Murphy. 1997. Report of the Marine Technology Foresight Panel Working Group on Exploitation of Non-Liv-

ing Marine Resources. *Journal of the Society for Underwater Technology* 22(3):103–22.

Swenson, M. 1999. The global drifter program. In *Proceedings of the International Conference on the Ocean Observing System for Climate, Volume 2*. OceanObs99, San Raphael, France, 18–22 October 1999. Toulouse, France: CNES.

Takahashi, P. 1992. *U.S. Ocean Resources 2000: Planning for Development and Management*. Reports for NSF/NOAA. Washington, D.C.: National Science Foundation and National Oceanic and Atmospheric Administration.

Takahashi, T., R. H. Wanninkof, R. A. Feely, R. F. Weiss, D. W. Chipman, N. Bates, J. Olafsson, C. Sabine, and S. C. Sutherland. 1999. Net sea-air CO_2 flux over the global oceans: An improved estimate based on the sea-air pCO_2 difference. In *Abstracts of the Second CO_2 in the Oceans Symposium*. Tsukuba, Japan. Maps available at http://ingrid.ldgo.columbia.edu/SOURCES/.LDEO/.Takahashi/.

Tamburri, M. N., E. T. Peltzer, G. E. Friederich, I. Aya, K. Yamane, and P. G. Brewer. 2000. A field study of the effects of ocean CO_2 disposal on mobile deep-sea animals. *Marine Chemistry* 72:95–101.

Tang, X. O., W. K. Stewart, L. Vincent, H. Huang, M. Marra, S. M. Gallager, and C. S. Davis. 1998. Automatic plankton image recognition. *Artificial Intelligence Review* 12(1–3):177–99.

Taylor, C. D., and K. W. Doherty. 1990. Submersible incubation device (SID), autonomous instrumentation for the in situ measurement of primary and other microbial rate processes. *Deep-Sea Research I* 37:343–58.

Taylor, L. 1988. Whale tourism: Look, don't touch. *Scanorama* 18:58–66.

Taylor, P. K., and E. C. Kent. 1999. *The Accuracy of Meteorological Observations from Voluntary Observing Ships: Present Status and Future Requirements*. First Session of the WMO Committee on Marine Meteorology Subgroup on Voluntary Observing Ships, Athens, 8–12 March 1999.

Taylor, P. K., E. F. Bradley, C. W. Fairall, D. Legler, J. Shulz, R. A. Weller, and G. H. White. 2001. Surface fluxes and surface reference sites. In C. J. Koblinsky and N. R. Smith (eds.), *Observing the Oceans in the 21st Century: A Strategy for Global Ocean Observations*. GODAE Project Office, Bureau of Meteorology, Melbourne, pp. 177–197.

Tett, S. F. B., P. A. Stott, M. R. Allen, W. J. Ingram, and J. F. B. Mitchell. 1999. Causes of twentieth-century temperature change near the Earth's surface. *Nature* 399:569–72.

Thiel, H., M. V. Angel, E. J. Foell, A. L. Rice, and G. Schriever. 1997. *Environmental Risks from Large-Scale Ecological Research in the Deep Sea: A Desk Study*. Luxembourg: Office for Official Publications of the European Communities.

Tokar, J. M., and T. D. Dickey. 2000. Chemical sensor technology: Current and future applications. In M. Varney (ed.), *Chemical Sensors in Oceanography*, 303–29. Amsterdam: Gordon and Breach Scientific Publishers.

Twardowski, M. S., J. M. Sullivan, P. L. Donaghay, and J. R. V. Zaneveld. 1999.

Microscale quantification of the absorption by dissolved and particulate material in coastal waters with an ac-9. *Journal of Atmospheric and Oceanic Technology* 16:691–707.

UNCED. 1987. *Our Common Future.* Report of the World Commission on Environment and Development. Oxford: Oxford University Press.

UNCTAD. 1997. *Review of Marine Trade.* Geneva: United Nations Conference on Trade and Development.

van der Veen, J. 1992. Land ice and climate. In K. Trenberth (ed.), *Climate System Modeling,* 437–50. Cambridge: Cambridge University Press.

Vinnikov, K. Y., A. Robock, R. J. Stouffer, J. E. Walsh, C. L. Parkinson, D. J. Cavalieri, J. F. B. Mitchell, D. Garrett, and V. F. Zakharov. 1999. Global warming and Northern Hemisphere sea ice extent. *Science* 286:1934–36.

Wang, P. 1999. Response of western Pacific marginal seas to glacial cycles: Paleoceanographic and sedimentological features. *Marine Geology* 156:5–39.

Warren, B. A., and C. Wunsch. 1981. Deep circulation of the world ocean. In B. A. Warren (ed.), *Evolution of Physical Oceanography,* 6–40. Cambridge: MIT Press.

Warrick, R. A., C. Le Provost, M. F. Meier, J. Oerlemans, and P. L. Woodworth. 1995. Changes in sea level. In J. Houghton (ed.), *Climate Change 1995: The Science of Climate Change,* 358–405. Cambridge: Intergovernmental Panel on Climate Change, Cambridge University Press.

Watson, A. J., and J. R. Ledwell. 2000. Oceanographic tracer release experiments using sulphur hexafluoride. *Journal of Geophysical Research* 105:14325–37.

Watson, R., and D. Pauly. 2001. Systematic distortions in world fisheries catch trends. 414 (Nov. 29):534–536.

Watson, R. T., J. A. Dixon, S. P. Hamburg, A. C. Janetos, and R. H. Moss. 1998. *Protecting Our Planet; Securing Our Future.* Nairobi: UNEP, NASA, World Bank.

Wattayakorn, G., B. King, E. Wolanski, and P. Suthanaruk. 1998. Seasonal dispersion of petroleum contaminants in the Gulf of Thailand. *Continental Shelf Research* 18:641–59.

Weaver, A. J., J. Marotzke, P. F. Cummins, and E. S. Sarachik. 1993. Stability and variability of the thermohaline circulation. *Journal of Physical Oceanography* 23:39–60.

Weber, M. L., and J. A. Gradwohl. 1995. *The Wealth of Oceans.* New York: W. W. Norton.

Weiher, R. F. 1999. *Improving El Niño Forecasting: The Potential Economic Benefits.* Washington, D.C.: NOAA.

Weller, R. A., M. F. Baumgartner, S. A. Josey, A. S. Fischer, and J. C. Kindle. 1998. Atmospheric forcing in the Arabian Sea during 1994–1995: Observations and comparisons with climatology and models. *Deep-Sea Research II* 45:1961–99.

Weller, R., J. Toole, M. McCartney, and N. Hogg. 1999. Outposts in the ocean. *Oceanus* 42(1):20–23.

Went, A. E. J. 1972. *Seventy Years A-growing: A History of the International Council for the Exploration of the Sea (1902–1972)*. ICES Marine Science Symposia 165. Copenhagen: International Council for the Exploration of the Sea.

Williams, A. J. III, J. S. Tochko, R. L. Koehler, W. D. Grant, T. F. Gross, and C. V. R. Dunn. 1987. Measurement of turbulence in the oceanic bottom boundary layer with an acoustic current meter array. *Journal of Atmospheric and Oceanic Technology* 4:312–27.

Willson, R. C., and H. S. Hudson. 1988. Solar luminosity variations in solar cycle 21. *Nature* 332:810–12.

Winn, C. D., Y.-H. Li, F. T. Mackenzie, and D. M. Karl. 1998. Rising surface ocean dissolved inorganic carbon at the Hawaii ocean time-series site. *Marine Chemistry* 60:33–47.

Woehler, E. J., R. L. Penney, S. M. Creet, and R. Burton. 1994. Impacts of human visitors on breeding success and long-term population trends in Adélie Penguins at Casey, Antarctica. *Polar Biology* 14:269–74.

Wolanski, E., P. Doherty, and J. Carleton. 1997. Directional swimming of fish larvae determines connectivity of fish populations on the Great Barrier Reef. *Naturwissenschaften* 84(6):262–268.

Wolanski, E., B. King, and S. Spagnol. 1999. The implication of oceanographic chaos for coastal management. In W. Salomons, R. K. Turner, L. D. de Lacerda, and S. Ramachandran (ed.), *Perspectives in Integrated Coastal Zone Management*, 129–41. Berlin: Springer-Verlag.

Wolanski, E., N. N. Huan, L. T. Dao, N. H. Nhan, and N. N. Thuy. 1996. Fine sediment dynamics in the Mekong River estuary, Vietnam. *Estuarine, Coastal, and Shelf Science* 4:565–82.

Wood, R. A., A. B. Keen, J. F. B. Mitchell, and J. M. Gregory. 1999. Changing spatial structure of the thermohaline circulation in response to atmospheric CO_2 forcing in a climate model. *Nature* 399:572–75.

Woods, J. D. 1997. The EuroGOOS Strategy. In J. D. Stel, H. W. A. Behrens, J. C. Borst, L. J. Droppert, and J. v. d. Meulen (eds.), *Operational Oceanography: The Challenge for European Co-operation*, 19–35. Proceedings of the First International Conference on EuroGOOS. Elsevier Oceanography Series 62. Amsterdam: Elsevier.

Woods, J. D. 2000. *Ocean Predictability*. Bruun Memorial Lecture, IOC 20th Assembly. IOC technical series 55. Paris: UNESCO.

Woods, J. D., H. Dahlin, L. Droppert, M. Glass, S. Vallerga, and N. C. Flemming. 1996. *The Strategy for EuroGOOS*. EuroGOOS Publication no. 1. Southampton, U.K.: Southampton Oceanography Centre.

Wu, J., and E. A. Boyle. 1997. Lead in the western North Atlantic Ocean: Completed response to leaded gasoline phase out. *Geochemica Cosmochemica Acta* 61:3279–83.

Young, D. K., and P. J. Valent. 1997. Is abyssal sea-floor isolation an environmentally sound waste management option? *Journal of the Society for Underwater Technology* 22(4):155–65.

ABOUT IOC, SCOR, AND SCOPE

About the Intergovernmental Oceanographic Commission

The Intergovernmental Oceanographic Commission (IOC) of UNESCO was established in 1960 as a specialized mechanism of the United Nations system to coordinate ocean scientific research and ocean services worldwide. The work of IOC focuses on: (1) development, promotion, and facilitation of international oceanographic research programs to improve understanding of critical global and regional ocean processes and their relationship to the sustainable development and stewardship of ocean resources; (2) planning, establishment, and coordination of an operational global ocean observing system to provide the information needed for oceanic and atmospheric forecasting, for oceans and coastal zone management by coastal nations, and for global environmental change research; (3) provision of international leadership for education and training programs and technical assistance essential for systematic observations of the global ocean and its coastal zone and related research; and (4) assurance that ocean data and information obtained through research, observation, and monitoring are efficiently handled and made widely available. IOC collaborates with other relevant organizations within and outside the UN system to pursue these objectives. The IOC is composed of its Member States (currently 129), an Assembly, an Executive Council, and a Secretariat. More information on IOC is available at hhttp://ioc.unesco.org.

About the Scientific Committee on Oceanic Research

The Scientific Committee on Oceanic Research (SCOR) was established in 1957 by the International Council for Science. It is the leading nongovernmental organization for the promotion and coordination of international oceanographic activities. SCOR's science activities focus on investigation of cutting edge topics in oceanography, promoting international cooperation in planning and conducting marine research, and solving methodological and conceptual problems that hinder research. Scientists from the thirty-seven SCOR member nations participate in working groups and scientific steering committees for large-scale ocean research programs. SCOR promotes capacity building for marine scientists in developing countries through special efforts to include such scientists in its activities, through travel grants, and through a new activity on Regional Graduate Schools of Oceanography and Marine Environmental Sciences. SCOR is a scientific advisory body to the Intergovernmental Oceanographic Commission (IOC) of UNESCO and works with the IOC and several other international organizations on cosponsored activities.

About the Scientific Committee on Problems of the Environment

The Scientific Committee on Problems of the Environment (SCOPE) was established by the International Council for Science (ICSU) in 1969. It brings together natural and social scientists to identify emerging or potential environmental issues and to address jointly the nature and solution of environmental problems on a global basis. Operating at an interface between the science and decision-making sectors, SCOPE's interdisciplinary and critical focus on available knowledge provides analytical and practical tools to promote further research and more sustainable management of the Earth's resources. SCOPE's members, forty national science academies and research councils and twenty-two international scientific unions, committees, and societies, guide and develop its scientific program.

LIST OF CONTRIBUTORS

MARY G. ALTALO	Science Applications International Corporation, La Jolla, CA, USA
RICHARD BALL	Suiderland Fishing Corporation, Cape Town, South Africa
PATRICO BERNAL	Intergovernmental Oceanographic Commission (UNESCO), Paris, France
KARL-HEINZ VAN BERNEM	GKSS-Research Center, Institute for Coastal Research, Geesthacht, Germany
PETER G. BREWER	Monterey Bay Aquarium Research Institute, Moss Landing, CA, USA
EDMO J.D. CAMPOS	Department of Physical Oceanography, University of Sao Paulo, Brazil
CHRIS CROSSLAND	LOICZ International Project Office, Netherlands Institute for Sea Research, Texel, The Netherlands
JOHN DELANEY	School of Oceanography, University of Washington, Seattle, WA, USA
TOMMY D. DICKEY	Ocean Physics Laboratory, University of California Santa Barbara, Santa Barbara, CA, USA
ROBERT A. DUCE	Department of Atmospheric Sciences, Texas A&M University, College Station, TX, USA
HUGH DUCKLOW	School of Marine Science, The College of William and Mary, Gloucester Point, VA, USA
JOHN G. FIELD	Department of Zoology, University of Cape Town, South Africa
RANA FINE	Rosenstiel School of Marine and Atmospheric Science, University of Miami, USA

List of Contributors

MIGUEL FORTES	Marine Science Institute, University of The Philippines, Diliman, Philippines
JAN HELGE FOSSÅ	Institute of Marine Research, Bergen, Norway
WOOI KHOON GONG	Center for Marine and Coastal Studies, University Sains Malaysia, Penang, Malaysia
W. JOHN GOULD	CLIVAR International Project Office, Southampton Oceanography Centre, Southampton, UK
GWYNN GRIFFITHS	Ocean Engineering Division, Southampton Oceanography Centre, Southampton, UK
JULIE A. HALL	National Institute for Water and Atmospheric Research, Hamilton, New Zealand
GOTTHILF HEMPEL	Scientific Adviser of the President of the Senate of the Free Hanseatic City of Bremen/Bremerhaven, Germany
GEOFFREY L. HOLLAND	2WE Associates Consulting Ltd, Victoria, B.C. Canada
NILS G. HOLM	Department of Geology and Geochemistry, Stockholm University, Sweden
RUSSELL HOWORTH	South Pacific Applied Geoscience Commission, Suva, Fiji Islands
MANUWADI HUNGSPREUGS	Department of Marine Science, Chulalongkorn University, Bangkok, Thailand
STJEPAN KECKES	Advisory Committee on Protection of the Sea, London, UK
HENK DE KRUIK	National Institute for Coastal and Marine Management, Ministry of Transport and Public Works, The Netherlands
PAULO DA CUNHA LANA	Center for Marine Research, Federal University of Paraná, Pontal do Sul, Brazil
HAN J. LINDEBOOM	Netherlands Institute for Sea Research, Texel, The Netherlands
KARIN LOCHTE	Institute for Marine Research, University of Kiel, Germany
EDUARDO MARONE	Center for Marine Research, Federal University of Paraná, Pontal do Sul, Brazil
JOHN A. MCGOWAN	Scripps Institution of Oceanography, University of California at San Diego, La Jolla, USA

List of Contributors | 353

FRANK VAN DER MEULEN	National Institute for Coastal and Marine Management, Ministry of Transport and Public Works, The Netherlands
GERALD R. NORTH	Department of Atmospheric Sciences, Texas A&M University, College Station, TX, USA
JIN EONG ONG	Center for Marine and Coastal Studies, University Sains Malaysia, Penang, Malaysia
ULRICH SAINT-PAUL	Center for Tropical Marine Ecology, University of Bremen, Germany
DANIEL PAULY	The University of British Columbia, Fisheries Centre, Vancouver, B.C., Canada
RALPH RAYNER	Fugro GEOS Limited, Swindon, UK
DAVID P. ROGERS	National Oceanic and Atmospheric Administration, Office of Oceanic and Atmospheric Research, Silver Spring, MD, USA
VLADIMIR RYABININ	World Climate Research Programme, World Meteorological Organization, Geneva, Switzerland
RAMIRO SANCHEZ	National Institute of Fisheries Research and Development, National University of Mar de Plata, Argentina
LISA R. SHAFFER	Scripps Institution of Oceanography, University of California San Diego, La Jolla, CA, USA
JOHN G. SHEPHERD	Southampton Oceanography Centre, Southampton, UK
LEONARDO F. SOUZA	PETROBRAS, Rio de Janeiro, Brazil
RICHARD W. SPINRAD	Office of the Oceanographer of the Navy, U.S. Naval Observatory, Washington, DC, USA
COLIN P. SUMMERHAYES	Intergovernmental Oceanographic Commission (UNESCO), Paris, France
DAVID SZABO	Fugro GEOS Limited, Swindon, UK
ILANA WAINER	Department of Physical Oceanography, University of Sao Paulo, Brazil
ERIC WOLANSKI	Australian Institute of Marine Science, Townsville, Queensland, Australia
CHRISTOPHER ZIMMERMANN	Institute for Sea Fisheries, Federal Research Centre for Fisheries, Hamburg, Germany

INDEX

Note: Page numbers in italics refer to illustrative material, tables, and charts.

Abbott, M. R., 216, 217, 219, 230
Academic oceanography, 10–11
Accidents, shipping, 146, 163–64, 171, 269
Acoustic Doppler current profilers (ADCPs), 214, 227, 228, 239
Acoustics and acoustic instruments, 8, 170, 218, 228, 229, 242
  bioacoustics, 237–41
Adger, W. N., 65
Advective feedback, 101
Aebischer, N. J., 32
Aeolian fallout, 28
Aerosols, anthropogenic, 93–94, 95, 103, 106
Afshan, A., 99
Agenda 21, 7–8, 50, 309
Agrawal and Pottsmith, 236
Altalo, Mary G., 165–85, 175
Alverson, D. L., 110, 118
American comb jelly, 78
Anderson, D. T., 21
Andreassen, C., 174
Angel, M. V., 138, 156
Anoxic conditions, 70
Antarctic Bottom Water, 13
Antarctic Circumpolar Current, 36
Antarctic ice sheet, 60, 100–101
Apparent optical properties (IOPs), 235, 236
Apt, J., 224
Aquaculture, 72–73, 78, 110, 269, 274
  mangroves and, 54
Arctic, persistent organic pollutants (POPs) in the, 258, 268
Arctic Ocean, 12
Arctic Oscillation (AO), 212–13

Arctic shores, 60
Argentina fisheries, 114–17
Argo project, xii, 203, *204*
Argo Science Team, 218
Army Corps of Engineers, U.S., 174
Asian clam, 78
ATESEPP, 153
Atkinson, A., 126
Atlantic Margin Metocean Project, 194–96
Atmosphere and the ocean, climate change and relationship between, *see* Climate and climate change
Atmospheric general circulation model (AGCM), 88
AURIS Decommissioning Report, 147
Autonomous temperature line acquisition system (ATLAS) environmental sensing buoys, *17*, 21
Autonomous underwater vehicles (AUVs), 217, 220–23
Azov Sea, 78

Backscattering, 229, 236, 239
Bakun, A., 126
Ball, Richard, 130–31
Ballast water, 78, 176–78, 269
Baranaov, F., 112
Barnett, T. P., 94
Bates, N. R., 26
Beaches:
  erosion of, 1, 75
  sandy, 60–62
Bell, N., 146
BENEFIT program, 304
Benfield, M. C., 239

355

Benthos, 70
Bergman, M. J. N., *52*, 53
Beverton, R. J. H., 112
Bioacoustics, 237–41
Biodiversity, 1, 237, 313–14
Biological oceanography, 10–11
  in situ sensors and systems for, 237–41
Biosphere Reserve Program, 50
Biotechnology, 154–56
Bishop, James, 219, 237
Bissett, Paul, 225
Black Sea, 78
Black Sea GOOS, 205
Blue whiting, 115, *116*
Bluth, G. J. S., 91
Bogucki, D., 215, 216
Bohnsack, J. A., 127
Bolch, C. J., 78, 177
Bond 1992, *16*
Bottom trawling, 118, 120
Bottom tripods, 215–16, 217
Boyle, E. A., 230
Brault, P., 216
Brazil, capacity building in marine sciences in, 290–91
Brazil Current, 88–89
Brent Spar oil facility, 146, 147
Brewer, P. G., 25, 158–59, 231
Brink, K. H., 50
Broeker, W. S., 101
Brook, E. J., *16*
Brown, M. G., 149
Brundtland Commission, 4
Bückmann, A., 112
Byrne, R. H., 231

Calanus finmarchicus, 30
California Cooperative Oceanic Fisheries Investigations (CalCOFI), 31
California Current, 31–32, *32*
California sardine, 32–33
Camphuysen, C. J., 76
Cane, M. A., 91
Capacity building, 283–319, 317–18
  conclusions and recommendations, 305–306
  determining needs, 284–90
  education and training, North/South and South/South assistance in, 289–94
  journals and the language barrier, 296–97
  regional and global networks, 300–305
  regional meetings, 297–300
  in technical and organizational infrastructure, 295–96
Caplan-Auerbach, J., 242
Carbon cycle, 29, 102–104
  JGOFS study of, *see* Joint Global Ocean Flux Study (JGOFS)
  ocean-atmosphere, 12–14, *13*
  water mass tracers, nutrients, and the, 22–28
Carbon dioxide, 6, 9, 13, 43–44
  deep sea disposal of, 157–59
  global warming and, *see* Global warming
  photosynthesis and, 12
  sequestration, 12, 157–59
  see also Carbon cycle
Carder, K. L., 226
Carneiro, Eulalia R., 143–44
Carpenter, S. R., 28
Caspian Sea, sea-level rise of, 64
Cavalieri, D. J., 98
CCAMLR (Commission for the Conservation of Antarctic Marine Living Resources), 125
Centre for Marine Studies (CEM), Federal University of Paraná, Brazil, 291
Chang, G. C., 216, 225, 228, 235, 264
Channel survey technology, 176
Charlson, R. J., 104
Charnock, H., *14*
Chave, A. D., 243
Chavez, F. P., 215, 219
Chemistry, in situ sensors and systems for marine, 229–33
Chloroflurocarbons (CFCs), 22–23, *24*, 86, 93
Cholera, 156
Christie, H., 51
Clark, J. R., 62, 73, 74
Clean Water Act, 178
Climate and climate change, 1, 9, 85–108, 237, 242, 284, 314
  atmospheric climate and sea surface temperature, 89–91
  attribution of causes of, 94–96
  carbon cycle and, 102–104
  conclusions and recommendations, 106
  consequences of interaction of climate and ocean, 96–104
  engineering of, 28
  global warming, *see* Global warming
  history of changes in, 33–35
  natural variability, 87–91
  overview, 85–87
  paleoceanography and, 104–106
  perturbations inducing, 91–94
  regime shifts in fisheries and variations in, 30–33
  research priorities, 107
  surface temperature, variability of Earth's, *92*
  thermohaline circulation and, 101–102

Index | 357

Climate sciences, 11
Climate Variability and Predictability (CLIVAR), 5, 87, 212
Cloern, J. E., 78
Coastal Zone Color Scanner, 236
Coastal zones, 49–84, 284, 285–86, 315
  conclusions and recommendations, 81–84
  defined, 49
  demographics, 65–66
  food production in, 72–74
  habitats, 53–63
  harmful algal bloom (HAB), *see* Harmful algal bloom (HAB)
  hazards, 63–65
  increasing human presence and pressures, 65–71
  information systems for shipping, 170–76
  instrumentation and equipment needs, special, 79–80
  invasive species, *see* Exotic species
  major challenges, 49–50
  mammals and birds, 76
  management tools, 80–81
  natural variability in, 51–53
  population growth and, 2, 276
  recreation and tourism in, 74–75, 274
  sustainable development of, 75–76
Coast Guard, U.S., 174
Coast Watch, 171–72
Cod, 111, 128, 132
Colwell, R. R., 154, 155, 156
Computered visualization, 251–52
Conference of the Parties to the UN Framework Convention on Climate Change, 200
Continental ice sheets, 99–100
Convective feedback, 101–102
Cook, P. J., 65
Cooperation, framework for, 257–81
  conclusions, 279
  evolving institutions in science, 264–66
  global efforts, organizing, 277–79
  growth of interdisciplinarity, 316
  institutional framework, 259–64
  overview, 257–59
  recommendations, 280–81
  regional perspective, 272–77
  societal context, 266–72
Cooperative Research Center for the GBRWHA, 57
Coral reefs, 2, 55–58, 104, 117, 138, 146
  deep-water coral ecosystems, 121–22
  river runoff and, 67
  structure of, 58
Coultier, M. C., 74

Coupled Ocean-Atmosphere Mesoscale Prediction System, 169, 170
Crosslandi, Chris, 56–57
Cruickshank, M. J., 148, 152
Csirke, J., 115
Curry, J. A., 99
Curtin, T. B., 220, 249
Cushing, D. H., 112
Cytometers, 237

Daan, N., 118
da Cunha Lana, Paulo, 290–91
Dansgaard-Oeschger events, 101
Davis, C. S., 238
Dawes, T. C., 220
Decadal shifts, 11, 132
Deep Sea Drilling Project, xii
Deep sea floor biosphere, 315
Deep water formation, 101
Defense, *see* Shipping and defense, marine information for
DeGrandpré, M. D., 216, 231, 233
Delaney, John, 243–47
Delcroix, T., 225
Department of Trade and Industry (DTI), 149, 164, 187, 197
Department of Transportation, 164, 179
Desiderio, R. A., 236
de Solla Price, Derek, 261, 264
Detrick, R. S., 218, 248
Developing countries, 273, 274, 284–85
Dickey, Tommy D., 209–54
Dikes, 60, 62, 77
DISCOL, 153, 154
Discovery Expeditions, 126
Distillation of seawater, 276
Doherty, K. W., 216
Done, Terry, 56
Dr. Fridtjof Nansen (vessel), 303
Drifters and floats, 218–19
Driver, Pressures, State Impact and Response approach (DPSIR), 80
Duce, Robert A., 85–108
Ducklow, Hugh, 26–27
DuRand, M. D., 236
Dushaw, B., 229
Dutch coast, lessons learned from case study of, 61–62

Earle, S. A., 1, 219
Earthquakes, 65
Earth system, 38–40
Ebert, E. E., 99
Ecosystems:
  deep-water coral, 121–22
  fisheries in, 123–26

Ecosystems (*continued*)
  large marine (LME), 285, 304
  ocean, 6, 30
Ecotourism, 74–75
Eddies, 18, *20*, 190
Education and training in the marine sciences, capacity building for, 289–94
Electronic chart display and information system (ECDIS), 174
Electronic nautical chart (ENC), 174
El Niño events, 5, 11, 21–22, 31, 43, 51–52, 87, 88–91, 103, 198, *199*, 212, 224, 225, 248, 250, 273–74, 284
  ocean-atmosphere coupling and, 19–22
Energy industry, 139–46
ENSO, *see* El Niño events
Environmental Protection Agency (EPA), 178
Erosion, coastal, 1, 75
Estuaries, 62–63
Euphausia superba, 30
EuroGOOS, 189, 203
European Centre for Medium Range Weather Forecasting, 167
European Space Agency, 173
European Union, 265–66, 300
Eutrophication of coastal waters, 68–70, 230
Evolution, 1, 9
Exclusive economic zones (EEZs), 7, 138, 284, 298, 303, 304
Exotic species, 2, 156, 176–78, 269
Exxon *Valdez*, 146, 164

Falkowski, P. G., 28, 216, 217, 233, 235, 236
FGGE project, xii
Fiberoptics, 232–33, 249
Fichefet, T., 99
Field, John G., xi–xvi, 1–8, 9–44, 309–19
Fine, R. A., 23, 228, 230
Fine Resolution Antarctic Model (FRAM), 193
Finlayson, A. C., 132
First International Conference for the Ocean Observing Systems for Climate, 200–201
Fischer, J., 189
Fisheries and fisheries science, 1, 73–74, 109–35, 237, 274–75, 298, 316, 317
  beyond 2000, 129–34
  bycatch, *110*
  co-management concept, 128
  conclusions and recommendations, 134–35
  conflicts between users, 119–23
  definitions, 109
  in ecosystems, 123–26
  environmental influences, 117
  fishing down marine food webs, 118–19
  globalization and, 111, 131–32
  management, 2
  maximum economic yield (MEY), *113*
  maximum sustainable yield (MSY), *113*, *116*, 310
  multispecies problem, 117–18
  origins of fisheries science, 111–12
  overfishing, *see* Overfishing
  Patagonian case, 114–17
  regime shifts, studying causes of, 30–33
  research priorities, 135
  single species fisheries, 123–25, 127
  statistics, 109–10
  sustainability of, 112–17, 126–29
Fitzroy, Vice-admiral, 166
Flather, R. A., 193
Flato, G. M., 99
Flemming, N. C., 189, 199, 203
Floats and drifters, 218–19
Floods, 1, 65
Foley, D., 215, 216
Food and Agricultural Organization, 114, 130
Forecasting ocean science, 4–7
Fortes, Miguel D., 283–319
Fosså, J. H., 121
Fossil fuels, 10, 25–27, 137–38, 146
  anthropogenic aerosols and, 93
  gas hydrates as alternative to, 148–49
  global warming and, *see* Global warming
Framework for cooperation, *see* Cooperation, framework for
Fraunhofer effect, 236
Friederich, G. E., 216, 231, 233
Frisian Barrier Islands, 62
Froese, *119*
Fromentin, *198*
Frye, Dan, 217, 218, 248
Future, the, 309–19

Ganopolski, A., 102
GARP project, xii
Garth, S., 76
Garzioli, S., 88
Gas hydrates, 148–49
Gattuso, J.-P., 138
Geographical information systems (GIS), 132
Geological Survey, U.S., 148
Geophysical Fluid Dynamics Laboratory, 98, 99
Gillman, B. C., 220

Glenn, S. M., 216, 218, 229
Global Ecology and Oceanography of
 Harmful Algal Blooms (GEOHAB), 6
Global Ocean Data Assimilation
 Experiment, 203, 313
Global Ocean Ecosystem Dynamics
 (GLOBEC) program, 6, 80, 212, 237
 mission of, 30
Global Ocean Observing System (GOOS),
 5, 79–80, 87, 165, 180, 197–205, 278,
 285, 298, 304–305, 313
 design of, 200–201
 development of, 198–99
 implementation of, 201–205
Global Ocean Science: Toward an Integrated
 Approach (NSC), 260
Global Positioning System (GPS), 129–30,
 174, 218, 219, 225, 310
Global Star, 168
Global warming, 2, 11, 25, 28, 87, 93, 95,
 102, 138, 229–30, 284
 coral reefs and, 55, 58, 138
 El Niño cycle and, 91
 polar areas and, 60, 100
 sea ice as indicator of, 98
 sea-level rise and, 99–100
 see also Climate and climate change
Gloersen, P., 97
Glomar Challenger expeditions, xii
Goldenfeld, N., 41
Gong, Wooi K., 54–55
Goolsby, D. A., 70
Gordon, H. B., 98, 101
Gould, W. John, 86–87
Gradwohl, J. A., 163
Granata, T., 216
Gray, J. S., 51, 99, 156
Great Barrier Reef, 56–57, 251–52, 316
Greenhouse warming, see Global warming
Greenland ice sheet, 60, 100–101
Greenland Sea, 101, 102
Griffiths, Gwyn, 218, 220, 221–23, 228
Groundwater discharge, coastal zones and,
 67–68
Groupers, 111
Group of Experts on the State of Marine
 Pollution, 4
Guild system, scientists and, 259–61, 264
Gulf Stream, 18, *19*, *20*, 170

Hadley Centre, 99
Hake, 111, 114–16
Hall, Julie, 177
Hallegraeff, G. M., 78
Hamblin, P. F., 252

Hansen, J. E., 99, 103
Harbison, G. R., 78
Harmful algal bloom (HAB), 6, 76–77, 226,
 239, 269
Haury, L., *42*
Hawaii Undersea Geo-Observatory,
 242
Hawken, P., 156
Hawkes, G. S., 219
He, C., 252
Hempel, Gotthilf, xi–xvi, 1–8, 109–35,
 283–319, 309–19
Henry, W., 219
Hibler, W. D., 99
Hierarchal ocean management, 271–72
H_2O (deep ocean observatory), 243
Holliday, D. V, 239
Holm, Nils, 36–37
Hopkins, H., 233
Houghton, J. T., 93, 94
Howarth, R. W., 69
Howorth, Russell, 298–99
Hoyt, D. V., 93
Hubold, G., 110
Hudson, H. S., 93
Hungspreugs, Manuwadi, 287
Hunt's Mercant Magazine, 166
Hurricanes, 63, 217
Hyde, W. T., 98
Hydrothermal activities, 36–37
Hydrothermal vents, 6, 230

Ianuzzi, R. A., 98
Ice Watch, 173
Independent World Commission on the
 Oceans, 309–10
Independent World Ocean Forum,
 proposed, 309–10
Indian Ocean, 12
Industries, ocean studies for offshore, *see*
 Offshore industry, ocean studies for
Inherent optical properties (IOPs), 235, 236
In situ sensors and systems for
 oceanographic applications, 226–49
 biology and bioacoustics, 237–41
 chemistry, 229–33
 interdisciplinary systems, 247–48
 marine geology and geophysics, 241–47
 optics and bio-optics, 233–37
 physics, 227–28
 telemetry, shore- and satellite-based,
  248–49
Instituto Nacional de Investigación y
 Desarrollo Pesquero (INIDEP), 114,
 115, *116*

Instrumentation, observation techniques, and technologies, 15, *17*, 310–11
autonomous underwater vehicles, 220–23
bottom tripods, 215–16, 217
challenges and recommendations for new technologies, 252–53
coastal zones, of special interest for, 79–80
data utilization, 249–52
drifters and floats, 218–19
fisheries science and, 129–30
in situ sensors and systems, *see* In situ sensors and systems for oceanographic applications
manned submersibles, 219–20
moorings, 215–16, 217
multiplatform approach, 212–14
observational challenges, 210–11
offshore and shore-based platforms, 215–18
platforms, 212–26
remotely operated vehicles (ROVs), 220
remote sensing data, 196–97, *198*, 224–26, 242, 312
ships and submarines, 214–15, 257–58
Intergovernmental Agreement on the Prevention of Marine Pollution from Land-Based Activities, 268–69
Intergovernmental Oceanographic Commission (IOC), xi, 3, 165, 169, 192, 199, 200, 201, 203, 205, 263, 294, 300
current study of, as basis of this book, 3–4
Global Oceanographic Data Archeology and Rescue Project, 196
previous assessments of marine science and technology, 3, 5
Intergovernmental Panel on Climate Change (IPCC), 63–64
International Convention for the Prevention of Pollution from Ships (MARPOL), 178
International Convention for the Prevention of the Pollution of the Sea, 178
International Council for Science (ICSU), xi, 3, 199
International Council for the Exploration of the Sea (ICES), 111, 125, 205, 317
International Geosphere-Biosphere Global Exchange Program (IGBP), 26, 30, 80, 264, 285, 314
International Hydrogrpahic Organisation, 174
International Marine Satellite Organization (Inmarsat), 168
International Maritime Organization (IMO), 2, 71, 144–46, 168, 178, 263, 305

Safety of Life at Sea (SOLAS) Convention, 174
International Ocean Colour Coordinating Group (IOCCG), 224, 225, 236
InterRIDGE Programme, 6
Invasive species, *see* Exotic species
IOCARIBE-GOOS, 205
IUCN, 201

Jackson, J. B. C., 120
Jaffe, 238
Jannasch, H. W., 216
Japanese Meteorological Agency, 167
Johannessen, O. M., 98, 171, 173
Johnston, A. K., 165
Joint Global Ocean Flux Study (JGOFS), 5–6, 25, 26–28, 80, 212, 260, 266, 312, 316
objectives of, 26–27
Joint Technical Commission for Oceanography and Marine Meteorology (JCOMM), 192, 202
Jones, P. D., 166
Jorgensen, B. B., 68
Journal of Geophysical Research, 176
Journals, international, 296–97
Junior, Waldemar T., 143–44

Kadanoff, L. P., 41
Kaku, M., 137, 154, 209, 310
Karl, T. R., 96
Kelp, 60, 155
Kemp, A. E. S., 52
Kent, E. C., 166, 181
Kleypas, J. A., 58, 104
Knox, R., 215
Koblentz-Mishke, O. J., 29
Koblinsky, C. J., 224
Kolber, Z. S., 235
Krill, 30
Krueger, A. J., 91
Kruik, Henk de, 61–62
Kuroshio, 18, *20*, 36
Kyoto Protocol, 149

Labrador Sea, 101, 102
Lagerloef, G., 225
Land-Ocean Interactions in the Coastal Zone (LOICZ) project, 5, 80
La Niña, 11, 31, 43, 91, 224
Large marine ecosystems (LME), 285, 304
Larkin, P., 310
Law of the Seas, United Nations Convention on the (UNCLOS), 7, 50, 111, 138, 267–68, 275, 276, 303

Lazier, J. R. N., 50
Lea, J., 93
Ledwell, J. R., 228
Leetma, A., 202
Le Traon, P. Y., 209
Libraries, 296
Lima, Jose Antonio M., 143–44
Lindeboom, Han, 49–84
Lloyds Register, 163, 164, 179
Lochte, Karin, 137–62
Longhurst, A., 132
Lophelia pertusa, 121–22, 146
Lorenz, E. N., 88
Lotti, R., *16*
Ludwig, D., 112

Mace, P. M., 112
McGillicuddy, D. J., 237
McGinn, A. P., 163
McGlade, J. M., 80
McGowan, John A., 9–44
McKenzie, D., *140*, 149
McNeil, J. D., 231, 237
McPhadden, M. J., 22, 203, 217, 248, 250
McPhee, M. G., 98
Macrobenthos, 51
MADAM (Mangrove Dynamics and Management), 55, 302–303
Malaysia, mangroves and aquaculture in, 54–55
Malvinas Current, 88–89
Mammals, marine, 76
Manabe, S., 96, 102
Mangrove Dynamics and Management (MADAM), 55, 302–303
Mangroves, 53–55, 302–303
Mann, K. H., 50, 132
Manned submersibles, 219–20
Maqueda, M. A. Morales, 99
Marchetti, E., 157
Mariculture, 132
Marine chemistry, 10
Marine geology and geophysics, 11
  in situ sensors and systems for, 241–47
Marine protected areas (MPAs), 315–16
Marine Stewardship Council, 134
Maritime transportation industry, 262–63
Marone, Eduardo, 290–91
MARPOL (International Convention for the Prevention of Pollution from Ships), 178
Marra, J., 217
Martin, J., 225
Martinson, D. G., 15, *42*, 98
Masson, 141, *141*
Masutani, S. M., 148

Matsuda, F., 149
Mauna Loa mountain observatory, 27–28
Maury, Dr. Mathew, 165–66
Meanders, 18–*19*
MedGOOS, 203–205
Medicine, marine resources and, 154–56, 275
Meier, M. F., 99
Merlivat, L., 216, 231
Metal mining, 138, 152–54, 275
Microelectromechanical system (MEMS), 233
Mining, seabed, 138, 152–54, 275
Mitchell, Greg, 219
Mitchell, J. P. B., 98, 99
Mitchum, G. T., 225
Modeling, 209
  globalization of modeling capacity, 313
  see also specific models and areas for modeling
MODE project, xii
Modular Oceanographic Data Assimilation System (MODAS), 170
Monastersky, R., 157, 158
Monterey Bay Aquarium Research Institute, *159*
Moore, C. C., 235
Moorings, 215–16, 217
Mullin, M. M., 132
Murphy, J. M., 98

Nakashima, T., 149
NAS 2000, 68, 69
National Oceanic and Atmospheric Administration (NOAA), 90, 98, 167, 169
  Physical Oceanographic Real-Time System (PORTS), 169–71, *172*, *173*, 179–80
National Oceanographic Data Center, 261
National Research Council, U.S., 127, 259–61
National Research Council of Thailand, 251
National Science Foundation, 4
National Weather Service, 167
NEAR-GOOS, 203
Neeland, J. D., 22
NEPTUNE program, 243–47
Nerem, R. S., 225
Nichols, F. H., 78
Niiler, P., 18, *20*
Nitrogen and eutrophication of coastal waters, 68–70
Nixon, S. W., 55, 68
Njoku, E. B., 225

Noctiluca scintillans, 51
Nonindigenous species, *see* Exotic species
Norris, R. D., 33
North, Gerald R., 85–108
North Atlantic Current, 102
North Atlantic Deep Water, 12–13, 15, *16*
   estimates of age of, *24*
North Atlantic Drift, 104
North Atlantic Oscillation, 198, *198*, 212
North Brazil Current Rings Experiment, 189, 190–91
Northern Atlantic, 12
Northern Hemisphere Glaciation, 36
North Pacific Ocean, 12
North Sea, 51, 62, 147
   oil pollution from drilling in the, 144, 145–46
North Sea herring, 123–25, 316
Nowlin, W. D., 187
Nuclear magnetic resonance (NMR) technology, 231–32
Nuclear weapons testing, 22, 86

Observation techniques, *see* Instrumentation, observation techniques, and technologies
Ocean Carbon Model Intercomparison Project (OCMIP), 27
Ocean Color and Temperature Scanner (OCTS), 236
Ocean Drilling Program (ODP), 33–35, 141–42, 148, 152, 155, 242, 265, 315
Ocean floor:
   deep sea floor biosphere, 315
   history of climate-ocean changes and, 33–35
Ocean Obs99, 200–201
Ocean Sciences at the New Millenium, 4
Ocean studies, 9–44
Ocean thermal energy conversion, 149–50, 298
Oceaonography, 237
O'Farrell, S. P., 98
Offshore and shore-based platforms, 215–18
Offshore industry, ocean studies for, 137–62
   biotechnology and medicine, 154–56, 275–76
   conclusions and recommendations, 159–60
   energy industry, 139–50, 275
   renewable marine energy sources, 149–50
   research priorities, 160–62
   submarine cables, 150–51
   waste disposal, *see* Waste disposal
Oil and gas exploration, 1–2, 139–49, 190, 275

   controls on tanker activity, 2
   disposal of offshore facilities, 146–47
   gas hydrates, 148–49
   pollution resulting from, 142–46
Oil spills, 146, 164
Olsgard, F., 142
Olson, Rob, 237
Ong, Jin E., 54–55
OOSDP, 187, 197
Operational oceanography, 187–206
   access to data, 196
   conclusions and recommendations, 205–206
   definition of, 188
   emerging trends, 191–92
   end uses and users, 187–91
   global ocean observing systems, *see* Global Ocean Observing System (GOOS)
   operational numerical models, 192–96
   public concerns about the oceans, 187, 189
   remote sensing data, 196–97
   research priorities, 206
Optics and bio-optics, in situ sensors and systems for marine, 233–37
Orbcomm, 168
Orcutt, J., 229, 242
Organization for Economic Co-operation and Development (OECD), 187
Overfishing, 1, 55, 73–74, 111–12, *113*, 120, 132–33, 284, 317
   variations in fish availability and, 31, 32
Owens, W. B., 218

Pacific Decadal Oscillation (PDO), 212
Pacific GOOS, 203, 298–99
Paduan, J., 228, 229
Paleoceanography, 11, 15, *16*
   climate change and, 104–106
Pap, J. M., 93
Parkes, R. J., 155
Parrish, J. S., 239, 240
Partnership for an Integrated Global Observing Strategy (IGOS), 205
Partnership for Observation of the Global Oceans (POGO), 278–79
Patagonia, 114–17
Pauly, Daniel, 109–35, *113*, 118, *119*, 128
PCBs, 70, 71
Peer review sciences, 259
Pérez, M. A., 115
Permafrost, 60
Petrae, Gary, 171, 179
Petrenko, A. A., 236
PETROBRAS, 142, 143–44

Phosphates, 24, 69–70
Photosynthesis, 12, 25, 67
Photosynthetically available radiation (PAR) measurements, 235
Physical oceanography, 10
Physics, in situ sensors and systems for marine, 227–28
Phytoplankton, 12, 51, *52*, 70, 103, 233, *234*, 236, 238
  iron and blooms of, 28
Pike, J., 52
PIRATA (Pilot Research Moored Array in the Tropical Atlantic), 203
Plankton, 10, 29
  phytoplankton, *see* Phytoplankton
Planque, *198*
Polar areas, global warming and, 60, 100
Pollard, B., 225
Pollution, 55, 230, 258, 268–69, 285
  from offshore oil and gas exploration, 142–46
  runoff and, 2
  water quality and, 70–71
POLYMODE project, xii
Porter, D. L., 225
Primary productivity, 25, 29, 233, 235
Provost, C., 217

Rabalais, N. N., 70
Rahmstorf, S., 12, 101, 102
Ramsar Convention on Wetlands, 50
Raney, R. K., 225
Rayner, Ralph, 187–206
Recreation, coastal zone, 74
Red tides, 78
Regime shifts, 30–33
Regional Graduate School of Oceanography in Southeast Asia, 294
Regional meetings, 297–300
Regional programs, cooperative, 272–77
Regional vessel data networks (RVDNs), 179–80
Regression analysis, 94
Reid, P. C., 32, 238
Remotely operated vehicles (ROVs), 220
Remote sensing data, 196–97, *198*, 224–26, 242, 312
Renewable marine energy sources, 149–50, 276
Research/Platform Instrument Platform (R/P FLIP), 215
Research Ship Sharing Scheme, European Union, 265–66
Research vessels, *see* Ships, oceanographic data and analysis from

Revelle, Roger, 25–27
Rice, A. L., 138, 156
Richardson, K., 68
Rind, D. R., 98
Ring Group, 18
RISKER project (Environmental Risks from Large-Scale Ecological Research in the Deep Sea: A Desk Study), 157
River runoff and load, effect on coastal zones of, 66–67
Roberts, C., 127
Robinson, A. R., 50, 209, 211, 241
Rocky shores, 59–60
Roe, H. S. J., 239
Roemmich, D. H., 31, *32*, 101, 203, 218
Rogers, A. D., 121
Rogers, David P., 165–85, 175
Root, T. L., 41
Rosenfeld, L. K., 228, 229
Rothschild, 112
Ruiz, G. M., 78
Russell, E. S., 112
Ryabinin, Vladimir E., 64

Salt marshes, 59
Sánchez, Ramiro, 114–17
Sand and gravel, near-shore mining of, 138, 151, 275
Sandy beaches, 60–62
San Francisco Bay, 78, 178
San Francisco Regional Water Quality Control Board, 178
Santer, B. D., 95
Sarmiento, J. L., 25, 103
Satellite-borne sensors, 196–97, *198*, 224–26, 312
Satellite communication, 168, 218
Satellite synthetic aperture radar (SAR), 173
Scatterometers, 86, *198*
Schaefer, M. B., 112
Schatten, R., 93
Schlesinger, W. H., 10
Schlosser, P., 22
Schmidt, G. A., 99
Schmitz, W. J., 101
Schneider, S. H., 41
Scholin, C., 239
Schopf, P., 90
Schwab, D. J., 252
Scialabba, N., 80
Scientific Committee of Ocean Research (SCOR), xi, 3, 26, 30, 300
Scientific Committee on Problems of the Environment (SCOPE), xi, 3
Sea, The (Brink and Robinson), 50

Seabirds, 76
Seagrass meadows, 58–59
Sea ice, 97–99
Sea-level rise, 1, 63–64, 99–100, 274
Seals, 62
SeaSat satellite, 86
Sea surface temperature (SST), 88–89, 89–91
Sea-viewing Wide Field-of-View Sensor (SeaWiFS), 236
SEAWATCH, 251
Sediment discharge, impact on coastal zones of, 66–67
Send, U., 216, 229
Severinghaus, J. P., *16*
Sewage treatment, 70
Shaffer, Lisa, 278–79
SHEBA (Surface Heat Budget of the Arctic), 98
Sheldon, R. W., *234*
Shepherd, John, 38–40, 147
Shipping and defense, marine information for, 165–85
  accidents and, 146, 163–64, 171, 269
  coastal information systems, 170–76
  conclusions and recommendations, 179–82
  costs and forms of communication, 167–68
  electronic charts and positioning systems, 174
  environmental impacts, 176–78
  growth in world shipping, 163, *164*
  intermodal transport, maritime ports and, 174–76
  international regulations, impact of advances in marine science on, 178–79
  research priorities, 182–83
  safety at sea and environmental security, 163–67, 262–63
Ships, oceanographic data and analysis from, xii–xiii, 214–15, 257–58, 265–66, 311
Sholkovitz, E., 231
Signell, R. P., 252
Significance of the oceans, 1–2, 9, 86
Sim Coast approach, 80
Skud, B. E., 118
Smethie, W. M., 22, 23, *24*
Smith, N., 224, 239
Solar luminosity, variability of, 93
Sole, 111
South Atlantic Ocean, climate and, 88–89
Southeast Asian countries, coastal management programs of, 288
Southern Ocean, 14

Southern Oscillation, 21
South Pacific, regional cooperation in the, 297–99
South Pacific Forum, 298
Souza, Leonardo F., 143–44
Spectral fluorescence, 236
Spinrada, Richard W., 165–85
Stammer, D., 209
START (System for Analysis Research and Training), 294
Stel, J. H., 187, 188
Stevens, M., 95
Steward, R. G., 226
Stewart, J. E., 78, 177
Stouffer, R., 96, 102
Stramska, M., 216
Suarez, M., 90
Submarine cables, 150–51
Submarine groundwater discharge (SGD), 67–68
Submarines, oceanographic data gathered by, 214–15
Sulfide mining, 151
Summerhayes, Colin P., xi–xvi, 1–8, 34–35, 137–62, 187–206, 309–19
Swenson, M., 218

Takahashi, T., 149–50
Tamburri, M. N., 158, 159
Tang, X. O., 238
Taylor, C. D., 216
Taylor, L., 74
Taylor, P. K., 166, 181, 182, 227
Telemetry, shore- and satellite-based, 248–49
Thailand, 286–87
Thermohaline circulation, 11–14, 97, 101–102, 104, 106
Thiel, H., 156
Tokar, J. M., 216, 230, 232, 233
Tourism, 120, 138
  coastal zone, 74–75, 274
  coral reefs and, 55
Tracers, water mass, 86–87
  nutrients, and the carbon cycle, 22–28
Trade, world, *see* Shipping and defense, marine information for
Trade winds, 21
Tropical Atmosphere Ocean, xii, 5, 202–203
Tropical Ocean Global Atmosphere (TOGA) experiment, xii, 5, *17*, 21, 22, 212, 250, 266
Tropical Pacific, 103, 202
Tsunamis, 63, *65*
Turtles, 62
Twardowski, M. S., 235
Typhoons, *65*, 217

UNCED, 4
UNCTAD, 163
United Kingdom:
  Marine Foresight Process, 261, 262
  Meteorological Office, 167
United Nations, 168
  Commission on Sustainable
    Development, 271, 279
  Conference on Environment and
    Development, 1992, 7, 198
  General Assembly, 310
United Nations Convention on the Law of
  the Seas (UNCLOS), 7, 50, 111, 138,
  267–68, 275, 276, 303
United Nations Development Program
  (UNDP), 300, 305
United Nations Education, Scientific and
  Cultural Organization (UNESCO), xi, 3,
  263, 294
United Nations Environmental Programme
  (UNEP), 199, 300
  Regional Seas Programme, 269–70, 285
U.S. Navy, 169, 170–71, 242
  Fleet Numerical Meteorology and
    Oceanography Center (FNMOC),
    167, 169
University of Concepción, Chile, 294
Unmanned aerial vehicles (UAVs), 225

Valent, T. J., 156
Van der Meulen, Frank, 61–62
van der Veen, J., 100
Varney, M., 216, 230, 231
Vertical velocity, 229, 239
Vessel monitoring systems (VMS), 179
Vinnikov, K. Y., 99
Vision to 2020, 309–19
Volcanic eruptions, 91, 95
Volovick, S. P., 78
Voluntary observing ships (VOS), 166, 180,
  181, 212

Wadden Sea, 51, 61–62
Wallace, J., 252
Wang, P., 36
Warren, B. A., 101

Warrick, R. A., 99, 100, 101
Waste disposal, 2, 138, 156–59, 314
Water quality, 276
Watson, A. J., 228
Watson, R. T., 2
Weather, *see* Climate and climate change
Weaver, A. J., 101
Weber, M. L., 163
Weddell Sea, 12
Weiher, R. F., 198, 199
Weller, R. A., 216, 227
Went, A. E. J., 112
Wetlands, 61
Whales, 76
Williams, A., 228
Williams, A. J., 216
Willson, R. C., 93
Wind-driven circulation, *14*, 14–15
Winn, C. D., 26
Woehler, E. J., 74
Wolanski, Eric, 251–52
Wood, R. A., 102
Woods, J. D., 188, 193, 196, 198, 203
World Bank, 300
World Climate Research Programme
  (WCRP), 5
World Meteorological Organization
  (WMO), 192, 199, 201, 263
  Voluntary Observing Ship (VOS)
    program, 166, 180, 181, 212
World Ocean Affairs Observatory, proposed,
  309
World Ocean Circulation Experiment
  (WOCE), xii, 5, 6, 22, 86–87, 212, 218,
  227, 260, 266, 312
World Summit on Sustainable Development,
  2002, 8
Wu, J., 230
Wu, Q., 94, 95
Wunsch, C., 101

Young, D. K., 156

Zibton, Paul, 245
Zimmerman, Christopher, 123–25
Zooplankton, 51, *52*, 238, 239, 240

ISLAND PRESS BOARD OF DIRECTORS

Chair
HENRY REATH
President, Collectors Reprints, Inc.

Vice-Chair
VICTOR M. SHER
Miller Sher & Sawyer

Secretary
DANE A. NICHOLS
Chair, The Natural Step

Treasurer
DRUMMOND PIKE
President, The Tides Foundation

ROBERT E. BAENSCH
Director, Center for Publishing,
New York University

SUSAN E. SECHLER
Senior Advisor on
Biotechnology Policy,
The Rockefeller Foundation

MABEL H. CABOT
President, MHC & Associates

PETER R. STEIN
Managing Partner,
The Lyme Timber Company

DAVID C. COLE
Owner, Sunnyside Farms

CATHERINE M. CONOVER

RICHARD TRUDELL
Executive Director,
American Indian Resources Institute

CAROLYN PEACHEY
Campbell, Peachey & Associates

DIANA WALL
Director and Professor, Natural
Resource Ecology Laboratory,
Colorado State University

WILL ROGERS
President, Trust for Public Land

CHARLES C. SAVITT
President, Center for
Resource Economics/Island Press

WREN WIRTH
President,
The Winslow Foundation